Cultures of Sustainability and Wellbeing

Cultures of Sustainability and Wellbeing: Theories, Histories and Policies examines and assesses the interdependence between sustainability and wellbeing by drawing attention to humans as producers and consumers in a post-human age. Why wellbeing ought to be regarded as essential to sustainable development is explored first from multifocal theoretical perspectives encompassing sociology, literary criticism and socioeconomics, second in relation to institutions and policies, and third with a focus on specific case studies across the world. Wellbeing and its sustainability are defined in terms of biological and cultural diversity; stages of advancement in science and technology; notions of citizenship and agency; geopolitical scenarios and environmental conditions.

Wellbeing and sustainability call for enquiries into human capacities in ontological, epistemological and practical terms. A view of sustainability that revolves around material and immaterial wellbeing is based on the assumption that life quality, comfort, happiness, security, safety always posit humans as both recipients and agents. Risk and resilience in contemporary societies define the intrinsically human ability to make and consume, to act and adapt, driving the search for and fruition of wellbeing. How to sustain the dual process of exploitation and regeneration is a task that requires integrated approaches from the sciences and the humanities, jointly tracing a worldwide cartography with clear localisations.

This book will be of great interest to students and researchers interested in sustainability through conceptual and empirical approaches including social theory, literary and cultural studies, environmental economics and human ecology, urbanism and cultural geography.

Paola Spinozzi is Associate Professor of English Literature at the University of Ferrara. Her research focuses on the relationships between literature and visual art, science, utopia as a genre and sustainability.

Massimiliano Mazzanti is Professor of Political Economy at the University of Ferrara. He specialises in environmental policy, economics of innovation and waste management.

Routledge Studies in Culture and Sustainable Development

Series Editors:

Katriina Soini

University of Jyväskylä, Finland, and Natural Resources Institute Finland

Joost Dessein

Institute for Agricultural and Fisheries Research (ILVO) and Ghent University, Belgium

Culture as an aspect of sustainability is a relatively new phenomenon but is beginning to attract attention among scholars and policy makers. This series opens up a forum for debate about the role of culture in sustainable development, treating culture and sustainability as a meta-narrative that will bring together diverse disciplines. Key questions explored in this series will include: how should culture be applied in sustainability policies; what should be sustained in culture; what should culture sustain; and what is the relationship of culture to other dimensions of sustainability?

Books in the series will have a variety of geographical foci and reflect different disciplinary approaches (for example, geography, sociology, sustainability science, environmental and political sciences, anthropology, history, archaeology and planning). The series will be addressed in particular to postgraduate students and researchers from a wide cross-section of disciplines.

Theory and Practice in Heritage and Sustainability
Between Past and Future
Edited by Elizabeth Auclair and Graham Fairclough

Cultural Sustainability and Regional Development
Theories and Practices of Territorialisation
Edited by Joost Dessein, Elena Battaglini and Lummina Horlings

The Politics of Cultural Development
Trade, Cultural Policy and the UNESCO Convention on Cultural Diversity
Ben Garner

Form, Art and the Environment
Engaging in Sustainability
Nathalie Blanc and Barbara Benish

Culturally Responsive Education
Reflections from the Global South and North
Elina Lehtomäki, Hille Janhonen-Abruquah, and George Kahangwa

Design for a Sustainable Culture
Perspectives, Practices and Education
Edited by Astrid Skjerven and Janne Beate Reitan

Cultures of Sustainability and Wellbeing
Theories, Histories and Policies
Edited by Paola Spinozzi and Massimiliano Mazzanti

Cultures of Sustainability and Wellbeing

Theories, Histories and Policies

Edited by Paola Spinozzi and Massimiliano Mazzanti

LONDON AND NEW YORK

First published 2018
by Routledge

2 Park Square, Milton Park, Abingdon, Oxfordshire OX14 4RN
52 Vanderbilt Avenue, New York, NY 10017

Routledge is an imprint of the Taylor & Francis Group, an informa business

First issued in paperback 2019

British Library Cataloguing in Publication Data
A catalogue record for this book is available from the British Library

Library of Congress Cataloging in Publication Data
A catalog record for this book has been requested

ISBN: 978-1-138-23454-3 (hbk)
ISBN: 978-0-367-27119-0 (pbk)

Typeset in Times New Roman
by Wearset Ltd, Boldon, Tyne and Wear

Contents

List of figures	viii
List of tables	ix
Notes on contributors	x
Series editor introduction	xviii

Pillars and circles: wellbeing at the core of sustainability   1
PAOLA SPINOZZI AND MASSIMILIANO MAZZANTI

PART I
Sustainable wellbeing in theory and history   17

Prologue   19
PAOLA SPINOZZI

I.1 **Creating capacities for human flourishing: an alternative approach to human development**   23
PAUL JAMES

I.2 **The incongruities of sustainability: an examination of the UN Earth Summit Declarations 1972–2012**   46
GONZALO SALAZAR

I.3 **Slow living and sustainability: the Victorian legacy**   63
WENDY PARKINS

I.4 **Innovation and consumption in the evolution of capitalist societies**   72
PIER PAOLO SAVIOTTI

I.5 In a prescient mode: (un)sustainable societies in the post/
apocalyptic genre   85
PAOLA SPINOZZI

PART II
Policies and institutions for wellbeing   105

Prologue   107
MASSIMILIANO MAZZANTI

II.1 Contextualising sustainability: socio-economic dynamics,
technology and policies   111
MASSIMILIANO MAZZANTI AND MARIANNA GILLI

II.2 Social equity and ecological sustainability through the lens
of degrowth   135
VALERIA ANDREONI

II.3 Institutionalist climate governance for pleasant cities and
the good life   146
GJALT HUPPES AND RUBEN HUELE

II.4 Assessing public awareness about biodiversity in Europe   167
ANNA KALINOWSKA

II.5 Is the current global role of English sustainable?   185
RICHARD CHAPMAN

PART III
Sustainable wellbeing via habitat and citizenship   201

Prologue   203
PAOLA SPINOZZI AND MASSIMILIANO MAZZANTI

III.1 The impact of settlements on urban river basins and the
case of the Belém River in Curitiba, Brazil   207
GILDA AMARAL CASSILHA,
MARTA MARIA BERTAN SELLA GABARDO,
SYLVIA RAMOS LEITÃO, AND
ZULMA DAS GRAÇAS LUCENA SCHUSSEL

III.2 Creative social innovation and urbanism: the case of
      Medellín 222
      ANA ELENA BUILES VÉLEZ AND
      MARÍA FLORENCIA GUIDOBONO

III.3 **Long-term visions and ordinary management:**
      post-earthquake reconstruction in the Italian region
      of Emilia 232
      GIANFRANCO FRANZ

III.4 **Urban life and climate change at the core of political**
      dialogue: a focus on Saint-Louis du Sénégal 244
      ADRIEN COLY, FATIMATOU SALL,
      MOHAMED B. C. C. DIATTA, AND
      CHÉRIF SAMSÉDINE SARR

III.5 **Is ethno-tourism a strategy of sustainable wellbeing?**
      A focus on Mapuche entrepreneurs 254
      GONZALO VALDIVIESO, ANDRÉS RIED AND SOFÍA ROJO

III.6 **Japanese castle towns as models for contemporary urban**
      planning 269
      SHIGERU SATOH

III.7 **Vietnam's pathway towards sustainability: stories**
      half-told 281
      NHAI PHAM AND YEN DAN TONG

      Index 305

Figures

I.1.1	Circles of social life: conducted as an urban assessment of the city of Melbourne	26
I.1.2	Circles of social capacities: conducted as an assessment of the contemporary global condition	43
II.1.1	Total patents filed under the PCT, 1977–2013, based on the inventor's country of residence	115
II.1.2	R&D expenditures in selected OECD countries	117
II.1.3	Total R&D expenditure variation in Europe (2005–2013)	118
II.1.4	Total patents filed under the PCT, 1977–2013, by inventor's country of residence	118
II.1.5	Green patents trends in selected countries	119
II.1.6	OECD environmental policy stringency indicator (EPS), 1990–2012	120
II.1.7	Variation in total environmental taxes in Europe (2005–2013)	121
II.1.8	Variation in per capita environmental taxes in Europe (2005–2013)	122
II.1.9	Production-based CO_2 emissions across income levels	123
II.1.10	Beta convergence in the development index in the EU	127
II.1.11	Sigma convergence in the development index in the EU	127
II.3.1	Different roles of decentral actors under planning and control and institutionalism	162
III.1.1	The Pinheirinho River Basin	215
III.1.2	The Guaíra River Sub-basin	215
III.4.1	Risk situation in Saint-Louis	245
III.6.1	Urban design of Tsuruoka	271
III.6.2	Urban design of Morioka	272
III.6.3	Compilation of design elements and techniques incorporating the natural environment into the structure of a castle town	278
III.7.1	GDP per capita in Vietnam, 1986–2015	284
III.7.2	Poverty rate in Vietnam, 1992–2014	285
III.7.3	Natural capital depletion in Vietnam	287

Tables

I.1.1	Core conditions for engaging in social life	27
I.1.2	Social capacities for positively engaging in social life	42
I.2.1	Two contradictory epistemologies in the International Sustainability Agenda	52
I.4.1	Trajectories of economic development	77
II.2.1	Wellbeing theories and hypothesis	138
III.4.1	Structures of the District Councils	247

Contributors

Valeria Andreoni is Senior Lecturer in Economics at Manchester Metropolitan University, UK. She is the author of peer-reviewed papers and reports for the European Commission, such as "Competitiveness and Sustainable Development Goals" (2016), "Mapping the Distribution of Wellbeing in Europe beyond National Borders" (2015), "How to Increase Wellbeing in a Context of Degrowth" (2014), "Global Resources Use and Pollution: Production, Consumption and Trade (1995–2008)" (2012). She is also involved in research activities related to energy economics, emissions and use of resources, among which "Drivers in CO2 Emissions Variation: A Decomposition Analysis for 33 World Countries" (in *Energy Journal*, 2016); "Global Use of Water Resources: A Multiregional Analysis for Water Use, Water Footprint and Water Trade Balance" (in *Water Resources and Economics*, 2016); "The Game of Trading Jobs for Emissions" (in *Energy Policy*, 2014).

Gilda Amaral Cassilha holds a degree in Architecture and Urban Planning from Universidade Federal do Paraná, Brazil, a Master's degree in Public Administration and Urban Planning from the São Paulo Business School at Fundação Getúlio Vargas-EAESP and a specialisation in Technical Management of the Urban Environment from FAE Centro Universitário in Curitiba, Paraná. Since 1997 she has worked as a Professor of Urban and Regional Planning at Pontifícia Universidade Católica do Paraná. Her main areas of interest are urban landscape and urban technical management, the cities in the state of Paraná, the borders of Brazil with other South American countries, sustainability and eco-socioeconomics. She is the author of *Planejamento Urbano e Medio Ambiente* (IESDE, 2005).

Richard Chapman is Researcher and Lecturer in English Language in the Department of Human Sciences at the University of Ferrara, Italy. A first degree in history from Cambridge University gives a cultural and anthropological flavour to his linguistic research, while extensive experience in teacher training results in a pragmatic bent in his approach to language. Publications include numerous course-books for English language learners (both teenagers and adults) and studies reflecting interest in developments in the language from a sociolinguistic, textual and pragmatic point of view. Language testing is another area of research interest, involving participation in the development of regional examinations for the certification of language teachers in Italy, and the theoretical and practical evaluation of current language tests. Recent publications include work on computer-assisted language testing, on the pragmatics of language test instruments and the possible roles of English as a Lingua Franca.

Adrien Coly holds a PhD in Geography from Université Cheikh Anta Diop de Dakar, Sénégal, and is a lecturer and researcher in the Department of Human Sciences at Université Gaston Berger – Saint-Louis du Sénégal. He is certified in Integrated Water Resources Management for developing countries by the Fondation Universitaire Luxembourgeoise in Belgium. He is a member of Laboratoire Leidi – Dynamique des Milieux et Développement and the Director of the water division Gouvernance des Territoires de l'Eau (Governance of Water Territories), www.watergov-senegal.org, which investigates the relationships between water and the territory by adopting a syndrome approach. Specialised in natural hazards and vulnerabilities, he co-ordinates research programmes on urban hazards, water safety, environmental pollution at Université Gaston Berger.

Mohamed B. C. C. Diatta is a Doctor in Geography and Management with specialisations in Water Management and Policy as well as in Environment and Rural Societies. He has co-ordinated scientific seminars and courses on Integrated Water Resources Management, Water Governance and Geography in Senegal and West Africa. He is an Associate Researcher at Laboratoire Société Environnement Territoire CNRS-UPPA, France, and a member of the National Executive Board of the Association des Jeunes Professionnels de l'Eau et de l'Assainissement du Sénégal – AJPEAS. He is currently the first Technical Advisor for the Ministry of Hydraulics and Sanitation of the Republic of Senegal.

Gianfranco Franz is Associate Professor of Policies for Sustainability and Local Development at the University of Ferrara, Italy, where he is the Co-coordinator of the International University Network *Routes towards Sustainability*. From 2012 to 2014 he was the co-ordinator of two special community programmes for recovering the area of the seismic crater in the Emilia-Romagna region. On this topic, he published *La ricostruzione in Emilia dopo il sisma del maggio 2012* (in *Urbanistica*, 2014). For many years he was a consultant on policies and programmes of urban regeneration for the Regional Government of Emilia-Romagna. He is the author of *La riqualificazione continua* (Alinea, 2005) and *Dieci anni di riqualificazione urbana in Emilia-Romagna. Processi, progetti e risultati* (Corbo Editore, 2010, with M. Zanelli). From 2003 to 2012 he was the co-ordinator of *Ecopolis*, an international MA course taught in collaboration with several universities in Latin America, Europe and Japan.

Marta Maria Bertan Sella Gabardo holds a degree in Architecture and Urban Planning from Universidade Federal do Paraná, Brazil, a Master's degree in

Geography from Universidade Federal de Santa Catarina and a PhD in Urban Management from Pontifícia Universidade Católica do Paraná. Since 2001 she has taught Urban and Regional Planning, Urban Design and Landscape Architecture at Pontifícia Universidade Católica do Paraná. Her research focuses on the development of sustainable cities. She is the author of "Metropolização, centralidades e periferias: Três cidades da Região Metropolitana de Curitiba" (Juruá Editora, 2004).

Marianna Gilli is a Post-Doctoral Research fellow at the University of Ferrara, Italy. She earned a PhD in Economics and Management of Innovation and Sustainability at the University of Ferrara and an MSc in Statistics and Econometrics at the University of Essex. Her main research interests are the environmental and economic implications of technological change and environmental behavioural economics.

María Florencia Guidobono earned a PhD in Economics and Management of Innovation and Sustainability at the University of Ferrara, Italy, and is an Assistant Professor of Urban Management and Public Policies at Universidad Católica de Córdoba, Argentina. She is also the co-ordinator of research programmes at the Technological and Postgraduate Department, Faculty of Architecture. Her areas of research encompass environmental and sustainable territorial policies, social economy, management and urban development. Her current research focuses on urban development and social economy. Recently she has published "Lukas Fusters' Plaza Ramansito" (in *Cuaderno LatinoAmericano de Arquitectura*, 2015).

Ruben Huele is affiliated to the Institute of Environmental Sciences at Leiden University, where he works on modelling, data analysis and strategy and has taught analysis of large datasets of greenhouse gases emission to Industrial Ecology students. He co-designed and contributed to an online course on metals in the circular economy, co-ordinated a European project on a photo database of cetaceans and contributed to building up an automated method of photo identification of sperm whales. He worked for Dutch ministries and the Dutch Museum of Natural History, specialising in strategy development. His publications focus on instrumentation strategies for climate policy and his latest project is aimed at producing a geological map of the built environment of Amsterdam.

Gjalt Huppes studied political science and economics at the University of Amsterdam. His PhD thesis *Macro-environmental Policy, Principles and Design* (Amsterdam: Elsevier, 1993) combines macro level institutional arrangements, such as substance deposits on toxic metals and carbon, with micro-level sustainability analysis. Micro-macro linkages are covered in "Environmental Impacts of Consumption in the European Union" (in *Journal of Industrial Ecology*, 2007), *Quantified Eco-Efficiency. An Introduction with Applications* (Springer, 2007, edited with M. Ishikawa), and "Eco-efficiency Guiding Micro-level Actions towards Sustainability" (in *Ecological Economics*, 2009). He was Head of the Department of Industrial Ecology at the Institute of Environmental Sciences – CML, Leiden University, from 1994 to 2011. The International Society for Industrial Ecology awarded him the Prize for lifetime contributions in 2011. Currently he is Senior Researcher at CML. Strategic design of long-term climate policy instrumentations is also the subject of a forthcoming Routledge book.

Paul James is Professor of Globalization and Cultural Diversity at the Western Sydney University, Australia, where he is Director of the Institute for Culture and Society. He is Scientific Advisor to the Senate Department for Urban Development, Berlin, and a Metropolis Ambassador of Urban Innovation. He is an editor of *Arena Journal* and author or editor of 34 books including *Nation Formation. Towards a Theory of Abstract Community* (Sage, 1996) and *Globalism, Nationalism, Tribalism* (Sage, 2006). His work for the Papua New Guinea Minister for Community Development became the basis for their Integrated Community Development Policy. From 2007 till 2014 he was Director of the United Nations agency, the Global Compact Cities Programme. His latest book is *Urban Sustainability in Theory and Practice: Circles of Sustainability* (Routledge, 2014).

Anna Kalinowska is a biologist with a PhD in Ecology and Professor of Environmental Education and Policy at the University of Warsaw, Poland, where since 1992 she has been Director of the Centre for Environmental Studies and Sustainable Development. Her research focuses on nature conservation, environmental policies, and education. She was an Advisor to the Ministry of the Environment from 1998 to 1999 and Director of the Bureau of Environmental Education at the Ministry of the Environment from 1999 to 2002. Actively involved in Polish and international institutions, such as CCM-NATO, Global Environment Facility, REC, she was an IUCN Regional Councillor for Europe from 1996 to 2004 and has been a Member of IUCN – International Union for Conservation of Nature at the Commission on Education and Communication since 1994. She is also a member of the Informal Advisory Committee for Communication, Education and Public Awareness at the Convention on Biological Diversity.

Sylvia Ramos Leitão holds a degree in Architecture and Urban Planning from Universidade Federal do Paraná, Brazil, and a Master's degree and a PhD in Urban and Regional Planning from Universidade de São Paulo. Since 1998 she has been Professor of Urban and Regional Planning and Urban Design at Pontifícia Universidade Católica do Paraná and since 2016 she has co-coordinated Laboratório de Cidades at the School of Architecture and Design. Her research revolves around the socio-environmental costs of planned cities, local housing plans in metropolitan regions, metropolitan integration polices and smart mobility. Her latest publications are "Gênese do discurso do planejamento urbano em Curitiba: bases políticas, filosóficas e técnicas" (in *Paranoá: Cadernos de Arquitetura e Urbanismo*, 2014), and "Política de mobilidade, mercado de terras e a nova lógica de expansão na Curitiba metrópole: inclusão do excluído?" (in *Oculum Ensaios*, 2015).

Massimiliano Mazzanti is Professor of Economic Policy and Lecturer in Macroeconomics, Environmental and Ecological Economics at the University of Ferrara, Italy. He is Director of the Interuniversity Research Centre SEEDS – Sustainability, Environmental Economics and Dynamics Studies and President of the Italian Association of Environmental and Natural Resource Economists – IAERE. He co-ordinates the research activities of two European projects: H2020 *Green.eu – European Global Transition Network on Eco-Innovation, Green Economy and Sustainable Development* and *European Topic Centre on Waste Material in a Green Economy – WMGE*. His publications concern environmental policy design and analysis, economics of innovation, economic evaluation of non-market goods, economic development and climate change, waste management and policy, cultural economics and policy. He is associate editor of *Journal of Environmental Planning and Management, Economia Politica – Journal of Analytical and Institutional Economics*, and *Italian Economic Journal*.

Wendy Parkins is Professor of English Literature and Director of the Centre of Victorian Literature and Culture at the University of Kent, UK. She is a member of the editorial boards of the *Journal of Victorian Culture* and *Victoriographies: A Journal of Nineteenth-Century Writing*. She was the President of the Australasian Victorian Studies Association from 2011 to 2013 and is currently a member of the Executive Committee of the British Association for Victorian Studies. She is also a member of the Peer Review College, AHRC. She is the author of *Slow Living* (University of New South Wales Press, 2006, with Geoffrey Craig) and has published widely on topics relating to gender and modernity in Victorian literature, such as *Mobility and Modernity in Women's Novels, 1850s–1930s* (Palgrave Macmillan, 2009) and *Jane Morris: The Burden of History* (Edinburgh University Press, 2013). Her latest book is *Victorian Sustainability in Literature and Culture* (Routledge, 2017).

Nhai Pham earned a Master's degree in Business Administration and a PhD in Economics from La Trobe University, Australia. She is currently a member of the Faculty of Management and Tourism at Hanoi University, Vietnam. Her research on the agricultural insurance market in Vietnam has been presented and well-received twice at the annual conferences held by the Australian Agricultural and Resources Economics Society (AARES). Thanks to an international scholarship for PhD students, in 2013 she participated in the Summer School of the European Association of Environmental and Resource Economists (EAERE). Her main research areas are sustainable development, environmental economics and the economics of climate change and disaster management.

Andrés Ried is Assistant Professor at Pontificia Universidad Católica de Chile – PUC, Campus Villarrica, where he teaches courses related to outdoor education and games as an educational resource. He is a researcher at the Centre for Local Development (CEDEL). His recent publications include "Leisure in Nature as a Space for Youth Development" (in *Journal of Sport Psychology*, 2016, with Joseba Doistua), "La experiencia de ocio al aire libre en contacto con la naturaleza, como vivencia restauradora de la relación ser humano-naturaleza" (in *Polis. Revista LatinoAmericana*, 2015), and "Ecotourism as a Humanistic Leisure Experience" (Institute of Geography, PUC, 2014, with Manuel Gedda). He is a member of OTIUM, the Ibero-American Leisure Studies Association. His main research topic is the experience of leisure in protected wild areas and in relation to the construction of the sense of the place.

Sofía Rojo is an agronomist working as General Coordinator of the Local Development Centre (CEDEL) at Pontificia Universidad Católica de Chile – PUC. She has worked as a research assistant for ICIIS and has been the co-ordinator of research and local development projects on education for sustainability and training programmes with communities and municipalities in the region of Araucanía.

Gonzalo Salazar holds a PhD in Philosophy from the Centre for the Study of Natural Design at the University of Dundee, UK. He is Assistant Professor at Pontificia Universidad Católica de Chile – PUC, Campus Villarrica and Institute of Urban and Regional Studies. He is Director of the Centre for Local Development (CEDEL) and a researcher at the Centre for Sustainable Urban Development (CEDEUS) and at the Centre for Intercultural and Indigenous Research (CIIR). His teaching and investigation focus on theories of sustainability, planning for sustainable local development, processes of collaborative governance, education for sustainability, human ecology and ecological design.

Fatimatou Sall holds a Master's degree in Management of Protected Areas from Senghor University of Alexandria, Egypt. She is a member of Laboratoire Leidi – Dynamique des Milieux et Développement at Université Gaston Berger – Saint-Louis du Sénégal and is in charge of research projects at the Institut Africain de Gestion Urbaine. She is the author of several papers and articles on environmental geography and has participated in various research programmes in the field of climate change, related to her PhD project on "Urbanity and Biodiversity: Study of the Resilience of a Socio Ecological System in an Estuarine environment (Saint Louis of Senegal)".

Chérif Samsédine Sarr is a PhD student at Université Gaston Berger – Saint-Louis du Sénégal as well as a member of the Laboratoire LEIDI and of the International Mixed Laboratory of Patrimoines et Territoires de l'Eau (Water Heritage and Territories). The focus of his research is on insularity and the vulnerability of rivers. As a geographer specialising on islands, he is in charge of research projects at the Municipality of Diembéring and the Coordinator of the Lower Casamance Islands Network for the Management of Integrated Coastal Zones.

Shigeru Satoh is Director of the Department of Urban Planning at Waseda University, Japan. He was Director of the Research Institute of Urban and Regional Studies and President of the Architectural Institute of Japan. His research on urban design and planning in relation to the cultural context and use of local potential and his study of *Joka-machi* (Japanese castle town) have established a methodology that explains the evolution of urban morphology in connection to historical ecological features, community traditional groups and other regional layers. He is the author of *Illustrated Japanese Castle-Town Cities* (Kajima Shuppan Co., 2015, in Japanese) and a representative of *Machizukuri*, a community-based comprehensive approach aimed at improving the built environment in theory and practice. In recent years, he has applied his concepts to a wide range of projects on pre- and post-disaster planning, rural and regional revitalisation, and public/private partnerships for regional management.

Pier Paolo Saviotti is a Research Fellow of the Innovation Studies Group at Copernicus Institute, Utrecht University, the Netherlands, where he works for the inter-faculty programme on Institutions, focusing on the sub-theme of innovation. He holds a PhD in Chemistry from McGill University, Montreal, and an MSc in the Social Studies of Science from the University of Manchester. From 2008 to 2012 he was Vice-President of the International Schumpeter Society. He is an Editor of the *Journal of Evolutionary Economics* and his research interests include economic development, evolutionary theories of economic and technological change, complexity and economy. His current projects address differentiation and economic development as well as the impact of innovation and education on the balance between growth and income distribution.

Zulma das Graças Lucena Schussel holds a degree in Architecture and Urban Planning from Universidade Federal do Paraná, Brazil, a Master of advanced studies in Socio-Economic Development at Université de Paris I (Pantheon-Sorbonne) and a PhD in Environment and Development from Universidade Federal do Paraná. She was a Professor of Urban Management, Architecture and Urbanism at Pontifícia Universidade Católica do Paraná till 2016. She has worked on sustainability, urban and regional design, urban and metropolitan planning, environment and urbanisation, and processes of metropolisation. More recently she has directed her research towards the impact of climate change on urban spaces and adaptation of cities to climatic variations. Her latest publications are "Gestão por Bacias Hidrográficas: do Debate Teórico à Gestão Municipal" (in *Ambiente and Sociedade*, 2015) and "Housing Policy: A Critical Analysis on the Brazilian Experience" (in *Journal of Land Use, Mobility and Environment*, 2012).

Paola Spinozzi is Associate Professor of English Literature at the University of Ferrara, Italy. Her research encompasses literature and the visual arts, literature and science, utopia as a literary genre and the sustainability of cultures and the humanities. She investigates the theories and methodologies of verbal-visual studies and is the author of *Sopra il reale. Osmosi interartistiche nel Pre-raffaellitismo e nel Simbolismo inglese* (Alinea, 2005) and of *The Germ. Origins and Progenies of Pre-Raphaelite Interart Aesthetics* (Peter Lang, 2012, with E. Bizzotto). She studies literary representations of scientific theories and is the editor of *Discourses and Narrations in the Biosciences* (V&R unipress, 2011, with B. Hurwitz). She researches utopia, art and aesthetics, imperialism, racism, Darwinism, post-apocalypse and is the editor of *Histoire transnationale de l'utopie littéraire et de l'utopisme* (Paris: Champion, 2008, with V. Fortunati and R. Trousson). She is the Co-coordinator of the International University Network *Routes towards Sustainability* and of the Erasmus+ project *Facing Europe in Crisis: Shakespeare's World and Present Challenges*.

Yen Dan Tong received a Master's degree in Resource and Development Economics from the University of Life Sciences at Ås, Norway, and a PhD in Environmental and Ecological Economics from La Trobe University, Australia. In 2012 she co-ordinated a research project entitled "Cost benefit analysis of dyke heightening in the Mekong Delta" supported by the Economy and Environment Program for Southeast Asia (EEPSEA) and is now a lecturer at Can Tho University, Vietnam. She investigates the economic implications of development policies with reference to large-scale water control infrastructures and rice intensification in the Mekong River Delta. Her latest article is "Rice Intensive Cropping and Balanced Cropping in the Mekong Delta, Vietnam – Economic and Ecological Considerations" (2017).

Gonzalo Valdivieso is Assistant Professor at Pontificia Universidad Católica de Chile – PUC, Campus Villarrica. He is a researcher at the Centre for Local Development (CEDEL) and at the Centre for Intercultural and Indigenous Research (CIIR). He was Director of the Municipality Programme (Puente UC) at the Public Policy Centre of PUC from 2006 to 2010 and has developed several projects on decentralisation, local planning and training at the municipality level. His research focuses on human capital, local development, public policies and decentralisation.

Ana Elena Builes Velez is a Product Design Engineer with a MA in Project Planning and Management. She works as an Assistant Professor at Universidad Pontificia Bolivariana, Medéllin, Colombia, where she is the Coordinator of Research and Postgraduate Studies at the School of Architecture and Design and the principal investigator of the research programme on Fashion, City and Economy. Her work on urban development focuses on social innovation, sustainability and the social, cultural and economic impacts of urban transformations in Latin-American cities. Her publications include "Nuevas Maneras de habitar la ciudad como resultado de las continuas migraciones intraurbanas y las transformaciones sociales y culturales" (in *Arquetipo Magazine*, 2014) and "Comprendiendo la innovación social sostenible. La necesidad de un lenguaje común" (UPB, 2013).

Series editor introduction

Katriina Soini and Joost Dessein

Finding pathways to ecological, social and economic sustainability is the biggest global challenge of the twenty-first century and new approaches are urgently needed. Scholars and policymakers have recognised the contribution of culture in sustainability work. "Cultural sustainability" is also being increasingly discussed in debates in various international, national and local arenas, and there are ample local actor-driven initiatives. Yet despite the growing attention there have only been very few attempts to consider culture in scientific and political discourses of sustainability in a more analytical and explicit way, probably as a consequence of the complex, normative and multidisciplinary character of both culture and sustainability. This difficulty should not, however, be any excuse for ignoring the cultural aspects of sustainability.

The series "Routledge Studies in Culture and Sustainable Development" aims to analyse the diverse and multiple roles that culture plays in sustainability. It takes as one of its starting points the idea that culture serves as a "meta-narrative" which will bring together ideas and standpoints from an extensive body of sustainability research, currently scattered among different disciplines and thematic fields. Moreover, the series responds to the strengthening call for inter- and transdisciplinary approaches which is being heard in many quarters, but in few fields more strongly than that of sustainability, with its complex and systemic problems. By combining and confronting the various approaches in both the sciences and the humanities and in dealing with social, cultural, environmental, political and aesthetic disciplines, the series offers a comprehensive contribution to the present-day sustainability sciences as well as related policies.

The books in the series take a broad approach to culture, giving space to all the possible understandings and forms of culture. Furthermore, culture is not only seen as an additional aspect of sustainability – as a "fourth pillar" – but rather as a mediator, a cross-cutting transversal framework or even as new set of guiding principles for sustainability research, policies and practices.

The essence of culture in, for and as sustainability is being explored through the series in various thematic contexts, representing a wide range of practices and processes (e.g. everyday life, livelihoods and lifestyles, landscape, artistic practices, aesthetic experiences, heritage, tourism). These contexts concern urban, peri-urban or rural settings, and regions with different socio-economic trajectories. The perspectives of the books will stretch from local to global and cover different temporal scales from past to present and future. These issues are valorised by theoretical or empirical analysis; their relationship to the ecological, social and economic dimensions of sustainability will be explored, when appropriate.

The idea for the series was derived from the European COST Action IS1007 "Investigating Cultural Sustainability", running between 2011 and 2015. This network was comprised of a group of around 100 researchers from 26 European countries, and represented many different disciplines. They brought together their expertise, knowledge and experience, and in doing so, they built up new inter- and transdisciplinary understanding and approaches that can enhance and enrich research into culture in sustainable development, and support the work of the practitioners in education, policy and beyond.

Wellbeing of nature and humans, and in particular their interrelationship, is at the core of sustainability. Yet, as a goal it is highly ambiguous and raises a number of questions such as, what is wellbeing, whose wellbeing is in question, how to enhance wellbeing, and how to determine its sustainability. In exploring these questions, it is revealed that sustainable wellbeing is, in the end, culturally embedded. This book advances our understanding of these questions by clarifying the historical and philosophical roots of the wellbeing discourse and by discussing various approaches ranging from socio-economic perspectives to the quality of life. It also leads us to learn about the complementary and sometimes contradictory policies that aim for enhancing wellbeing. The conceptual and political aspects presented in this book are explored through cases from interdisciplinary perspectives, and show that wellbeing is a result of various local and global processes in different geographical and cultural settings. Although sustainable wellbeing is, indeed, a complex issue, the conceptual frameworks and empirical examples provided in this book convince readers that pathways for ecologically sound and socially just wellbeing can be found.

Pillars and circles

Wellbeing at the core of sustainability

Paola Spinozzi and Massimiliano Mazzanti[1]

I A methodological manifesto

In the last few decades, the notion of sustainable development has acquired a wide assortment of meanings. Appropriated and exploited, it has become ubiquitous, high-sounding and formulaic. Mentioned de rigueur as the primary goal for ensuring the future of humankind, it serves rhetorical and instrumental purposes as well as arouses criticism. Definitions of sustainability have been affected by constraints deriving from a tripartite model, in which economics stands out as the chief domain, while the environment and society are forever engaged in confrontation.

The Circles of Sustainability have introduced a broader and more balanced view by presupposing a quadripartite structure that includes economics, ecology, politics and culture (Hawkes, *The Fourth Pillar of Sustainability*; James *et al.*). While the production, use, and management of resources is still a major object of enquiry, over the last decade increasing attention has been directed towards the role of humans as producers, users and custodians of cultures which demand to be sustained (UNESCO).

The claim that sustainability and wellbeing should be integrated within a cultural discourse presupposes a view of humans as makers and consumers in a post-human age. This very view has generated an interdisciplinary dialogue between the authors, a literary scholar and an ecological economist who share the assumption that sustainable development and wellbeing are interrelated because they comprehend the duties as well as the rights of humankind to sustain and enjoy nature and culture. We support cross-fertilisations between positivist and interpretivist methods of enquiry and propose a methodological framework formed by three parts: theoretical and historical approaches; discussion of policies and institutions; presentation of a wide range of case studies across the world and from diverse disciplinary fields.

Inequality of access to resources and capabilities has been rediscovered as a primary barrier for a balanced development of societies (Neumayer; United Nations Development Programme; Piketty). We believe that a focus on social welfare is an alternative to socio-economic approaches to sustainable development based on the pursuit of growth per se. The goal is to increase the quality of life, to be assessed according to the quality of the environment (air, soil, water), the level of security, safety, health, the availability of essential social services, the possibility to appreciate intangible values and spiritual aspects of life. A more advanced view of sustainable development thus considers investments in natural, technological and human resources as the main pillar of enhanced welfare possibilities for future generations. The "human development" approach emphasises the role of education, health, environment and gender as factors of sustained capabilities to generate wellbeing. Thus, we argue that sustainable wellbeing can be achieved only by ensuring that technologies develop and, concurrently, cultural diversity, heritage and history are cultivated in different geopolitical scenarios.

The complex range of interactions between sustainability and wellbeing evoke a Proteus-like and Janus-like object of investigation that incorporates diverse aspects and requires historical insight into models adopted in previous epochs of civilisation as well as a predictive ability to anticipate possible outcomes. In order to understand sustainable wellbeing as a dynamic concept that involves material as well as immaterial aspects, it is necessary to define and strengthen the idea that sustainability is situated within cultural discourses (Benhabib) and its public understanding requires efficient modes of communication and narratives. Far from being clearly defined wholes with permanent sets of values, cultures nurture always new assumptions and expectations about life quality, comfort, welfare, happiness, security, safety.

A culture of wellbeing presupposes that economic health, social equity and cultural vitality are seen as interdependent, encompassing tangible and intangible manifestations. In order to situate biological and cultural diversity in a broader vision of sustainable development, it is necessary to understand that wellbeing is multifaceted. It derives from health and security, economic capabilities, productivity, commodities. While requiring objective factors, it involves subjective, emotive components, such as the ones poignantly defined by Arjun Appadurai as "the spirit of participation, the enthusiasm of empowerment, the joys of recognition and the pleasures of aspiration" ("Diversity and Sustainable Development" 16; Sennett).

Owing to its multiple facets, sustainable development has raised questions as to the domains that should be sustained. Critical revisions and redefinitions have broadened frameworks, programmes, and good practices. However, the pursuit of inclusivity has backfired by increasing the level of imprecision and ambiguity of the term. The choice of elaborating an innovative view of sustainability that revolves around material and immaterial wellbeing is based on the assumption that life quality, comfort, happiness, security, and safety always posit humans as both recipients and agents. Risk and resilience in contemporary societies define the intrinsically human ability to make and consume, to act and adapt, driving the search for and fruition of wellbeing. How to sustain the dual process of exploitation and regeneration is a task that requires integrated approaches from the sciences and the humanities tracing a worldwide cartography with clear localisations.

Wellbeing and sustainability call for further conceptual and political work and advocate the importance of philosophical enquiries into human capacities in ontological and epistemological terms. Identifying and examining different theoretical perspectives on sustainability helps assess how they affect local praxis. Definitions of sustainability related to wellbeing need to consider specific contexts from the nineteenth and twentieth centuries to the second millennium. Multiple meanings and ambiguities of the two key terms emerge by comparing sustainable lifestyles theorised in nineteenth-century Great Britain, where industrialisation first began, and contemporary outcomes. A diachronic approach is necessary to retrieve the roots of social, political, and cultural mindsets attached to good life and its sustainability at the beginning of the Industrial Revolution, to understand how they have changed owing to the development of capitalist societies and to speculate on how they will evolve along with scientific and technological progress and economic innovations. Speculative literary genres such as utopia, dystopia and steampunk fiction can be powerfully predictive by envisioning new forms of wellbeing in modern and contemporary novels that conjure up vivid post-apocalyptic scenarios. Viewed through the lens of environmental and ecological economics, sustainable wellbeing invites more sustainable and human-focused economic practices promoted through bottom-up initiatives such as voluntary reduction of consumption, sustainable and ethical productions, voluntary work, reciprocity activities. Basic choices on climate policy instrumentation can support, or be detrimental, to the development of pleasant and healthy cities and ultimately the good life for citizens in the broadest sense (Syse and Mueller). Sociological research and reports on environmental awareness indicate that the level of education and self-employment strongly influence public awareness and knowledge of biodiversity issues. Civic engineering and urban planning show how urbanisation and the costs of building technology and housing transform the environment, alter life conditions and affect wellbeing. Political programmes informed by strong cultural policies are needed to address national identities and ethnicities, and enhance social cohesion, above all in turbulent countries, and improve the citizens' sense of belonging. Urban sociology, planning and cultural geography must strengthen research on the protection of ancient symbolic systems and the preservation of cultural identities and memories, with a focus on post-disaster reconstruction. They must also study how the material and cultural wellbeing of local communities can be enhanced by investing in forms of sustainable tourism that shun commodification and preserve the cultural heritage.

Speculative and historical approaches should be complemented by outstanding case studies delving into social structures, political agendas and cultural contexts across the continents. We shun a top-down EU-centric approach by claiming that attention must be directed towards sustainability thinking and policies in local case studies connected to international discourses and narrations. As developing countries and emerging economies are strongly involved in the pursuit of sustainability targets set by governments and international institutions, global-local integrated approaches are crucial. Interdisciplinary projects have gained a wide circulation within numerous funding programmes (such as Horizon 2020); interdisciplinary journals devoted to sustainability have flourished and increased in number and quality; English-speaking countries have first paved the way to fully interdisciplinary sustainability courses from various angles, such as environmental economics and ecology, humanities and urban planning. Such experiences, above all in emerging countries, need focusing on frameworks that are more specific.

The humanities, architecture, urban planning, the social sciences and life sciences can share their methodologies while investigating green economy, resilience and urban creativity, (un)sustainable societies of the future, sustainable cultural heritage and tourism. In particular, environmental economics, literary and cultural studies, ecology, chemistry, bioethics, engineering, urbanism, cultural geography, sociology and anthropology are well equipped to develop multifocal, hybrid discourses of sustainable wellbeing.[2]

II Wellbeing and economics

Economics and more specifically political economy and economic policy can contribute to the development of a more synergic scientific environment. First, political economy derives from a branch of moral philosophy. Adam Smith is a transdisciplinary thinker, as are other classical social scientists. In addition, classical economists, many of whom are no longer included in university courses,[3] examine social phenomena with eyes wide open. The writings of Thomas Robert Malthus, John Stuart Mill, and Karl Marx connect economic disciplines, sociology, and anthropology. Mid-nineteenth century utilitarianism and mathematical reasoning gave rise to economics, or better microeconomics, which studies the (rational?) behaviour of people and firms, and the public agent which intervenes with policies in order to correct market failures such as overproduction of pollution or underprovision of innovation and training. The analysis of behaviour, aimed at understanding how individual and social welfare can be enhanced, involves sustainability insofar as it tries to understand why the welfare of consumers and firms may not parallel the achievement of social welfare (Koch and Mont). The influence of Benthamism is evident.

The ways in which economists represent reality as a bi-dimensional world, using simple curves in a graph, have been criticised as naive. It is indeed fundamental to check whether the simple assumptions chosen to support models are empirically relevant. This applies to many other non-formal ways of modelling reality, such as the ones adopted in sociology and philosophy. Simplicity, in terms of Occam's razor, is often useful. Instead of claiming to represent the complexity of reality, models should provide environments where real world facts are consistently analysed in a variety of non-specific settings. Interpretations of reality do need various models and yet in contingent and extreme situations they can produce fallacies that contribute to engender undesirable policies. Limits arise because models derive from assumptions and develop in a specific historical time. How models respond to dynamic change can be better understood by embracing multi-, inter- and trans-disciplinary projects. By recognising the scope and constraints of their disciplines, scholars are stimulated to explore new avenues of research, reopen locked gateways and discover unused properties.

Microeconomics and macroeconomics work on different assumptions and adopt separate tools to interpret socio-economic issues. An integration between the two would facilitate interactions with other disciplines and elude the risk of a partial and misleading dialogue with a part of the whole. Two examples can elucidate this point. Within the microeconomics field, welfare economics, if broadly intended, can dialogue with psychology, anthropology, medicine, and engineering, as far as the analysis of people's behaviour, technology and public goods production is concerned. Welfare economics is about setting policies to increase and also measure social welfare, including goods that are not usually priced in markets such as biodiversity, CO_2, health, et cetera. The typical monetisation of non-market goods is useful for planning, project appraisal and policymaking, but should be implemented when possible and practicable. Within interdisciplinary projects, monetisation can be avoided if it is not possible or not well received. As Ece Ozdemiroglu and Rosie Hails point out in *Demystifying Economic Evaluation*:

> Economic analysis is not a replacement for social or political debate. The best practice should be to use all sorts of high quality evidence to support better decisions – including different interpretations of 'value' of resources and our choices. 'Value' has different meanings in different contexts as defined by different disciplines.
>
> (1)

> Do not aim for a single number that claims to answer all questions. Be open about uncertainties and assumptions. Use sensitivity analysis to show how sensitive the results are to key factors and assumptions. If there are no quantitative data or methods available, describe impacts qualitatively rather than leaving them out of the analysis.
>
> (11)

> Present economic value evidence as part of the three-stage process, together with qualitative and quantitative assessments of change. Also remember economic value evidence is only one of many inputs to the decision about which option might be the best.
>
> (12)

This is an example of how economics as a social science can set a broad, flexible and open framework, which recognises the strengths and limits of disciplinary toolkits mainly aimed at quantitatively monetising non-market values (Mazzanti).

Macroeconomics can productively interact with other social sciences and the hard sciences by broadening a paradigm of economic growth measured by technology, human capital and man-made capital in order to encompass "capital forms" which include social capital and institutional quality, renewable and non-renewable natural resources, and capture wellbeing components as far as possible. Growth thus becomes an instrument to achieve a better and more inclusive life "for everyone" (United Nations Development Programme). Poverty and all forms of social inequality, gender and ethnicity issues, health and life expectation, literacy and the quality of the environment are the major concerns[4] of human societies, above all considering the alarming population increase, soon heading towards nine billion, while richer areas witness growth decrease and a transition towards an ageing society, the management of which depends on migration issues.

Stefan Speck has recently declared that there is a need to analyse:

> the future potential for environmental tax reform in view of the need to integrate the economic and environmental demands while taking into account social inclusiveness. Demographic change will have significant budgetary implications for some EU Member States, creating pressures for increased expenditure (e.g. on pensions and health care) while undermining revenues due to a reduction in the labour force.
>
> (1)

This exemplifies how macro-issues could be scrutinised through multi-dimensional and heterodox "models".

A pivotal interaction involves economic policy, planning and communication. Communication sciences and the humanities are essential to strengthening the feasibility and acceptability of public projects and policies. First, it is necessary to draw attention to welfare enhancements, which may be hidden behind other factors; second, although the "change" is overall desirable from a social point of view, the allocation of costs and benefits among stakeholders is often uneven. In addition, most wellbeing-oriented public policies act over a long-term scenario, which should be "explained" to current generations bearing substantial investment costs. Identifying who the winners and losers are, clarifying how governments intend to compensate the losers, defining what kind of shared resources are targeted by policies and projects are the most important issues an effective social-oriented communication must tackle. Environmental policies present many criticalities connected to long-term, diversified national and global factors. "Optimality" becomes a meaningful term only when environmental effectiveness, cost-effectiveness and feasibility are jointly discussed and integrated within policy narratives.[5]

III   Socio-economic discourses of sustainable wellbeing

Economic methodologies offer econometric analyses, formal modelling, mathematical analysis, case studies, and narrative analysis to share with other disciplines.

Empirically speaking, what the macroeconomics of sustainability has shown is that intangible assets are more important than other forms of capital (e.g. man-made or natural) across income levels. Along with drivers of growth and welfare, human, social and institutional capitals play a key role. Thus, even from an economic angle it is relevant to place culture and sustainability as pillars of integrated methods, approaches and policy thinking. As David Throsby strongly argues:

> Culture is seen to have both an instrumental role in promoting economic progress and a constituent role as a desirable end in itself, the characteristic of civilisation that gives meaning to existence. These two roles for culture are construed as consistent with, respectively, a narrow view of economic development (in which culture may play an instrumental role, assisting or impeding development) and a more all-encompassing view of human development (in which culture plays a constituent role, being valued for its own sake).
>
> (9)

Sustainability calls for diversity: the diversity of capital forms behind growth and their complementary nature, the diversity of sustainability definitions that constantly evolve, and the diversity of policy models. The value of biodiversity and ecosystem services, crucial for analysing environmental sustainability, can be used as a general social and cultural metaphor capturing the abundance of methods, critical perspectives and stakeholders. Without that abundance, the long-term implementation and enforcement of sustainability in globalised, yet heterogeneous and regionalised societies is at risk (Werlen).

Knowledge and technology are historical drivers of wellbeing. They bring together intangible and man-made capital, with knowledge being an intangible component concretely represented by research and development expenditures, which often cannot be separated from human capital investments. Since the beginning of industrialisation, technology has been perceived as a creative and destructive force. Technological revolutions drive change (2010): after the fifth technological revolution, characterised by the introduction of ICT, the transition towards resource efficiency and future climate change targets could induce a sort of "sixth technological revolution" an "age of low carbon – resource efficient economy". All in all, the evolution of capitalist society has so far generated a net benefit out of technological development, possibly including the last wave of radical changes such as digitalisation and information technology, in terms of welfare and employment. The effects of technology on welfare are always an open research question. With regards to the green economy transition, for instance, technological development is a key force in reducing emissions and use of resources (Montini and Mazzanti; Costantini and Mazzanti). In addition, the skill-biased technology-driven effects of the green transition are also potentially strong (Consoli *et al.*). Technology thus continues to create the need to assess its human, cultural, economic and environmental implications, at present as well as in the past and future. Along with the green economy transition, new issues are radically emerging while technological development accelerates, supported by knowledge spillovers and international market development. Melanie Arntz *et al.* examine the new phase of robot-wise automatisation and its effects on skills contents and labour markets:

> First, the utilisation of new technologies is a slow process, due to economic, legal and societal hurdles, so that technological substitution often does not take place as expected. Second, even if new technologies are introduced, workers can adjust to changing technological endowments by switching tasks, thus preventing technological unemployment. Third, technological change also generates additional jobs through demand for new technologies and through higher competitiveness. The main conclusion from our paper is that automation and digitalisation are unlikely to destroy large numbers of jobs. However, low qualified workers are likely to bear the brunt of the adjustment costs as the automability of their jobs is higher compared to highly qualified workers. Therefore, the likely challenge for the future lies in coping with rising inequality and ensuring sufficient (re-)training especially for low qualified workers.
>
> (4)

Sustainability is all about infra- and intergenerational inequalities. The disruptive effects of technological development must be anticipated by sustaining a continuous investment in education and training. This strategy intrinsically aims at sustainability, given that societies tend to accumulate more capital assets year by year and reduce inequality while they develop. Evidence shows that economic development is associated with lower inequality and better environmental performances (United Nations Industrial Development Organization). Far from being part of a "natural" evolution, growth, inequality and the environment are dependent on factors such a policies and institutional quality, which are likely to vary across countries (Kasuga and Takaya). Considering the current economic stagnation in many high-income parts of the world and the associated rising inequalities especially within countries, it is worth looking at the evolutions that characterise growth, inequality and green economy with an eye to the role of regions and municipalities. In the globalised world, countries might paradoxically end up being less relevant. First, it is increasingly difficult to set up international agreements that satisfy all countries. Second, most social and environmental issues are dealt with by local institutions through policy competence. Third, differences across regions are increasing. The decentralised levels of governance are regaining momentum and shaping the global picture. Well-being is linked to effective and participated policy efforts at a local level and well-designed international agreements that set the scene, with countries operating to enhance coherence across actions and observing whether the various local development paths converge or diverge.

GDP indicators, though relevant for accounting research and iconic in our societies, are merely an instrument towards the goal of achieving higher social welfare. Inequality is surprisingly still a broadly overlooked issue in the policy agenda, even within environmental and ecological economics. Jamie Morgan argues that, while Piketty's work has certainly been able to bring inequality back to the foreground, his critical statements remain detached from any ecological considerations.

Tim Jackson and Peter A. Victor contrast Piketty's classical hypothesis that slow growth is related to increasing inequality. The correlation is certainly complex. They find and propose some conditions under which slow growth (degrowth) may reduce inequality. This debate again draws attention to the "quality" of growth. On the one hand, it involves diverse socio-economic components such as welfare, happiness, human development, and inequality, on the other it depends on policy/institutional factors such as social innovation and ecological policies that could bring about different growth and development paths.

James K. Boyce explains how, in addition to income distribution, other inequalities in terms of capability and power can affect the quality of the environment. Environmental degradation and inequality may well be interrelated. Mariano Torras' hypothesis is that greater (income and power) inequality may generate inferior environmental and health outcomes. The environment, innovation and inequality are among the main policy themes through which socio-economic discourses can fruitfully engage with other narratives. As definitions and goals pose environmental, social, economic and cultural challenges, it is imperative to build up sound operational discourses of sustainability and wellbeing.

IV Cultural footprints of sustainability and wellbeing

Research on sustainability and wellbeing invites focusing on footprints, intended as multiple layers to be defined with regard to diverse, interconnected aspects of life on Earth. The notion of multiple footprint relates to biological diversity and cultural diversity, which are interrelated and also distinct. Gianfranco Franz suggests that multiple footprints should be defined as a latticework in which the Anthropocene, Planetary Boundaries, climate change, and resilience form an intertwined discourse ("Multiple Footprints"). In *Geologia stratigrafica*, the second volume of *Corso di Geologia* (1871–1873), the Italian Catholic priest, geologist and palaeontologist Antonio Stoppani defined the origin of humankind in terms of its impact on Earth:

> non dubito di proclamare l'era antropozoica. La creazione dell'uomo è l'introduzione di un elemento nuovo nella natura, di una forza affatto sconosciuta ai mondi antichi … questa creatura veramente nuova in sè stessa, è anche pel mondo fisico un nuovo elemento: è una nuova forza tellurica che, per la sua potenza o universalità, non sviene in faccia alle maggiori potenze del globo.… L'era antropozoica è un'era incominciata … Quando diciamo era antropozoica non guardiamo allo scarso numero dei secoli che furono, ma a quelli che saranno. Nulla al certo ci fa sorgere nemmeno il sospetto che il seme d'Adamo sia vicino a spegnersi; ché l'umanità è troppo bambina

a fronte di quell'ideale di perfetta civiltà di cui non invano il primogenito degli uomini avrà piantato il germe sulla terra ... la terra non uscirà dalle mani dell'uomo, se prima non sia tutta profondamente istoriata dalle sue orme.

(732, 740)[6]

The idea of a new geological epoch marked by the appearance and intervention of humans on Earth has evolved over 150 years, as Nobel Prize Paul Crutzen explains in "Geology of Mankind". He acknowledges the contribution of Antonio Stoppani and points to the term noösphere introduced by Pierre Teilhard de Chardin in *L'Hominisation* (1923) and Vladimir I. Vernadski in *Biosfera* (1926) to define the ability of the human brain to shape its own future and environment. The magnitude of the impact has been sharply marked and in the past three centuries its acceleration has been exponential owing to a relentless exploitation of resources (23).

It is the role of humans as makers that requires closer inspection in discourses of sustainability. Sustainable wellbeing presupposes understanding how the behaviours and decisions of social groups as well as individuals can produce visible or invisible environmental, social and ecological effects; it requires raising awareness about "unsustainable consumption patterns, destructive production practices, malfunctioning governance structures, and financial systems that prioritize short-term returns" (World Wildlife Fund, 88). While anthropogenic impacts are measured, monitored and mitigated, the Planetary Boundaries framework must circulate, be taught and understood. A global paradigm shift depends on effective communication and public understanding. The humanities and the social sciences are thus expected to share the paramount task of shedding light on mental and societal models, value systems and individual behaviours, which are at the core of change. To repurpose economic goals, contrasting short-term profit; to ensure that companies accept social and environmental responsibility; to shift the emphasis from material goods to immaterial benefits, from consumer to intangible culture; to promote more sustainable leisure and food habits: these are the agents of change allowing humans and nature to coexist and thrive.

Since 2005 the United Nations have defined the cultural environment, shaped by linguistic and cultural diversity, as a fourth pillar of sustainable development. Since 2007 the Forum d'Avignon has defined cultural investment as a dimension of the wellbeing of citizens and investigated its impact on territorial attractiveness. The Forum d'Avignon has proposed that the cultural footprint should be measured according to four parameters: cultural intensity, academic intensity, economic performance and pride of belonging. Main objects of enquiry are the relationships between economic performance of a territory and its culture and academic intensity, the lever effect of public cultural expenses on GDP, the strategies of cities in cultural matters with a focus on jobs and training courses, and the impact of culture on local economies (Forum d'Avignon 1).

The 2013 meeting of the Asia–Europe Foundation – ASEF in Hanoi offered Asian and European cultural actors a significant opportunity to discuss the dynamics of creative economy and its limitations, bringing out the need of "creative ecology". Based on the belief that culture and creative industries have a positive effect on economic sectors, creative economy fosters synergies between culture and its industries such as fashion, gastronomy, tourism and sports. Creative economy requires specific cultural policies, as clarified by Anh Tuan HO, the Minister of Culture, Sports and Tourism of Vietnam, who declared that the Socialist Republic is willing to move from a logic of manufacturing to a service economy based on creative economy, including tourism. However, creative economy involves public regulation as it requires defining legal frameworks, allocating funds to support cultural diversity, setting up plans for education, identifying the recipients among states, regions, multinationals, consumers, and measuring to what extent it actually becomes a sustainable economic reality for the people. A creative economy that incorporates social cohesion and individual responsibility should coincide with creative ecology, aimed at developing the diversity and plurality of cultural ecosystems (Kaltenbach).

In *Approaches to a Cultural Footprint. Proposal for the Concept and Ways to Measure it* Jordi Baltà Portolés and Elna Roig Madorran discuss political implications, especially with regards to protecting and promoting cultural and linguistic diversity in Europe. Intangible components, such as beliefs, languages and ways of life, and tangible elements, such as the arts, point to a dynamic notion of culture which encompasses the individual and collective dimension. In particular they refer to linguistic diversity, intangible heritage, creative expressions and examine how each of these factors contributes to defining the contemporary notion of sustainable development. The transversal effects of culture and its interrelationship with other strategies and policies that have a bearing on individual and collective development call for tools aimed at measuring their impact on economy, spatial planning, education and communication. On the one hand, cultural footprints often show their impact in developing countries, on the other hand European minority languages and cultures also offer an extremely diverse internal landscape. Globalisation engenders cultural fragility but also creates new resources and opportunities to give expression and visibility to linguistic and cultural diversity (8). Significantly, the authors do not situate the concept of cultural diversity within an exclusively conservative territory. Nor do they magnify cultural niches:

> interaction also offers the possibility of a mutual learning process. We must therefore consider the possibility that, in certain contexts, an external 'cultural footprint' can have a positive effect on a certain region's resources and expressive and creative capacities, and give rise to a win-win situation.
>
> Finally, there is one last difference between diversity in the biological sense and diversity in the cultural sense, which relates to their measurement mechanisms. The "ecological footprint" paradigm has succeeded in proposing an assessment methodology based on clear and easily quantifiable identification of the raw materials that make up a region's biological capacity or

biocapacity. In contrast, the "raw materials" that determine a region's or community's cultural and linguistic capacity are multidimensional and often difficult to measure.

(15)

The definition of cultural footprint by Baltà Portolés and Roig Madorran lends itself to further elaboration. Culture is pervasive and its centrality in human, economic and social development calls for an assertion of its fundamental role within sustainability. A holistic and interrelated view of the various dimensions of development is essential. The need to assess the cultural dimension of development while critiquing the process of globalisation that began in the 1990s is a reaction to the tensions caused by globalisation itself. It is also, and more importantly, an intellectual process aimed at redefining and sharing worldwide scenarios from which it is no longer possible to step back.

The ecological footprint has become a tool particularly effective and advanced in terms of representation and communication; in comparison, the cultural footprint has not yet reached a similar elaboration of theories and methods. It does require further conceptualisation and debate. In order to develop cultures of sustainability and wellbeing it is necessary to achieve a number of goals: research on cultural footprints requires comparative perspectives encompassing urban studies, environmental economics and literary studies; footprints in medium-sized European towns and sites with high density of cultural heritage should be thoroughly mapped; how multiple footprints can interact in order to produce regeneration and refunctionalisation must be investigated; interdisciplinary courses and training programmes for scholars and students of cultural heritage are needed.

Articulated in theoretical ideas and along a historical continuum, strengthened by insights into environmental economics and policies, and confirmed by emblematic case studies, cultures of sustainable wellbeing find fertile ground for development in novel forms of cultural creativity and resilience, preservation as well as creolisation, management and adaptation.

Wellbeing, quality of life, slow living, diversity, interplay of individuals, participation, citizenship, nature and the built environment, risk society, vulnerability, post-apocalyptic societies, post-disaster planning, resilience, cultural memories, green economy, cultural and ethnic tourism invite scientists and scholars to rethink their roles, starting from the authors' exploration of the environmental humanities and ecological socioeconomics.

Notes

1 This chapter arises from an exchange of critical perspectives and ideas. Paola Spinozzi is the author of sections I and IV. Massimiliano Mazzanti is the author of sections II and III.

2 The elaboration of interdisciplinary, intercultural, transnational frameworks and practices is the major aim of *Routes towards Sustainability*, www.routesnetwork.net, founded by the University of Ferrara in 2012 and formed by a broad range of universities across five continents. The members of the network have co-ordinated annual

symposia, general meetings, conferences, workshops, seminars; set up exchange and double degree programmes at MA and PhD level; interacted with policymakers and developed Third Mission activities in the areas of technology transfer and innovation, continuing education and social engagement.

3 While the history of economic thought has lost ground, mainstream theories have become predominant. Paradoxically it seems easier to teach and learn maths-oriented economics than direct multi-skills efforts towards more heterogeneous discourses involving maths, philosophy, and history.

4 It is worth noting that the UN Sustainable Development Goals, sustainabledevelop ment.un.org/?menu=1300, indicate a slight "return to the past", as the focus is on separate sustainability goals, while the human development approach integrates and measures diverse approaches.

5 See Cecilia 2015. Optimal EU Climate Policy, cecilia2050.eu/publications/171.

6 For the English translation see Etienne Turpin and Valeria Federighi (2012):

> I do not hesitate in proclaiming the Anthropozoic era. The creation of man constitutes the introduction into nature of a new element with a strength by no means known to ancient worlds … this creature, absolutely new in itself, is, to the physical world, a new element, a new telluric force that for its strength and universality does not pale in the face of the greatest forces of the globe…. The Anthropozoic era has begun…. When we say Anthropozoic, we do not look to the handful of centuries that have been, but to those that will be. Nothing makes us suspect that Adam's seed might be close to extinguishing; for humanity is too young if compared to that ideal of perfect civilization of which mankind's first-born has planted the seed, surely not in vain … the earth will never escape the hands of man if not thoroughly and deeply carved by his prints. The first trace of man marks the beginning of the Anthropozoic era.
>
> (36, 41)

Bibliography

Appadurai, Arjun, editor. *Globalization.* Duke University Press, 2001.

Appadurai, Arjun. "Diversity and Sustainable Development". *Cultural Diversity and Bio-diversity for Sustainable Development.* A jointly convened UNESCO and UNEP high-level Roundtable held on 3 September 2002 in Johannesburg, South Africa during the World Summit on Sustainable Development. UNEP, 2003.

Arntz, Melanie, Terry Gregory, and Ulrich Zierahn. "The Risk of Automation for Jobs in OECD Countries: A Comparative Analysis". *OECD Social, Employment and Migration Working Papers*, no. 189. OECD Publishing, 2016, doi.org/10.1787/5jlz9h56dvq7-en.

Baltà Portolés, Jordi, and Elna Roig Madorran. *Approaches to a Cultural Footprint. Proposal for the Concept and Ways to Measure It.* Centre Maurits Coppieters, 2011, http://ideasforeurope.eu/wp-content/uploads/2013/01/CMC_1557_book_culturefootprint.pdf.

Beck, Urich. Risikogesellschaft. *Auf dem Weg in eine andere Moderne.* Suhrkamp, 1986. English translation *Risk Society. Towards a New Modernity.* Sage Publications, 1992.

Benhabib, Seyla. *The Claims of Culture*: *Equality and Diversity in the Global Era.* Princeton University Press, 2003.

Boyce, James K. "Inequality as a Cause of Environmental Degradation". *Ecological Economics*, vol. 11, no. 3, 1994, pp. 169–178.

Burford, Gemma, Elona Hoover, Ismael Velasco, Svatara Janoušková, Alicia Jimenez, Georgia Piggot, Dimitry Podger, and Martin K. Harder. "Bringing the 'Missing Pillar'

into Sustainable Development Goals: Towards Intersubjective Values-Based Indicators".
Sustainability, n. 5, 2013, pp. 3035–3059, http://mdpi.com/2071-1050/5/7/3035/pdf.

Consoli, Davide, Giovanni Martin, Alberto Marzucchi, and Francesco Vona. "Do Green Jobs Differ from Non-green Jobs in Terms of Skills and Human Capital?" *Research Policy*, vol. 45, no. 5, 2016, pp. 1046–1060.

Costantini, Valeria, and Massimiliano Mazzanti, editors. *The Dynamics of Economic and Environmental Systems. Innovation, Policy and Competitiveness.* Springer, 2013.

Crutzen, Paul J. "Geology of Mankind". *Nature*, vol. 415, 2002, p. 23.

Culture: Key to Sustainable Development. How Does Culture Drive and Enable Social Cohesion and Inclusion? International Congress on Culture and Sustainable Development, Hangzhou, People's Republic of China, 15–17 May 2013, http://cdc-ccd.org/IMG/pdf/Culture_and_social_inclusion_Hangzhou_papers_Revised.pdf.

European Environment Agency. *Resource-Efficient Green Economy and EU Policies.* European Environment Agency, 2014.

Forum d'Avignon. *Exclusive Report: Cultural Footprint*, October 2014, http://forum-avignon.org/sites/default/files/editeur/CULTURAL_FOOTPRINT_update_V4_1707.pdf.

Franz, Gianfranco. "Multiple Footprints". Meeting of *Routes towards Sustainability*, Universidad Pontificia Bolivariana, Medellín, Colombia, 27–29 April 2017, www.routesnetwork.net. Paper.

Hamilton, Clive, and Jacques Grinevald. "Was the Anthropocene anticipated?" *The Anthropocene Review*, 2015, vol. 2, no. 1, pp. 59–72, http://journals.sagepub.com/doi/pdf/10.1177/2053019614567155.

Hamilton, Clive, Christopher Bonneuil, and François Gemenne. *The Anthropocene and the Global Environmental Crisis. Rethinking Modernity in a New Epoch.* Routledge, 2015.

Harris, Graham. *Seeking Sustainability in an Age of Complexity.* Cambridge University Press, 2007.

Hawkes, Jon. "Why Should I Care?" *Museums and Social Issues*, vol. 1, no. 2, 2006, pp. 239–246, doi.org/10.1179/msi.2006.1.2.239.

Hawkes, Jon. *The Fourth Pillar of Sustainability. Culture's Essential Role in Public Planning.* Common Ground Publishing, 2001.

Islam, S. Nazrul. "Inequality and Environmental Sustainability". DESA Working Paper, no. 145, 2015, http://un.org/esa/desa/papers/2015/wp145_2015.pdf.

Jackson, Tim, and Peter A. Victor. "Does Slow Growth Lead to Rising Inequality? Some Theoretical Reflections and Numerical Simulations". *Ecological Economics*, vol. 121, 2016, pp. 206–219.

James, Paul, with Liam Magee, Andy Scerri, and Manfred Steger. *Urban Sustainability in Theory and Practice: Circles of Sustainability.* Routledge, 2014.

Kaltenbach, Laure. "After Creative Economy, Here Comes Creative Ecology". 12 October 2013, http://forum-avignon.org/en/after-creative-economy-here-comes-creative-ecology.

Kasuga, Hidefumi, and Masaki Takaya. "Does Inequality Affect Environmental Quality? Evidence from Major Japanese Cities". *Journal of Cleaner Production*, vol. 142, Part 4, 2017, pp. 3689–3701.

Koch, Max, and Oksana Mont, editors. *Sustainability and the Political Economy of Welfare: Perspectives, Policies and Emerging Practices.* Routledge, 2016.

Mazzanti, Massimiliano. "Cultural Heritage as Multi-Dimensional, Multi-Value and Multi-Attribute Economic Good: Toward a New Framework for Economic Analysis and Valuation". *Journal of Socio-Economics*, vol. 31, no. 5, 2002, pp. 529–558.

Mira, Ricardo García, José Manuel Sabucedo Cameselle, and José Romay Martínez. *Culture, Environmental Action and Sustainability*. Hogrefe & Huber, 2003.

Montini, Anna, and Massimiliano Mazzanti, editors. *Environmental Efficiency, Innovation and Economic Performances*. Routledge, 2010.

Morgan, Jamie. "Piketty and the Growth Dilemma Revisited in the Context of Ecological Economics". *Ecological Economics*, vol. 136, 2017, pp. 169–177, doi.org/10.1016/j.ecolecon.2017.02.024.

Neumayer, Eric. *Weak Versus Strong Sustainability: Exploring the Limits of Two Opposing Paradigms*. Edward Elgar, 2003.

Ozdemiroglu, Ece, and Rosie Hails, editors. *Demystifying Economic Valuation*. Valuing Nature Paper VNP04. Valuing Nature Programme, June 2016, pp. 1–12, http://valuing-nature.net/demystifying-economic-valuation-paper.

Perez, Carlota. "Technological Revolutions and Techno-economic Paradigms". *Cambridge Journal of Economics*, vol. 34, no. 1, 2010, pp. 185–202, doi.org/10.1093/cje/bep051.

Piketty, Thomas. "Putting Distribution Back at the Center of Economics: Reflections on Capital in the Twenty-First Century". *Journal of Economic Perspectives*, vol. 29, no. 1, 2015, pp. 67–88.

Salisbury, Frank B., editor. *Geochemistry and the Biosphere: Essays by Vladimir Vernadsky*, Synergetic Press, 2007.

Sennett, Richard. *Together: The Rituals, Pleasures, and Politics of Cooperation*. Yale University Press, 2012.

Speck, Stefan. "Environmental Tax Reform and the Potential Implications of Tax Base Erosions in the Context of Emission Reduction Targets and Demographic Change". *Economia Politica*, i-first, 2017, pp. 1–17, doi:10.1007/s40888-017-0060-8.

Stoppani, Antonio. *Corso di Geologia*. G. Bernardoni and G. Brigola, 1871–1873. 3 vols.

Syse, Karen Luke, and Martin Lee Mueller, editors. *Sustainable Consumption and the Good Life. Interdisciplinary Perspectives*. Routledge, 2015.

Teilhard de Chardin, Pierre. *L'Hominisation (1923)*, in *Oeuvres de Teilhard de Chardin. La vision du passé*, vol. 3. Éditions du Seuil, 1955.

Throsby, David. "Sustainability and Culture: Some Theoretical Issues". *International Journal of Cultural Policy*, vol. 4, no. 1, 1997, pp. 7–20.

Torras, Mariano. "Income and Power Inequality as Determinants of Environmental and Health Outcomes: Some Findings". *Social Science Quarterly*, vol. 86, no. 1, 2005, pp. 1354–1376, http://jstor.org/stable/42956040.

Turpin, Etienne, and Valeria Federighi. "A New Element, a New Force, a New Input: Antonio Stoppani's Anthropozoic". *Making the Geologic Now. Responses to Material Conditions of Contemporary Life*, edited by Elizabeth Ellsworth and Jamie Kruse, Punctum Books, 2012, pp. 34–41.

UNESCO, "Towards a Culture of Sustainable Diversity". *Cultural Diversity and Biodiversity for Sustainable Development*. A jointly convened UNESCO and UNEP high-level Roundtable held on 3 September 2002 in Johannesburg, South Africa during the World Summit on Sustainable Development. UNEP, 2003, pp. 8–11.

United Nations Development Programme. Human Development Report 2016: Human Development for Everyone. UNDP, 2017.

United Nations Industrial Development Organization. *Industrial Development Report 2016*. UNIDO, 2015.

Van den Bergh, and Jeroen, C. J. M. "Environment versus Growth – A Criticism of 'Degrowth' and a Plea for 'A-growth'". *Ecological Economics*, vol. 70, no. 5, March 2011, pp. 881–890.

Vernadsky, Vladímir. *Biosfera*. Nauch. khimiko-tekhn. izd-vo Nauch. tekhn. otdel V.S.N. Kh., 1926.

Werlen, Benno, editor. *Global Sustainability. Cultural Perspectives and Challenges for Transdisciplinary Integrated Research*. Springer, 2015.

World Commission on Environment and Development. *Our Common Future*. Oxford University Press, 1987.

World Wildlife Fund, "Chapter 3. Exploring Root Causes". *Living Planet Report 2016. Risk and Resilience in a New Era*. WWF International, 2016.

Part I

Sustainable wellbeing in theory and history

Prologue

Paola Spinozzi

Far from being defined as straightforward, unambiguous concepts, sustainable development and wellbeing are presented as contested fields. Their porosity has allowed for a proliferation of definitions and hermeneutic approaches. Why wellbeing is essential to sustainable development is explored here from interconnected theoretical and historical perspectives.

What wellbeing is, for whom, and how it can be sustained requires considering expectations, degrees of power and agency expressed by various social cross-sections, political parties, economic interest groups, and exponents of diverse cultural and religious backgrounds. Reaching an agreement about a fair sharing of benefits and hazards associated with the ever-increasing potentials of science-driven technologies and industries presupposes a complex synergy of concepts, methodological approaches and historical awareness. Environmental decision-making requires integrated paradigms for the sciences and the humanities, allowing the ethical implications of scientific research and its applications to be studied along with the sustainability of cultural histories and memories. Sustainability derives from an understanding of culture as the domain in which people's everyday lives intersect with the various structures of social formation – from traditions to class, politics to the personal. Whether our concerns are with the impact of climate change, global agriculture, or fossil-fuel economies, our response is often enacted in a local context and every single day. The food we buy, cook and eat, the means of transport we choose to travel from home to work, the ways in which we spend time with colleagues, in the family and in the community are framed within specific cultural beliefs about production and enjoyment, fruition, consumption and new production.

The goals of sustainability and wellbeing are experienced at the micro-level while simultaneously featuring as priorities in national and international agendas. How these two dimensions interact is at the core of Paul James' and Gonzalo Salazar's chapters. Theories variously informed by positivism or relativism, invoking functionalist agendas or ontological attributes, are examined in order to assess ways in which wellbeing could and should be sustainable.

"Creating Capacities for Human Flourishing: An Alternative Approach to Human Development" develops the idea that social wellbeing and sustainability have become interwoven developmental aspirations but still require adequate theorisation. Being well means being healthy and vital but can also involve individualistic and consumerist pursuits. Dialoguing with Amartya Sen and Martha Nussbaum, Paul James explains why he believes that their notion of capabilities is attached to a singular notion of freedom, posited as intrinsic and universal but also ambiguously portrayed as contingent and thus negotiable. Absorbing the early fifteenth-century term *capacité*, "ability to hold", and the Latin term *capacitas*, "to have breadth", and *capere*, "to take", James's definition of capacity draws a clear distinction from the concept of capability developing from a liberal approach. He proposes an integrated framework based on an ontological understanding of capacities as essential components of the human condition. While offering a broad range of exemplifications as to why capacity is always-already social, James also articulates a lucid meta-discourse, reflecting upon catch-phrases, clichés, and overused concepts of sustainability and wellbeing.

"The Incongruities of Sustainability: An Examination of the UN Earth Summit Declarations 1972–2012" by Gonzalo Salazar tackles sustainability in relation to economic and technocratic objectives as well as environmental and holistic goals. The conceptual indeterminacy of the term adjusts diverse perspectives, actors and scales, but it can also generate political exploitation and disorientation in local planning processes. By examining the UN International Agendas over the past three decades, particularly the Earth Summit Declarations of 1972, 1992, 2002 and 2012 and the role attributed to local planning and practice in a global scenario, Salazar identifies two contradictory perspectives on sustainability in the global agenda: an economic and pro-growth vision committed to the epistemology of progress versus a socio-ecological view committed to a relational epistemology. His arguments, elucidating the benefits of socio-ecologically effective sustainability practices, are connected to Anna Kalinowska's analysis in "Assessing Public Awareness about Biodiversity in Europe" in Part II. As socio-demographic surveys reveal that the level of education and self-employment strongly influence the knowledge of biodiversity issues and the *Natura 2000* network, it is clear that biodiversity can only be preserved through effective communication strategies and public understanding.

The chapters by Wendy Parkins, Pier-Paolo Saviotti and Paola Spinozzi move along a historical continuum spanning from the beginning of the Industrial Revolution in about 1780 to the contemporary age. Industrialisation has led to the greatest increase in productivity in the history of humankind and the challenges of sustainability faced by Great Britain in the nineteenth century still strongly resonate worldwide in the twenty-first century.

In "Slow Living and Sustainability: The Victorian Legacy" Parkins explains how the critique of modernity associated with "slow living" originated in the Victorian period, when a preference for a slower mode of life began to be articulated as an alternative to the dislocation and mechanisation of modernity. William Morris and Edward Carpenter began from the micro-level of everyday life in thinking about how the deleterious global consequences of industrial modernity could be challenged and, they hoped, reversed. They advocated an alternative approach to daily life and work in which the values of nature, beauty, and creativity were posed against the alienation, distraction and dissatisfaction attributed to the growth of Victorian consumer society. Celebrating the handmade and the artisanal, Morris and Carpenter variously began to formulate ideals of sustainability that they believed would enhance everyday life and restore the natural environment. Though permeated by strong ideological and idealistic views, their writings sound pragmatic and feasible; indeed, they show how addressing the problem of sustainability and wellbeing requires connecting vision and context, radical thought and practical agendas. The elements of complexity and simplicity these two Victorian writers associate with a good life resonate in theories of degrowth illustrated by Valeria Andreoni's "Social Equity and Ecological Sustainability through the Lens of Degrowth" in Part II. The DIY (Do It Yourself) lifestyle and the practice of reusing and recycling as a form of ecological intervention indicate how nowadays individuals can take direct responsibility towards sustainable wellbeing.

"Innovation and Consumption in the Evolution of Capitalist Societies" by Saviotti illustrates how the evolution of consumption, work and leisure can be traced back to the rapidly rising productivity and slow progress of wellbeing for working-class people, triggered by the Industrial Revolution. Till the end of the nineteenth century consumption was limited to basic necessities and it was only after the First World War that a wider range of goods and services could be consumed by a growing amount of people. Working hours decreased at the beginning of the twentieth century and continued to decrease, but at a much lower rate, after the Second World War. In 1928 J. M. Keynes foresaw that 15 hours of work per week could cover all future needs and people would devote their free time to "higher" activities. Unlike his predictions, the observed behaviour shows consumption of higher quality and more differentiated goods, but the time for leisure activities is unevenly distributed. Saviotti persuasively explains the apparent paradox of the rapidly rising productivity and slow progress of wellbeing.

Speculations about humanity, the human environment, and its sustainability have been at the core of utopia as a literary genre since its origin in 1516. From the second half of the eighteenth century the focus of utopian writers began to shift from another place to another time. As the temporal dimension became central, classical utopia became uchronia, expressing concern about history and human development. "In a Prescient Mode. (Un)Sustainable Societies in the Post/Apocalyptic Genre" examines modes of foretelling sustainable or unsustainable futures for humankind. The demise of civilisation caused by a global catastrophe and the coming of a new era characterise the ancient, universal topos of apocalypse. Futuristic novels portraying the coming of the new era after a catastrophe are inspired by the apocalypse. Natural or anthropogenic disasters – floods, droughts, fire, earthquakes, total eclipses, climate change, pandemics, famine, nuclear explosions, and collisions of planets – all mark the end of declining societies. By undermining mankind's confidence in eternal progress and dominion of nature, catastrophe in the contemporary age prompts a radical revision of the expectations about scientific and technological advancement. Perception of risk as a defining component of existence has proliferated in present times. Apocalypse and palingenesis in literary texts can be related to the categories of disaster and rebuilding in order to understand how human beings negotiate risk existentially and socially, within the private and public sphere, and how they make sense of, and respond to, catastrophic events. Adopting the categories of risk and disaster to interpret post-apocalyptic fiction shows how literary criticism can enhance our understanding of risk in present-day societies and strengthen our capacity building. Preparedness, management of risk and post-disaster reconstruction, tackled by Spinozzi, are further explored by Gianfranco Franz in Part III. In "Long-term Visions and Ordinary Management: Post-Earthquake Reconstruction in the Italian Region of Emilia" he contends that urban planning can be efficient and effective only if the areas hit by earthquakes are regarded as spaces in which individuals and communities have lived generating specific social and cultural histories. Post-disaster reconstruction can never reproduce what is lost: the culture, the traditions and daily habits of a community are forever changed and the next generations must find ways of metabolising the catastrophic event. Post-disaster sustainability is achieved if tangible and intangible capital assets are recognised as being equally relevant. Mature societies present a large stock of manmade, cultural, social capital. The larger the stock the more intense its potential erosion due to exogenous shocks and lack of investments: diversified investments in all complementary forms of capital are necessary, before and after disasters occur, to create a resilient social system.

I.1   Creating capacities for human flourishing

An alternative approach to human development

Paul James[1]

I   Core capacities

What are the core capacities that make for a flourishing life? It is an incredibly difficult question to answer. Every philosopher, public commentator and back-yard critic seems to have a different view on the matter. Occasionally the terms of what makes for a good life are developed explicitly, but mostly the grounding of such claims is either left implicit or undeveloped. It is as if we all agree on what is good – something that is clearly not true – and spelling out the terms of a good life is unnecessary.

In the Global North the most common appeals assume some variation on the capacities for freedom, connectivity, democracy, and inclusion, with the ideology of freedom usually prevailing. The initiating questions differ. What makes a life worth living? What capacities does a person need to lead a good life? Or what digital capacities should a person ideally have? These questions orient toward the personal and tend to stay focused upon the individual. They are very different from more socially expansive questions. What makes a city liveable? What capacities make for conditions of human flourishing? The first set of questions emphasises individual capacities as the basis of the enquiry; the second set begins with the social as their basis, and includes individual capacities but extends those to the form of social habitation or the conditions of human flourishing as they are lived both variably and relationally.

Our approach begins with the last question: What capacities make for conditions of human flourishing? It suggests that if we can give a working answer to that question, then we have the foundation for answering all those other more narrowly framed or precisely oriented questions. Put the other way around, if we want to know the answers to practical and policy issues such as what makes for a liveable city, what constitutes good digital engagement or what capacities we need to learn in order to live a good life, we need to go back to the basics concerning human flourishing in general. This move will not give us one-to-one or complete answers concerning what should be done – which in any case would partly depend upon differences in time and across place. But at least it will slow down the current tendency towards falsely connected fashion-statements about what constitutes good ways of doing things. Wherever such statement chains begin, they all tend to end in the same ideologically condensed place: the dominance of a certain regime of economics. For example, one recent chain instrumentalises mindfulness: "a good life is mindful", "mindfulness is smart", "smart cities are better cities", "better cities require fast connectivity", "connectivity brings growth", and "economic growth is the only way to increase the quality of life". Providing a different basis to understanding human capacities is the task of the present chapter.

Philosophically, the long history of an interest in what constitutes the conditions of human flourishing has been one of considerable contestation. The written enquiry goes back to the Greek philosophers of the fifth century BCE arguing over the meaning of the polis and the household. Aristotle used the concept of *eudaimonia* as the highest human good – from *eu*, meaning "good", and *daimon*, meaning "spirit". Across the mid-to-late twentieth century, *eudemonia* tended to be translated as "happiness", but more recently the term is equated with "flourishing". A publication that marked a turning point was the special issue of *Social Philosophy and Policy* (Rasmussen) called "Human Flourishing and the Appeal to Human Nature".

Here we use the term *eudaimonia* with a much deeper sense of ontological wellbeing than the term "happiness", but also a broader sense of flourishing than Aristotle's emphasis on the development of an ethos or spirit of wellbeing. For the framework being developed here, it is as much about human embodiment as it is about spirit. Here, the dialectic of wellbeing/adversity replaces the singular search for wellbeing (adversity is not treated as a bad thing). And here, the bases for human flourishing, from spiritualties and subjectivities to practicalities are treated as a way of grounding an alternative *social* approach to human development. This move allows us to bring wellbeing and sustainability, two very different categories of being, into the same integrated framework. Usually these two terms just rub against each other without being theorised in explicit connection.

In short, we begin with social capacities. Other possible names were considered for headlining the approach. "Life-skills" sounded too instrumental. "Life-capabilities" sounded too much as if it came out of a self-help book, and "life-ways" sounded too folkloric. We also wanted to distinguish our framework from the primary writings in this field that congregate around the concept of "capabilities" – namely, the Capabilities approach or liberal Human Development approach (Nussbaum 2011, "Creating Capabilities"; Sen). The concept of "capacity" is used here in the sense of its early progenitors – the early fifteenth-century term *capacité* or "ability to hold" (not necessarily to control), and the Latin term *capacitas*, "to have breadth", and *capere*, "to take". This seemed most usefully able to engage with contemporary debates and to distinguish the approach from the liberal Capabilities approach.

The Social Capacities framework being developed here is situated within a more encompassing approach – the *Circles of Social Life* approach, itself the empirical layer of a form of *engaged theory* (James *et al.*, *Urban Sustainability*; Magee; James, *Globalism*). Therefore, before developing an argument about the basic capacities needed to produce and reproduce the conditions of human flourishing, some background to that approach is first required.

II The *Circles of Social Life* approach

The Circles approach works at the level of empirical generalisation as part of a larger methodology that works simultaneously at different levels of epistemological abstraction. It begins by making the claim that it is heuristically useful to consider all questions of social life holistically across integrated domains of ecology, economics, politics and culture (see Figure I.1.1). The approach offers an integrated method for practically responding to complex issues of sustainability, resilience, adaptation, liveability and vibrancy. One version, for example, takes an urban area, city, community or organisation through the difficult process of responding to complex or seemingly intractable problems and challenges, defined as sets of critical issues.

The Circles approach to cities provides a way of responding to a series of questions. First, how are we best to understand and map the vibrancy of social life in our cities, communities and organisations, treating this question in all its complexity across those four domains – economics, ecology, politics and culture? Second, what are the central critical issues that relate to making the city or community more vibrant? Third, what should be measured and how? Instead of designating a pre-given set of indicators, the approach provides a process for deciding upon indicators and analysing the relationship between them. Fourth, it provides a process pathway for showing how positive responses can be planned. The approach provides a series of pathways for achieving complex main objectives. It offers a deliberative process for negotiation over contested or contradictory critical objectives and multiple driving issues in relation to those main objectives. Finally, it supports a monitoring and evaluation process and a reporting process.

III Conditions, capabilities and capacities

In *Urban Sustainability in Theory and Practice* we began by developing a set of "core conditions" for engaging positively in social life (Table I.1.1). This list was originally developed to be complementary to the *Capabilities* or *Human Development* approach of Sen and Nussbaum. However, in now moving to explore the question of what capacities are fundamental to the human condition, we find that this step now presents us with an awkward closure. While our seven chosen concepts – adaptability, learning, liveability, reconciliation, relationality, resilience and sustainability – can be comfortably stretched to name core *conditions of* human flourishing and development, they can only variably be translated into *capacities for* human flourishing. For example, liveability is indeed a contextual condition of good sociability, but it cannot be translated as a capability or capacity. That is, rather than being a capacity in itself, liveability is an outcome or a condition produced by positive capabilities used well. And,

CIRCLES of SOCIAL LIFE

and beyond

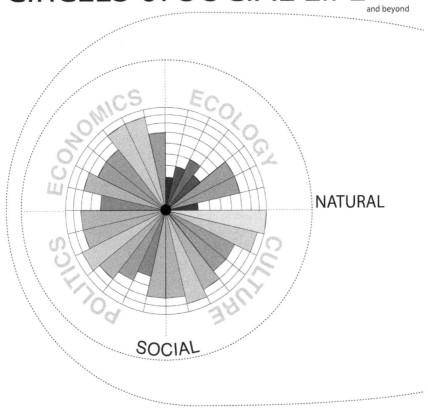

DOMAINS OF THE SOCIAL

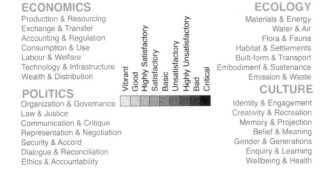

ECONOMICS
Production & Resourcing
Exchange & Transfer
Accounting & Regulation
Consumption & Use
Labour & Welfare
Technology & Infrastructure
Wealth & Distribution

POLITICS
Organization & Governance
Law & Justice
Communication & Critique
Representation & Negotiation
Security & Accord
Dialogue & Reconciliation
Ethics & Accountability

ECOLOGY
Materials & Energy
Water & Air
Flora & Fauna
Habitat & Settlements
Built-form & Transport
Embodiment & Sustenance
Emission & Waste

CULTURE
Identity & Engagement
Creativity & Recreation
Memory & Projection
Belief & Meaning
Gender & Generations
Enquiry & Learning
Wellbeing & Health

Figure I.1.1 Circles of social life: conducted as an urban assessment of the city of Melbourne.

Source: James *et al.*, *Urban Sustainability in Theory and Practice: Circles of Sustainability.* Routledge, 2015.

Table I.1.1 Core conditions for engaging in social life

Core conditions	Definitions of the positive side of these core conditions
1 Adaptability	The ability to adapt to change, including adapting to changes brought about by external forces that threaten the sustainability of conditions of liveability and security.
2 Learning	The capacity to seek knowledge, learn and use that understanding for enhancing social life. When learning becomes reflexive understanding, the highest form of learning, it includes the possibility of acknowledging the profound limits of one's knowledge.
3 Liveability	The life-skills and milieu that allow for living in ways that enhance wellbeing. Liveability includes having the resources to secure social life for all across the various aspects of human security, both in an embodied sense and an existential sense. One of the capacities here is the possibility of debating and planning possible alternative ways of living.
4 Reconciliation	The capability to reconcile destructive or *negative* differences across the boundaries of continuing and flourishing *positive* social differences.
5 Relationality	The capacity to relate to others and to nature in a meaningful way. This includes the capacity to love, to feel compassion, to reconcile.
6 Resilience	The flexibility to recover and flourish in the face of social forces that threaten basic conditions of social life.
7 Sustainability	The capacity to endure over time, through enhancing the conditions of social and natural flourishing.

therefore, we are still faced with the quandary of delineating what are those capacities that produce positive liveability and, more broadly, human flourishing.

The second problem in translating the original set of core conditions from the *Circles of Social Life* approach into capacities is that one of the core conditions in that original list, "relationality", is actually a general condition/capacity that encompasses at least one of the other conditions/capacities listed in our core group. Reconciliation, for example, is a relational capacity: the capacity to reconcile potentially negative differences across social and natural boundaries of continuing and flourishing positive differences. It names, in other words, a particular form of *relation* to others where Otherness is a potential source of conflict. The problem turns on that way that *relationality* is a significantly more abstract and generalising category than *reconciliation*. Relationality encompasses relations of reconciliation and more.

In these terms, the capacity to relate to others and to nature is a first-order core capacity. It includes the capacity to relate to the ecological conditions that surround us and, as embodied beings, makes us human. Reconciliation is a second-order core capacity. This point will become important to how we reorganise our *Social Capacities* framework into first-order capacities and second-order capacities – their subdomains. For such a listing, the categories need to operate at comparable levels of generality. This becomes our *first* consideration in constructing the capacities list. Moreover, instead of having two lists, "core conditions" and "central capabilities", and treating them as independent but complementary, this chapter begins the task of rewriting both conditions and capacities in terms of each other.

While our original list of "core conditions" had problems – problems that we now intend to address – we are now also forced to conclude that Martha Nussbaum's list is critically flawed because of its profound liberal bias. *The* fatal problem comes from her insistence that the *Capabilities* approach is founded upon freedom: "It is *focussed on choice or freedom* [her emphasis], holding that the crucial good [that] societies should be promoting for their people is a set of opportunities, or substantial freedoms" (*Creating Capabilities* 18). Later in the same book she writes: "it is unclear whether the idea of promoting freedom is even a coherent political project. Some freedoms limit others" (71). Hence some freedoms are more important in her approach that others. However, if this is the case – namely, freedom needs to be qualified by just negotiation over what are the important freedoms, then the negotiation and justice are as an important basis for thinking capabilities as freedom.

This is a fundamentally reductive view of the human condition. Moreover, the emphasis on freedom betrays a profoundly modern orientation. The compound problem is that freedom in Nussbaum's hands is both given an intrinsic and primary value (a reductive claim), and, at the same time, the list is treated as a contingent negotiated relation in tension with other virtues such as justice, equality and rights. Both propositions cannot hold. This question of normative grounding then becomes a *second* consideration in constructing the capacities list. By comparison, the *Circles of Social Life* approach works with a delineated series of negotiated tensions or antimonies (James *et al.*, *Urban Sustainability* 77–83):

1 Accumulation/Distribution (currently a dominant theme of contention in the domain of economics)
2 Security/Risk
3 Needs/Limits (currently a dominant theme of contention in the domain of ecology)
4 Freedom/Obligation
5 Participation/Authority (currently a dominant theme of contention in the domain of politics)
6 Inclusion/Exclusion
7 Difference/Identity (currently a dominant theme of contention in the domain of culture)

These couplets are expressed as two-way tensions, but they could be three-way tensions or otherwise, just by adding a third of fourth term: for example, "accumulation/distribution/equality". However, it does not much matter for this purpose, except perhaps for those who for some reason hate dualism. The emphasis here is on tension. The list includes "freedom", sometimes written as the more general and embracing term "autonomy", but the key point is that no concept is left to stand alone (to be individualised) or to become *the* singular basis of human development. By contrast, Amartya Sen treats freedom "as both the primary end and the principle means of development". Therefore, conversely for him "the removal of substantial unfreedoms … is constitutive of development" (xii). His position is not contradictory. Nevertheless, like Nussbaum's, it is profoundly reductive. The *Circles of Social Life* approach suggests rather that there are many negotiated pathways to human development, not just the removal of constraints to freedom. In the era of the Anthropocene, for example the nineteenth-century fetish for freedom is beginning to look more than a little problematic. It could be said that the removal of constraints to freedom on a "flourishing" economy (a tenet of neoliberalism) is contributing to destroying the planet.

This leads to a *third* consideration for constructing the capacities list: the constitutive grounding of the approach. This consideration refers questions about where and how capacities are produced. By comparison with the Capabilities approach, which begins with the individual, our starting point, and therefore our entire approach, is organised in social or inter-relational terms. It includes the personal embodied capabilities that Sen and Nussbaum's approach emphasises but it treats them socially, and as only one layer of the full expression of capacities relevant to persons in social life. It is our argument that human flourishing does not reside predominantly in the *personally* embodied and rational qualities (practical reason) of the individual, but rather depends upon socially framed capacities held by persons and other agents in social engagement: families, communities, institutions, organisations, corporations, states, and so on. That is, persons can have capacities, communities can have capacities, and institutions or organisations can have capacities. But in each case the grounding of those capacities is social rather than just individual. In other words, rather than a methodological individualism or even institutional individualism that focuses on individual persons as the carriers of capacities (Hodgson), the Circles approach takes persons-in-interrelation as its starting point. All capacities are always-already social.

Examination of Martha Nussbaum's list of fundamental capabilities leads us to further considerations for choosing the core capacities. She chooses ten crucial capabilities (*Creating Capabilities* 33–34):

1 Life
2 Bodily health
3 Bodily integrity
4 Senses, imagination and thought
5 Emotions
6 Practical reason
7 Affiliation
   a   Being able to live with and toward others
   b   Having the social bases of self-respect and non-humiliation

8   Other species
9   Play
10   Control over one's environment
   a   Political. Being able to participate effectively in political choices
   b   Material. Being able to hold property

Are ten capacities the right number? Should more capacities be added? Could the schema be organised differently? What are the most important social capacities of the innumerable possibilities? What constitutes a workable list? How are they to be chosen? The Circles approach begins by suggesting that there is no right number, no essential list, and no perfect balance or structure to such a taxonomy. Creating taxonomies is itself a cross-cultural social activity (Bowker and Star).

The author who developed the Capabilities approach did so by adding and subtracting capability-domains until the list seemed "right" and "balanced". Some capabilities have sub-domains; some do not. Our approach also depends upon an author in intense dialogue with others, but, to counter the problem of there being no right number, we begin the other way around by positing the constraining numbered structure before we start. That is, we begin with an empty template of four (unnamed) primary domains of capacities, each with seven (unnamed) subdomains. Starting with four domains, we argue, gives a number that allows sufficient range and complexity without the list of capacities becoming too long and unwieldy at the top level. This is a technique of the Circles approach. Dividing each of those four domains into seven subdomains then gives added complexity while keeping the structure simple enough to operationalise. These subdomains are used in a way that allows for resolution into related elements or constituent parts.

This is precisely the modern definition of *analysis* as a process – the breaking down of an object of enquiry into its elements. However, this is also a reflexive analysis, tempered by the postmodern critique of certainty. By choosing such a contingent restraint as a numbered set, it accentuates the contingency of the chosen number in the first place. This means that there is nothing wrong with a smaller set of primary domains. The Human Development Index, currently the pre-eminent index for measuring human capacities, has three primary indicator-sets. Problems arise here not because of the number three, but because there is no systematic relation between this set of three indices and the ten capabilities posited by the supposedly connected Capabilities approach. Why were these three domains chosen and how do they relate to the longer list of ten capabilities?

This question can only really be answered by talking of the messiness of the political process. And it introduces a *fourth* consideration that needs to be added to our list of considerations for choosing capacities: namely, the capacities need to be chosen *and* ordered to allow the structure of the framework and the structure of its operationalisation to consistently be mapped onto each other. If it is necessary for operational viability that the primary set is small, then the primary set needs be coherent, with each of its elements both necessary to the primary set and representative of a larger secondary set.

The United Nations Human Development Index suggests that the three domains of human development are having a "long and healthy life", "being knowledge-able" and having a "decent standard of living". In fact, the indices that are chosen for each of the domains suggest something much narrower. In the case of the Human Development Index, the indices actually only measure life expectancy at birth, years of formal schooling, and gross national income per capita. Certainly, these could be argued to be proxies for a more complex set of capabilities. But just as the mismatch accentuates the problem of unstructured mapping, it also opens up the issue of arbitrary or reductive assessment and operationalisation. This suggests a *fifth* consideration. Ideally, the chosen first-order capacities – *taken together* as a full list and standing in for the longer list of second-order capacities – need to provide a minimal basis for human flourishing. The three chosen domains of the Human Development Index arguably do not. Years and years of formal education, for example, does not necessarily give one a good capacity for being knowledge-able, let alone for experiencing wellbeing.

Going into the detail of Martha Nussbaum's set, the first six elements in the Capabilities approach emphasise body and mind, focusing on the individual. The sixth domain of practical reason takes only one of the many formations of know-ledge and gives it priority over others. And then with the seventh element, "Affiliation", the social is added on. Finally, the approach moves out to its broadest category, "[individual] control over one's environment". There are many issues here of which we only have the space to discuss a couple. Watene and Robeyns provide other critiques. One issue is that the list does not include basic capabilities such as being able to communicate, or build a shelter, or grow food. A core list of capabilities would not necessarily list such particularities, their importance notwithstanding, but it should take very seriously the technical and technological capacities necessary for producing basic existence. This is then a *sixth* consideration. Technical capacities are only one dimension of the many intersecting capacities that underpin human flourishing that are left out of Nussbaum's list. A *social* capacities framework should provide guiding prin-ciples to projects that involve what has been called "capacity-building". It should do so in a way that includes but goes beyond the usual emphasis on technique or training, but it cannot leave them out.

Still on the content of the Capabilities list, the emphasis in Nussbaum's list on *control over* one's environment suggests a particular kind of instrumentalism. In the Age of the Anthropocene that is neither necessarily positive nor norma-tively defensible – remember the earlier definitional point that *capacité* is the "ability to hold", not necessarily to control. By comparison, the Circles approach describes capacities in ways that puts a dual emphasis on the sociality of the pro-cesses and the making possible of good ecologically embedded life without pre-suming that there is a politically right way. This becomes a *seventh* consideration. The capacities need to be chosen *in a way* that allows for the pos-sibility of arguing about and planning possible alternative ways of living (and at the same time, because it is a lived list, the list of chosen capacities needs to include the capacity to develop such a framework including through negotiating and reconciling social contention). It is here in the content of the list that the modern liberal bias of the Capabilities approach becomes quite stark, including putting what seem to be modern property rights and choice-based participation at the centre of its claim to basic human development.

Collating all the problems and issues considered across the course of this chapter, we can now enumerate seven key considerations that we need to take into account in structuring and choosing the capacities:

1   The first-order categories chosen as the most critical ways of describing the complex range of capacities – operating at the same level of generality as each other and acting to name a constellation of second-order capacities – need to operate in a consistent way in relation to the related sets of second-order categories, also themselves operating at a common level of generality;
2   The normative grounding of all the chosen capacities needs to avoid a reductive emphasis on a singular normative value such as freedom;
3   The constitutive grounding of capacities needs to be understood as always-already social rather than intrinsic to individuals;
4   The domain structure and the assessment structure need to be consistent, allowing the named core capacities to be mapped consistently onto a non-reductive and non-distorting set of indicators of positive human development;
5   The chosen first-order capacities, *taken together*, need to provide a minimal basis for human flourishing;
6   The chosen capacities need to encompass the full range of human capacities from creative play and imagination to those technical and technological capacities needed to reproduce the basic conditions of existence;
7   The chosen capacities need to be able to be mapped onto positive outcomes or conditions without presuming a single blueprint for living or a set-politics.

When completed, we suggest, the overall framework needs in addition to meet four methodological criteria: (1) practical usefulness, (2) analytical coherence, (3) simple complexity and, (4) normative reflexivity (Nussbaum, *Creating Capabilities* ch. 3; McKinney).

Based on these four methodological criteria in relation to the seven listed considerations, our process of choosing the basic domains and their subdomains was long and tortuous. It was based on reading the classics, consulting and arguing, reading the contemporary literature on capabilities and capacities, and more arguing and consulting, and finally testing with experts in different fields. The chosen list presented below remains contingent and open to negotiation. On the basis of this method, the first area that we suggest is basic to a flourishing human condition is the capacity for *vitality*. This domain names the various aspects of the social that Sen and Nussbaum emphasise – embodied, emotional, and mindful wellbeing. The second is *relationality*, the constellation of capacities for relating to other and to nature, from the capacity to communicate to the capacity to reconcile difference and negotiate hospitality to friends and strangers. The third is *productivity*, the set of capacities that allow us to produce the conditions of existence. And the fourth is *sustainability*, capacities for reproducing those conditions in an enduring way that project into the future. Without all of these capacities, at least available in some variable measure, individually/socially, our lives would be solitary, poor, nasty, brutish, and short (Hobbes 65). The remainder of the chapter elaborates on these four inter-related constellations.

IV Vitality as a basic constellation of capacities for human flourishing

The first constellation names those capacities necessary for enjoying a flourishing embodied, emotional, human life. Here the concept of "enjoy" does not depend on the contemporary thin concept of "happiness". Happiness is an important emotion, but only one of the many emotions necessary to a full life. It is also primarily a feeling, understood colloquially as a particular state of mind. By comparison, the domain of vitality sets out a threshold set of mental *and* embodied capabilities that are basic to human flourishing. Within this domain we have identified seven key subdomains.

1.1. The first critical subdomain is *health and wellbeing*. Without at least basic embodied health and a basic sense of wellbeing maintaining relations with others and developing a flourishing life-world is put under considerable strain. This cluster of capacities is self-evidently important. This is not to suggest that a person with ill-health cannot positively experience other elements of the good life – adversity negotiated well is an important and productive part of the human condition – but chronic and consequential ill-health certainly qualifies that potential in a significant way. This encompasses Martha Nussbaum's first and second capacities: (1) life, and (2) bodily health.

1.2. Adding the capacities for *strength and vigour* underlines the way that *vitality*, as one of the first-order capacities necessary to a flourishing world, requires more than minimal capacities for bodily health and mental wellbeing. It requires the vigour to engage physically in relating to others, producing the means of existence and sustaining social and environmental life. This is not in any way to imply that a person who has disabilities in specific areas of embodied strength has a lesser life. In many cases, an incapacity in one area leads to focusing on alternative ways of being in the world. However, neither is it to ignore embodied or mental incapacities. As disabilities and incapacities compound, it becomes imperative that other socially supported capacities are enhanced to counter limits of strength and vigour.

1.3. The cluster of capacities for *emotion and feeling* is fundamental to being human and is one of the capacities on which relating to others (*relationality*) is most intimately connected. The capacity to have and express emotions, including the so-called negative emotions such as anger and sadness, is one of the bases for responding in complex ways to others, events, things and processes.

1.4. *Dignity and recognition* is a set of capacities that is relevant to all social situations. In her debate with Axel Honneth, Nancy Fraser emphasises the political dimension of recognition, but the cluster of capacities associated with *dignity and recognition* also has economic, cultural and ecological dimensions. On the other hand, to recognise the multiple dimensions of recognition is not to agree with Honneth that it therefore is a singular overarching category that can encompass all others.

1.5. The capacity to maintain bodily *integrity and consonance* names a further critical dimension of *vitality*. Rather than referring to ethical integrity or consistency or ethics (which is covered in our framework under the heading "Relationality: Justice and Truth"), *integrity* here refers to embodied and mental integrity. For example, it presumes the capacity – individual and collective – to impose clear limitations on interference, penetration or violation by others. This inclusion parallels Nussbaum's emphasis on "bodily integrity". The concept of *consonance*, that is, consonance of identity, adds to this the capacity to act as if one's identity is relatively continuous in relation to self and others. This is an anti-*Anti-Oedipus* argument, countering the politics of post-structuralists such as Deleuze and Guattari. Despite postmodern romanticism, schizophrenia or being a stranger to oneself is not a positive way of living. Negotiated and relative consonance of identity and difference – without on the other hand asserting the importance of easy comfortable and closed *certainty* about one's identity – is fundamental to being a functioning person and relating to others in a meaningful way. It is also basic to an ongoing community, polity or organisation.

1.6. *Security and safety* as a cluster of capacities that cannot be left out of the primary list. As with all our clusters it can be taken either as an individually held set of capacities (albeit, always understood as always-already social) or as a socially extended set of capacities dependent upon the practices of communities, polities and institutions. This is another of the many capabilities that Martha Nussbaum leaves out.

1.7. Capacities for *sensuality and sexuality* are included as a basic set of capacities under the heading of *vitality* because without them humanity would cease to be viable. Like all the other capacities in this list, they are relational before they are personal, and they are both individual and collective. Without a social capacity for sexuality, for example, there would be no viable reproduction of the species (see also *productivity* below), and without sensuality and the capacity to enjoy sensory experience, including in relation to sexuality, reproduction of social life or reproduction of the species would be reduced to a technical or empty post-human activity.

V Relationality as a basic constellation of capacities for human flourishing

The second constellation of capacities concerns *relationality* – relations to others and to nature. This domain of capacities is so rich and complex that it is profoundly difficult to decide on the seven primary subdomains. For relationality to be meaningful it requires that we have the capacity to establish regimes of mutual care, affinity, reciprocity and so on. At the same time, it is important to recognise the complexities of social difference. Therefore, positive relationality also requires capacities for reconciliation and negotiation across the boundaries of that difference. It also requires capacities for basic communication, a capability not included in Nussbaum's list. Proceeding on this basis, the following seven sets of capacities are taken to be fundamental.

2.1. *Communication and dialogue* are fundamental to all questions of relationality. Capacities for *communication and dialogue* include the capacity on the one hand to share ideas with others in a way that is understandable and expressive, and, on the other hand, to listen, take in the ideas of others, and respond. While there are some writers who make communications *the* basis of social life, here, while it is treated as basic, it is not a master category. This is therefore to go against Niklas Luhmann's systems theory, which treats communications, however contingent, as the autopoetic basis of society.

2.2. A second basic set of capacities for relating is *affinity and reciprocity*. There is a vast literature on this area, with *affinity* naming the capacity to develop ongoing affiliations as families, friends, groups and communities. The capacity for affinity, also extending to objects and the natural world, makes it possible for us to feel close to things, animals and places. Practices of embodied reciprocity are associated with the dominant form of exchange in customary communities, but also in the more abstract form of generalised co-operation they are also necessary to well-functioning modern social systems. Richard Sennett's *Together: The Rituals, Pleasures and Politics of Cooperation* is instructive on this capacity, except in relation to his treatment of tribalism as an internally directed affinity essentially involving aggression to those who are different. This ontological insensitivity is all too common in the non-anthropological literature, and mars an otherwise searching book.

As one important way of relating, *reciprocity* is a difficult concept. It cannot be reduced to the process of exacting *quid pro quo* through mutual self-interest (a reductive form of modern understanding). And it means more than equal giving and taking. In our terms, *reciprocity* is defined broadly as exchange relations of negotiated mutuality that emphasises exchange relations based upon circles of return. The importance of the principle of reciprocity therefore cannot be reduced to liberal considerations of balance or fairness.

2.3. The cluster of capacities for *care and trust* in and for others involves a stronger claim about a social inter-relation than the acceptance of others or tolerance of difference. This cluster brings in Carol Gilligan's embodied notion of an "ethics of care", developed by others including Joan Tronto. But it also includes, with critically qualification, more abstract notions of trust in co-operative activities from communal relations to market exchange processes discussed by writers as different as Barbara Misztal and Francis Fukuyama.

2.4. Capacities for *justice and truth* also need to be included as a set of capacities that are basic to good relationality. It is interesting that Martha Nussbaum's list of fundamental capabilities does not include this cluster directly even though the Capabilities approach is directed towards developing a theory of justice (*Frontiers of Justice*). This is a self-defeating omission. Without the capacity for understanding and enacting the good that an approach is attempting to achieve, the approach itself falls over.

2.5. The capacity to reconcile potentially destructive or negative differences across social and natural boundaries of continuing and flourishing positive differences, including through positive friction, is named under the subdomain of *reconciliation and negotiation*. The possibility of embodied encounter and social friction is important here (Sorkin). A flourishing world is not one in which all differences have dissolved into empty harmony or slide past each other because lines of "neutral" infrastructure facilitate easy flows. This highlights the need to reconcile across *continuing* (and positively tension-producing) boundaries of difference caused by human-to-human encounter and beyond: including the relationship to nature, to objects in the world, and to entities that some believe exist beyond the immanent world, from gods such as God, Yahweh, and Allah to ancestors, spirits and animated nature. *Continuing* boundaries of difference are emphasised here because there is a postmodern tendency to argue for the dissolution of boundaries into differences that, in our terms, do not make a difference.

2.6. The cluster of capacities concerning *faith and love* includes the capacity to have faith in others, to love them, but it also opens the way to including the capacity for faith in forces and beings beyond other humans or human-created things and processes. This suggests a capacity either for some kind of spirituality or at least for an understanding of the limits of human rationality. To have faith includes the capacity to step beyond modern scientific reasoning, whatever that faith may be directed towards. To have the capacity for love ranges from the capacity for interpersonal intimacy to the capacity for relating to nature (or the transcendental beyond nature) as it envelops the self (Milton).

2.7. Practices of *conviviality and hospitality* express another important set of capacities for relating to others. The concept of *conviviality* from the Latin *con* and *vivium*, meaning to come together in live-affirming ways: to eat, to celebrate, or to enjoy social engagement. *Hospitality* in relation to others can include both hospitality to intimate others and to strangers (Derrida).

VI Productivity as a basic constellation of capacities for human flourishing

The third constellation of capacities is the most difficult of all to name. Here various terms were considered as possible ways of naming the general capacity to reproduce the conditions of existence. For a time, we settled on using an older Greek term, *poesis*, meaning *to make*. However, the problem with the term *poesis* was its archaic heaviness and the contemporary tendency to emphasise its poetic dimension. We then moved back and forth between the concepts of *making*, *creativity*, and *productivity*. However, the constellation of capacities that we are trying to get at is broader than either physical production or *making*. As Henri Lefèbvre (528) writes: "Making reduces social practice to individual operations of the artisan kind on a given material which is relatively pliant or resistant". The constellation is also broader than what can be contained by the concept *creativity*. It also includes the capacity for basic practical technique.

We finally settled on the last of those broad concepts: *productivity*. This is still a dangerous choice of terms. The concept of *productivity* can be easily misunderstood, particularly given the contemporary narrowing of its meaning by productivity commissions and the like as they measure efficiency and output. *Productivity* is used here with all the nuanced complexity entailed in describing the creative process of reproducing the conditions of existence. In this sense then, it cannot be reduced to the capacity to produce a measured number of objects. Rather it refers to the capacity for creating the means for living, including practices ranging from practical technique to creative play.

The theme of productivity brings in Adam Smith's work on the division of labour, and the even richer work of Karl Marx on production of the means of existence. However, it significantly widens the meaning of theoretical discussions of "the means of production". Creative play can be as productive of social life as can structural engineering. (We would happily use Donald Winnicott's concept of "creative living" here, except that, following Freud, he accepts the ugly idea that creative living is not available to adults until, through monotheistic religion and science, they are lifted out as individual integrated units rather than embedded in nature as tribal aggregates. Perhaps children, he implies, might have inklings of this before the weight of social life comes down upon them (Winnicott 93–94). This appalling ethnocentrism, endemic also in Freud, relates to what was earlier criticised as the dominance of liberal modernism as a framing ontology in thinking about human capacities.

On the other hand, productivity, as used in this sense, is not as all-encompassing as Marx's notion of *praxis* – at least to the extent that praxis is understood as the production of social life in general, including the social relations that ground human being (relationality). Productivity is the practice of working on the world, either practically or through ideas. Its scope ranges from the intimate world of immediate personal relations to the global world of extended communities, materialities, and processes – and beyond. It is the capacity to bring something into existence, the power to produce. It includes the capacity to procreate, to give birth. In this sense it is complementary to the capacity for vitality and its life-affirming emphasis on embodiment, including *sensuality and sexuality* (Capacities 1.7) as described in the first constellation of capacities for *vitality*.

3.1. Capacities for *learning and teaching* are fundamental for achieving positive *productivity*. They require receptivity to events, processes, and meaning across time as well as capacities for *communication and dialogue* (Capacities 2.1 above). Unlike the emphasis of the UN Sustainable Development Goals on formal and institutionalised practices of reading and writing, this cluster of capacities has an equal emphasis on informal processes. A child obviously learns, but can also teach. And teaching occurs not just by inference where an adult learns from the unintended consequences of a child's actions. Under conditions of flourishing productivity, a child learns positively to teach quite early in life, even if this begins as imitation.

3.2. *Learning and teaching* in turn have an object – knowledge – that requires certain capacities for acquisition and elaboration. This brings us to the second subdomain of *productivity*: *knowing and comprehending*. It is worth elaborating here because Martha Nussbaum's list emphasises a single form of knowing: practical reason. There are however many different forms of knowledge. There is a massive literature on knowing and epistemology. Where, for example, in Nussbaum's list of capabilities is reflexive knowledge: the form of knowing that gives one the capacity to write, criticise or actively respond to a capacities framework in the first place? Where is the capacity to learn? Here, to exemplify how a Social Capacities framework might go deeper and deeper, we further distinguish four main ways of knowing (James and Verrest):

Sensory experience (feeling)

- Sensate knowing: knowing based on being attuned to one's senses: sight, hearing, touch, smell, and taste.
- Perceptive knowing: the cognitive apprehension of having experienced a sensation.
- Emotional knowing: the somatic feeling of affect, including the feeling for someone else's situation; for example, the blush of shame; the clenched fists of anger.
- Revelatory knowing: a visceral response to a particular scene or sound paradoxically experienced as "out of body": for example, the experience of the sublime or "being touched" by the transcendental.

Practical consciousness (pragmatics)

- Experiential knowing: knowledge based on doing things many times: for example, craft knowledge.
- Intuitive knowing: knowing through projecting possibilities; "conscious embodiment" before it comes to find reflective or articulated understanding; sometimes called "being savvy".
- Tacit knowing: knowledge that cannot be articulated or translated into written form.
- Situated knowing: knowledge that is specific to a particular place or time.

Reflective consciousness (reflection)

- Trained knowing: knowledge based on learning supported by teachers and/or curriculum.
- Contemplative knowing: knowledge that emerges in the saying or the thinking. For example, knowing that comes through linguistic consciousness, such as in the moment of saying "I love you" and realising in the act that it is true or otherwise; or knowing that comes through trying out ideas and seeing if they sound right.

- Analytical knowing: knowledge based on breaking things down into their constituent parts: deductive knowledge.
- Theoretical knowing: theoretical work that makes a claim about the determination, framing or meaning of something.

Reflexive knowing (reflexivity)

- Recursive knowing: knowledge that bears back upon itself and constantly interrogates the basis of its own knowledge.
- Epistemological knowing: knowledge about the different forms of knowledge; that is, classic epistemology understood in the sense of the study of knowledge.
- Meta-analytical knowing: analysis that reflects back on the basis of its analysis. For example, methodology studies, which work through the way in which we make claims about things. Another example is psychoanalysis of the kind that entails its practitioners reflecting on their own reflectiveness as they do their work. In other words, this is a kind knowing in which the subject and the object are brought into constant dialogue.
- Meta-theoretical knowing: theoretical work that seeks to understand the world while theorising the possibilities of its own theorising.

3.3. The other side of *knowing and comprehending* (Capabilities 3.2) is the capacity to act upon that knowing: that is, capacities for *practicality and technique – for means and ways. Practicality* is the capacity to adopt different means to an end, from *praktos*, to be done. The associated concept of *technique* or *techne* refers to the capacity to use craft-knowledge, technical proficiency, and so on, in adopting practical ways to chosen ends.

3.4. Without capacities for vocation and labour all the practicality in the world amounts to little. This cluster of capacities is almost self-evident in naming the capacities for work and developing a bounded, committed and renewing set of productive technical skills – a vocation.

3.5. Imagination and creativity is also critical to flourishing social life. This cluster brings together a whole range of capacities and allows us to express in perhaps a more abstract way the capacity to play, one of the ten important capabilities in Martha Nussbaum's list. However, imagination and creativity is, of course, much broader than play and enters into every aspect of vital, relational, productive and sustainable life. Without imagination and creativity, life would be instrumental, brutish and curt.

3.6. Capacities for *enquiry and vision* take imagination (Capacities 3.5) in an interrogative direction that seeks to project social possibilities into the future. They build upon the capacities for *knowing and comprehending* (Capacities 3.2) and give knowing a self-active dimension, addressing the world.

3.7. Finally, in this constellation of capacities for productivity, we need to recognise the importance of capacities for *innovation and change*. It is now dominant in contemporary thinking about development. *Innovation* as a concept has been used consistently in the English global literature across the post-1800 period, but it began to enter more general use as a concept in the 1920s to 1940s alongside other terms such as *technology* and *economy*. This set of capacities is included as basic not because it is fashionable, but because it is one side of the innovation–conservation dialectic and both sides need be included (see Capacities 4. *Sustainability* below for the other side of this dialectic).

VII Sustainability as a basic constellation of capacities for human flourishing

Finally, there is an important fourth constellation of capacities that enable us to sustain the conditions of social and natural flourishing. For all of the capacities for bringing about change (Capacities 3.7) we also need capacities to respond to change and to affect continuity and positive conservation. This entails having the capacity to adapt in relation to rapid external change, to recover from social forces that threaten basic conditions of social life, and to resolve to continue on in the face of adversity. This is the domain that we have called *sustainability*.

4.1. Resilience and flexibility are those important capacities that enable us to respond positively to changes brought about by external forces that threaten basic liveability. It includes the capacity to bounce back from adversity. "Resilience" has become a fashionable term these days in the sustainability lexicon, lifted out as a singular and primary good. Nevertheless, while we agree with the critiques of the contemporary turn to over-emphasising resilience (that is, to the extent that the proponents of such a turn use it as a mechanism to place overriding responsibility back on those experiencing adversity), the capacity for resilience considered as one capacity among others, remains critical to human flourishing.

4.2. The capacities for adaptation and limitation have always been salient across human history, but in the Age of the Anthropocene, and with the intensifying structural pressures of climate change, these capacities have come to the fore and become central to the conditions of our survival. The necessity for climate change adaptation (an instance of the broader capacity of adaptation) is now prevalent in environmental discourses and practices and rightly linked to carbon-emission mitigation (an instance of limitation).

4.3. Receptiveness and responsiveness are necessary for being resilient and adapting. Without receptiveness to the world around, to both social and natural relations, and to the patterns of pressures, changes, forces and critical issues, then adaption and resilience can become self-defeating and unthinking practices of mere survival. One of our critics rightly expressed a concern here that this subdomain sounds like it should be in the domain of inter-relationality, and perhaps it should. We will work though this and other issues in further refinements of the approach.

4.4. Similarly, without the capacities for endurance and patience, the prior capacities for receptiveness and responsiveness do not have a temporal purchase.

Both are social capacities with endurance being a characteristic that is enhanced by institutional or community embeddedness, while patience tends to be a characteristic of persons.

4.5. The capacities for commitment and purpose name the possibilities of directing the various capacities that enhance the long-run sustainability of the human condition towards chosen ends. Such capacities are complementary to and potentially enhancing of the productive capacities for enquiry and vision discussed earlier (Productivity 3.6). They give social practice the possibility for purposive orientation, often enhanced by good leadership.

4.6. Capacities for *stewardship and custodianship* are two sides of the same coin, at least in the way that we use them here. They build upon the capacities for *commitment and purpose* (Capacities 4.5 above), directing that commitment to an object. The capacity for *stewardship* is to commit oneself or one's community to care for an entrusted object (linked to Capacities 2.3 above). It is a commitment from above. It is bestowed and bestowing. By comparison, the capacity for *custodianship*, using this concept less in the conventional modern sense of the word and more as the indigenous literature uses it, is commitment from within or from below. The capacity, for example, to be a custodian of one's land is to belong to that land, to be ontologically embedded in that land, to care for it as one cares for a consonant and inalienable self-other: "The land is my mother".

4.7. In the contemporary world of constant flux – and dominant arguments that constant flux is a good and necessary thing – we conclude our list of capacities with the opposite: capacities for stability and continuity. Not all change is good. While under the heading of innovation and change, we discussed the important capacity to bring about change, change becomes a problem when it is an ideologically charged injunction: thou must change or life will be stagnant, static and bad.

VII Conclusion

In summary, the central capacities for a flourishing social life range from *vitality*, the capacity to enjoy embodied life to the full; to *relationality*, the capacity to relate to others and to nature in a meaningful way; *productivity*, the capacity to reproduce the conditions of existence; and *sustainability*, the capacity to set up the conditions for enduring, and therefore good vitality, relationality and productivity that extends over time. Table I.1.2 summarises this discussion.

This approach, first, has consequences for how discussions of human flourishing are conducted. We would go halfway with Douglas Rasmussen when he writes: "Thus instead of trying to launder ethical reasoning through such devices as 'veils of ignorance', 'impartial ideal observers', or agent-neutral conceptions of practical reason, practical wisdom remains concerned with the temporal and the individual" (19). Yes, veils of ignorance privilege abstract modernist considerations of the good. Yes, a broader conception of knowledge is needed that includes but goes beyond practical reason. Our difference from his conclusion, however, is that even practical wisdom is one capacity among many, and concerns about human flourishing need to be wider than the temporal and individual.

Table 1.1.2 Social capacities for positively engaging in social life

Social capacities	Positive definitions	Subdomains of the social capacities	Outcomes: For example, aspirational urban forms
1 Vitality	The capacity to *enjoy* embodied life to the full, where the concept of "enjoyment" does not depend on the contemporary thin concept of "happiness".	1 Health and wellbeing 2 Strength and vigour 3 Emotion and feeling 4 Dignity and recognition 5 Integrity and consonance 6 Security and safety 7 Sensuality and sexuality	• Healthy cities • Liveable cities
2 Relationality	The capacity to relate to others and to nature in a meaningful way, recognising the complexity of difference; to negotiate and establish regimes of mutual care, trust, and reciprocity.	1 Communication and dialogue 2 Affinity and reciprocity 3 Care and trust 4 Justice and truth 5 Reconciliation and negotiation 6 Faith and love 7 Conviviality and hospitality	• Caring cities • Inclusive cities • Just cities • Peaceful cities • Information cities • Networked cities
3 Productivity	The capacity to bring things into existence, including objects, ideas, processes and events – that is, the capacity to reproduce the basic conditions of a flourishing existence.	1 Learning and teaching 2 Knowing and comprehending 3 Practicality and technique 4 Vocation and labour 5 Imagination and creativity 6 Enquiry and vision 7 Innovation and change	• Prosperous cities • Learning cities • Smart cities • Innovative cities • Knowledge cities
4 Sustainability	The capacity to adapt to change, recover and flourish in an enduring way, particularly in the face of social forces that threaten basic conditions of social life.	1 Resilience and flexibility 2 Adaptation and limitation 3 Receptiveness and responsiveness 4 Resolution and endurance 5 Commitment and purpose 6 Stewardship and custodianship 7 Stability and continuity	• Sustainable cities • Resilient cities • Adapting cities • Carbon-neutral cities

Expressed in terms of the Social Capacities framework that has been unfolded across this chapter, such practical wisdom entails ongoing social negotiation as part of a manifold set of capacities for vitality, relationality, productivity, and sustainability (Figure I.1.2). Instead of being left as an abstracted set of general claims, any framework – including the one we have just presented – needs also to *be able* to be enacted by temporally specific *individuals*-in-social-relation. The *social* layering of ontological differences, for example, cannot, in the end, be handled by an advocacy of liberal pluralism with its emphasis on individual differences. Social capacities need to be enacted, reworked and lived as part of a *transitional practice* by individuals in communities and organisations. They need to work from a level of generality down to differences that include not just temporal differences, but also spatial, corporeal, performative, and epistemological differences (and then back again to inform that level of generality).

Second, the framework has consequences for the Human Development Index. Instead of a series of disconnected indicators that only have a marginal connection to the liberal Capabilities approach, proxy indices could be developed at two

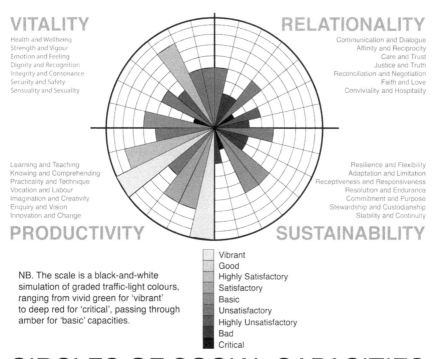

CIRCLES OF SOCIAL CAPACITIES

Figure I.1.2 Circles of social capacities: conducted as an assessment of the contemporary global condition.

Source: original by the author.

levels: (1) a set of four aggregate proxy indicators for each of the four domains, vitality, relationality, productivity, and sustainability; and (2) a set of 28 more directed indicators for each of the subdomains. At the second level, operational-ising the index could be variable across different states, regions and cities, depending upon available statistics and capacity to collect data.

Third, the framework has implications for the classic definition of sustainable development. The now classic text *Our Common Future* (World Commission on Environment and Development), more commonly known as the Brundtland Report, defined sustainable development as "development that meets the needs of the present without compromising the ability of future generations to meet their own needs" (8). Post-September 2015, as the world's nation-states and international organisations now set out on the 2030 Agenda for Sustainable Development Goals "to end poverty, protect the planet and ensure prosperity for all", this definition continues to inform most thinking on the subject.

The definition still works in a superficial sense; however, it has many prob-lems. It was written before the entry of ecological considerations into the heart of development thinking. Its meaning turns on the undefined implications of the word "needs", and it leaves unspecified the assumed importance of specifying economic-material needs as well as social and environment needs (the usual Triple Bottom Line grouping of categories). Moreover, and most remarkably, the Brundtland and post-Brundtland definitions of sustainable development do not actually define development at all. They actually only define the *sustainable* part of *sustainable development*, and then only in a minimal sense.

In terms of the Social Capacities framework, sustainable development would be redefined as a particular kind of social change – with all its intended or unin-tended outcomes – that brings about a significant and patterned shift in the tech-nologies, techniques, infrastructure, and the associated life-forms of a place or people that enhances capacities for human flourishing.

Note

1 The context for this chapter was a project to develop a Digital Capacities Index for measuring the capacities different people have for engaging in a positive digital life. The team compromises Delphine Bellerose, Philippa Collin, Louise Crabtree, Justine Humphry Emma Kearney, Liam Magee, Tanya Notley, Amanda Third and myself. This chapter could not have been written without this collaborative setting. Liam Magee and I worked on the terms of the social capacities framework, and I wrote it up. Thanks also to Stephanie Trigg, Paola Spinozzi, and Massimiliano Mazzanti.

Bibliography

Bowker, Geoffrey C., and Susan Leigh Star. *Sorting Things Out: Classification and its Consequences*. MIT Press, 2000.

Deleuze, Giles, and Félix Guattari. *Anti-Oedipus: Capitalism and Schizophrenia*. Viking Press, 1977.

Derrida, Jacques. *Of Hospitality*. Stanford University Press, 2000.

Fraser, Nancy, and Axel Honnett. *Redistribution or Recognition? A Political Philosophical Exchange*. Verso, 2003.

Fukuyama, Francis. *Trust: The Social Virtues and the Creation of Prosperity*. Hamish Hamilton, 1995.

Gilligan, Carol. *Psychological Theory and Women's Development*. 2nd edn, Harvard University Press, 1993.

Hobbes, Thomas. *Leviathan*. 1651. Dent, 1973.

Hodgson, Geoffrey M. "Meanings of Methodological Individualism". *Journal of Economic Methodology*, vol. 14, no. 2, 2007, pp. 2011–2026.

James, Paul. *Globalism, Nationalism, Tribalism*. Sage Publications, 2006.

James, Paul, and Hebe Verrest. "Beyond the Network Effect: Towards an Alternative Understanding of Global Urban Organizations". *Geographies of Urban Governance: Advanced Theories, Methods and Practices*, edited by Joyeeta Gupta, Karin Pfeffer, Hebe Verrest, and Mirjam Ross-Tonen, Springer, 2015, pp. 65–84.

James, Paul, with Liam Magee, Andy Scerri, and Manfred Steger. *Urban Sustainability in Theory and Practice: Circles of Sustainability*. Routledge, 2015.

Lefèbvre, Henri. *Critique of Everyday Life*. Verso, 2014. 3 vols.

Luhmann, Niklas. *Social Systems*. Stanford University Press, 1995.

Magee, Liam. *Interwoven Cities*. Palgrave Macmillan, 2016.

McKinney, John C. *Constructive Typology and Social Theory*. Meredith Publishing, 1966.

Milton, Kay. *Loving Nature: Towards an Ecology of Emotion*. Routledge, 2002.

Misztal, Barbara A. *Trust in Modern Societies*. Polity Press, 1996.

Nussbaum, Martha C. *Creating Capabilities: The Human Development Approach*. Harvard University Press, 2011.

Nussbaum, Martha C. *Frontiers of Justice: Disability, Nationality, Species Membership*. Harvard University Press, 2006.

Rasmussen, Douglas B. "Human Flourishing and the Appeal to Human Nature". *Social Philosophy and Policy*, vol. 16, no. 1, 1999, pp. 1–43.

Robeyns, Ingrid. "The Capability Approach: A Theoretical Survey". *Journal of Human Development*, vol. 6, no. 1, 2005, pp. 93–114.

Sen, Amartya. *Development as Freedom*. Oxford University Press, 1999.

Sennett, Richard. *Together: The Rituals, Pleasures and Politics of Cooperation*. Yale University Press, 2012.

Sorkin, Michael. "Introduction. Traffic in Democracy". *Giving Ground: The Politics of Propinquity*, edited by Joan Copjec and Michael Sorkin, Verso, 1999, pp. 1–15.

Tronto, Joan C. *Moral Boundaries: A Political Argument for an Ethic of Care*. Routledge, 1993.

United Nations Human Development Index, http://hdr.undp.org/en/content/human-development-index-hdi. Accessed 4 March 2017.

United Nations Sustainable Development Goals, www.un.org/sustainabledevelopment/sustainable-development-goals/. Accessed 4 March 2017.

Watene, Krushil. "Beyond Nussbaum's Capability Approach: Future Generations and the Need for Ways Forward". *New Waves in Global Justice*, edited by Thom Brooks, Palgrave Macmillan, 2014, pp. 128–148.

Winnicott, Donald W. *Playing and Reality*. Routledge, 1971.

World Commission on Environment and Development. *Our Common Future*. Oxford University Press, 1987.

I.2 The incongruities of sustainability

An examination of the UN Earth Summit Declarations 1972–2012

Gonzalo Salazar

I Introduction

In recent decades, the notion of sustainability has gained social and communicational prominence. It has gone from being an idea espoused by small environmental groups to a global social movement. As pointed out by Caradonna, "we might not live in a sustainable age, but we're living in the age of sustainability" (176). In a surprisingly short period of time, sustainability, as a social and public policy construct, has: (1) globally disseminated and solidified the notion of an unprecedented socio-ecological crisis; (2) joined and connected actors and institutions of various scales and dimensions; and (3) contributed to the need to produce changes in the way in which global society is organised.

However, this global movement has not come about free of conceptual contradictions, which have contributed to minimising the possibility of living in more sustainable ways. Although nowadays countless organisations claim to work for sustainable development, it is increasingly unclear what this refers to, both from an epistemological and practical point of view. As different perspectives on sustainability have emerged over the past 40 years, so has a scenario marked by practical-theoretical imprecision and ambiguity with regards to the concept of sustainability. As such, the integrity of the concept as an agent of political change has been seemingly decimated (Bosselmann; Foster), along with its practical application in various spheres of the global era.

This chapter delves into the rationales behind the global agenda on sustainability as defined by the United Nations over the past four decades, and the role that these agendas have played in defining local sustainability practices. Specifically, two contradictory sustainability rationales are examined: an economic and pro-growth rationale committed to the epistemology of progress versus a socio-ecological rationale committed to a relational epistemology. The theoretical discussion is followed by a critical analysis of how the former rationale has exerted a dominant and hegemonic role throughout the past 40 years. This not only denies the ethical and socio-ecological essence of sustainability that has been promoted over the years, but also maximises practical-theoretical ambiguities and incongruences, which end up misguiding the role of local practices in the transition to sustainability.

From a theoretical perspective, the first two sections explore respectively the essential contradiction of sustainability in the global agenda and the need to understand sustainability practices in a local dimension, and how they have been affected by existing incongruences within the global sustainability agenda. The third section examines two opposing sustainability rationales in the international agenda and with particular reference to the Earth Summit Declarations of 1972, 1992, 2002, 2012, the Brundtland Report of 1987, and *Agenda 21*. The fourth section provides a critical discussion of the way in which each of these rationales leads to local sustainability practices. Finally, emphasis is laid on aspects to be addressed in order to conceive of a more socio-ecologically appropriate sustainability practice.

II The incongruities of sustainability: epistemological change or the usual drive to progress?

To a certain degree, the concept of sustainability might be historically understood as a reaction to the socio-ecological crisis of the second half of the twentieth century. Particularly, the ethical and epistemological ideas propelled by environmental movements of the 1960s and 1970s can be considered as the prologue to the concept of sustainability. However, in this same period the essential contradictions of the concept of sustainability were also generated.

On one hand, two ideas associated with a certain understanding of sustainability emerged out of the environmental movements of the 1960s and 1970s: (1) the socio-ecological crisis originates in an epistemological and ethical dimension – i.e. in the way we understand and relate to our contemporaries, future generations and the rest of nature; (2) profound changes are required to the way in which societies in the global era are organised.

The environmental movement was key to culturally disseminating the need to transcend the Cartesian outlook on the world. Contrary to the mechanistic and reductionist fashion of this tradition, the environmental movement intended to put into practice a more holistic and ecological way of thinking as proposed by different disciplines such as cybernetics, phenomenology and ecology (Capra). This realisation not only implies moving to a relational epistemology, in which the world is created from moment to moment through the process of living with others (Ingold). Environmentalism in the 1960s and 1970s also meant that this new paradigm was inherently intertwined with a fundamental ethical challenge: re-evaluating the fact that human wellbeing is complementary to the protection of nature (Rozzi). From this standpoint, the environmental movement contributed to the realisation that complex social problems of the twentieth century – such as poverty and global resource distribution – are intricately related to an ecological crisis, and above all to the way in which societies in the global era understand their relation to nature.

Emphasis was laid on a socio-ecological rationale to be expressed in a practical and political dimension. Environmentalism was not only a political movement per se; it also advocated that the basic organisation of society should be changed, leading to a new global *polis* thriving on the interdependence of contemporary society, future generations and nature. Thus, the invitation to produce an ethical and epistemological shift of the relationship between man and the environment proposed by various thinkers during the 1960s and 1970s triggered concrete actions towards a more sustainable way of life. The concept of Land Ethic expressed by Aldo Leopold in *A Sand County Almanac* and *Una Ética para la Tierra* laid the foundations of environmental ethics as a philosophical discipline and inspired the formation of prominent environmental groups in the following decades. Rachel Carson envisioned the destruction of the habitat and the insurgence of human disease as a result of the indiscriminate use of pesticides in agriculture, triggering governmental discussions in various Western countries. Lewis Mumford developed a comprehensive vision of urban development by maintaining that social, natural and technological dynamics are parts of a whole urban phenomenon. Finally, in *Design with Nature* Ian McHarg introduced the term "urban ecological planning" and circulated it among British and North-American academics.

However, the need to integrate the epistemological and ethical components has not been regarded as a primary goal. The predominant concepts of sustainability have not stemmed from a new ecological epistemology or land ethic proposed by environmentalists of the past decades, but rather from economic arguments (Castro; Singer), which also took shape during the 1960s and 1970s. Environmental literature at that time focused on the increasingly evident contradiction between global growth-based economic policy and the ecological conservation of a planet with finite resources. Several reports from the 1970s, among which *The Limits to Growth* (Meadows *et al.*) by the Club of Rome was seminal, identified modern industrial capitalism as the primary cause of environmental decay. Such discourse limited the conversation on sustainability to an economistic dimension, which ultimately became marked by the maintenance of neoliberal capitalism.

The Club of Rome predicted a global economic and environmental collapse during the second half of the twenty-first century as a result of humanity extending beyond the planet's capacity to satisfy the constantly growing "consumer economy" (Princen). Despite this dire prediction, the limits to growth argument has not led to any significant inquiry into the founding principles of global neoliberalism: continuous growth, free market, and technological innovation. To the contrary, the Club of Rome has suggested that sustainable development practice is to be found in the development of "new measurements for growth that place a real economic value on our natural capital, including carbon, water, forests, biodiversity and our marine environment. This will ensure that the real costs of environmental degradation are reflected in economic decision-making" (Singer 129). In other words, rather than question the epistemological and ethical basis of the concept of growth, such conclusions suggest incorporating social and environmental variables into the prevailing economic regime.

In a deeper sense, such a pro-growth posture regarding sustainability is another example of a long-standing cultural attachment to the epistemology of progress. As Foster suggests:

willed environmental optimism … is an aspect or manifestation of the same destructive dynamics against which it superficially sets itself, it is thoroughly implicated in the basic commitment of the modern world that drives the destruction. That is in shorthand, the commitment to progress, the continuous overall improvement of the human condition.

(43–44)

The conceptual nexus between sustainability and the paradigm of progress results in several problematic issues, beyond the previously explained economistic reductionism. First, the idea that sustainability involves a culminating and then a complete stage is hardly tenable. From this perspective, sustainability practices are limited to the establishment of (almost always quantifiable) goals. "Entirely unsurprisingly, the numbers gain no real purchase. Either they become methodologically manipulated … to produce the results we first thought of, or they are presented as duly drastic and then ignored" (Foster 37). As an example, the dominant environmental policy and strategy delivered by the UN regarding climate change has been based on establishing goals for CO2 emissions reductions, measured according to specific percentage points and within a certain timespan (for example, 20 or 50 years from now). Yet, as already recognised at the UN Earth Summit in 2012, after 20 years this kind of strategy has not even gotten close to the emissions reductions that governments had previously discussed. In such a scenario, instead of relating to daily life, sustainability goals (in this case, reduction of CO_2 emissions) are perceived as disassociated phenomena pertaining to an abstract domain.

Second, the definition of sustainability in universalistic terms is controversial. The paradigm of progress implies the search for standardised and universal solutions that fail to attend to and appropriately manage highly diverse and complex socio-ecological systems (Agrawal; Bruckmeier). A method epistemologically founded on detachment from the living ecology does not only establish and cultivate a profound sense of ecological illiteracy within society (Orr) but is also the bedrock of domineering forms of public policies. As persuasively argued by scholars of phenomenology and the theory of cognition (Bateson; Husserl; Maturana; Merleau-Ponty), a universalistic approach to constructing reality implies a political will to invalidate and even discredit certain kinds of experiences and perspectives.

Thus, the notion of sustainability leads to a fundamental contradiction that goes beyond the typical distinction between weak and strong sustainability, as suggested by Costanza and Daly, or the dichotomies of deep and shallow ecology, as argued by Naess in "The Shallow and the Deep, Long-range Ecology Movement. A Summary". If we do not pay real attention to the rationales behind sustainability trends, we are at a point in which many of the actions that we perform in the name of sustainability can in fact be quite unsustainable. However, we also have the opportunity to seek out more profound insights into ethics and epistemology, which may be spatially associated with a process of localisation.

III Sustainability practice and localisation

As a multi-scalar phenomenon, the global socio-ecological crisis has taken on a high level of complexity. Although the effects of the crisis are now expressed globally, as in the case of climate change, peak oil, financial and economic crises, or global social inequity, it is locally that such complexity is felt the most. One of the main consequences is the weakening and degradation of the local scale in environmental, social, economic and political terms (Douthwaite; Hess). In the face of this multi-level crisis, the movement for localisation in a globalised world has not only been reinvigorated, but is also one of the greatest sustainability planning challenges of the twenty-first century, having been recognised within the UN *Agenda 21* and UN-Habitat programme.

The call for localisation is not a new phenomenon. On the contrary, it seems to have emerged hand in hand with a more ecological and holistic outlook from within the social sciences and humanities since the beginning of the twentieth century. Generally, two stages in the evolution of localisation can be identified. The first refers to the call for localisation as the best solution to face complex social and ecological problems produced by modernity. This stage is most strongly identified by the slogan "Think global, act local", extracted from Patrick Geddes' writings at the beginning of the twentieth century and further developed by several authors during the second half of the century (Aberley; Berry; Sale; Schumacher). This concept has influenced sustainability practice during the past decade, in which it is suggested that although the global socio-ecological crisis demands collaborative planning and governance processes on an international level, practical actions must be carried out within all socio-political systems and adapted to local territorial conditions (Barton; Guimaraes). As such, it is argued that sustainability practice must emerge from territorial and socio-cultural diversity, rather than as a universal model (UN-Habitat).

The second stage began at the end of the 1990s with a new group of actors who warned that localisation was no longer simply an option, but rather an inevitable phenomenon (Douthwaite; Hess; Hopkins; Shuman). As phenomena such as peak oil, climate change and global financial and economic crises are all intertwined, the exhaustion of the epistemological system that underlies the prevailing global economy becomes clear. Such a system is no longer capable of sustaining industrialisation and economic growth as the primary engines of development. In this context, there is an obligation to design a global socio-economic system, based on networking for the exchange of goods and services on a far more reduced and decentralised scale. If the modern era is characterised by systems that have taken us beyond our human and ecological means, the new era of sustainability foreseen is organised within local limits, as argued in "Localización del Diseño Ecológico: del Paisaje al Hacer-del-hogar" (Salazar).

Essentially, the localisation of sustainability requires a new cognitive form and new forms of agency (or institutional practices). Such processes should be unaffected by universalistic visions of sustainability, as a transition towards local sustainability is inherently shaped by heterogeneity, as well as by multiple "systems of knowledge" (Bruckmeier). Regarding a new cognitive form needed to deal with complex tasks of sustainability, it is imperative to progress towards the validation of diversified sources of knowledge that emerge from the experience of diverse places (Ingold; Massey; Thrift). This is what Haraway refers to as "situated knowledge" (Bruckmeier 1391). Similarly, Orr has called the same notion "ecological literacy", defined as a process in which not only do people learn to read the ecosystemic patterns and limits of the particular places in which they live, but also come to understand the human element as an integral part of a network of socio-ecological relations. Similarly, from a more philosophical standpoint, in *Ecology, Community, and Lifestyle* Naess introduces the term ecosophy, defined as a code of values and a way of seeing the personal world that guides one's decisions. Haraway, Orr and Naess emphasise the importance of the individual's and/or organisation's learning experience as developing through an intimate dialogue with places. However, situated knowledge is disrupted when the hegemonic concept of development associated with unidirectional progress requires (among other things) the homogenisation of knowledge production as well as the systematic epistemological disconnection between knowledge and place.

Regarding new forms of sustainability agency, there is a seeming consensus that global problems ought to be addressed by considering participatory local efforts (Das and Takahashi; Fung and Wright). This implies focusing on the institutional dimension of sustainability and enhancing the fact that sustainability practice "is all about deliberate decision making in order to direct global development and system evolution towards a more sustainable route" (Spangenberg *et al.* 70). From this perspective, the articulation of a variety of local actors and institutions involved in planning and governance processes plays a central role. However, it is important to determine the extent to which local sustainability agency is possible, considering that a homogenising structure that depends on a top-down approach has been used to define sustainability practice for at least the past 40 years.

IV Contradictory rationales in the global sustainability agenda

The contradictions illustrated above provide a new opportunity to examine the global sustainability agenda more exhaustively and in relation to various dimensions. Specifically, this section reviews the UN Earth Summit Declarations (Stockholm, Rio, Johannesburg, and Rio+20) regarding their sustainability rationales, the way in which they are applied, and the role of the local dimension in sustainability practice. As illustrated in Table I.2.1, two main trends can be identified within the UN sustainability agenda since 1972. These two trends do not correlate and one is not an effect of the other. To the contrary, there has been a mutual tension within each Earth Summit Declaration. Together these rationales reveal the essential contradiction of sustainability agendas over the past four decades. By presupposing opposite outlooks on the same phenomenon, they

Table I.2.1 Two contradictory epistemologies in the International Sustainability Agenda

	Trend based on the epistemology of progress	*Trend based on a socio-ecological epistemology*
Illustrated by	Three pillars of sustainability (1992) and green economy (2012)	*Agenda 21* (1992)
Rationale	Economistic and pro-growth	Socio-ecological
Model of application	Universalist	Contextual
Direction	Top-down	Bottom-up
Role of local	Negation of local agency	Based on local agency
Effects	Points towards sustainability rhetoric	Points towards sustainability practice

Source: author, based on UN Earth Summit Declarations of 1972, 1992, 2002 and 2012.

work as bipolar political forces that seek to appropriate the concept of sustainability. It is important to highlight the political nature of the Earth Summit Declarations, which have to a great extent formed the global sustainability agenda and are the result of political negotiations on an inter-governmental scale. Although a variety of conceptual agreements and practical commitments by hundreds of signatory states have resulted from these summits, they have also been key platforms for political persuasion exerted by various public and private actors in order to impose a certain vision regarding the concept of development on a global scale.

The description of these two trends provided here does not seek to increment their inherent dualistic relationship. To the contrary, the present work attempts to clarify and transcend this dichotomy and its underlying ambiguities and incongruences, in order to generate a new scenario for the notion of sustainability as well as an agenda with clearer directives.

IV.1 Sustainability trend based on the epistemology of progress

The first sustainability trend is clearly represented by the popular "three pillars of sustainability", also known as the "sustainability triangle", or the "three E's of sustainability" (environment, economy and equity), or what in *Cannibals with Forks* Elkington defined as the Triple Bottom Line (TBL) of sustainability. The three pillars of sustainability took shape in various ways in the Brundtland Report of 1987 and the Rio Declaration of 1992, being explicitly consecrated in the Johannesburg Declaration of 2002: we assume a collective responsibility to advance and strengthen the interdependent and mutually reinforcing pillars of sustainable development – economic development, social development and environmental protection – at the local, national, regional and global levels (Chapter 1, Art. 5). This view presupposes a multidimensional understanding of the complex problems caused by the global crisis. In general, the triangle invites an integration of economic, social and environmental variables in order to conceive of development on local and global scales. Typically represented by three intersecting circles, sustainable development emerges where the various dimensions intersect. This has derived mainly from the recognition that the diverse problems are interconnected and thus must be addressed systematically from the three angles. The 1987 Brundtland Report introduces this idea very clearly:

> There has been a growing realization in national governments and multilateral institutions that it is impossible to separate economic development issues from environment issues; many forms of development erode the environmental resources upon which they must be based, and environmental degradation can undermine economic development. Poverty is a major cause and effect of global environmental problems. It is therefore futile to attempt to deal with environmental problems without a broader perspective that encompasses the factors underlying world poverty and international inequality.
>
> (Art. 8)

The primary challenges addressed by the UN sustainability agenda emerged from this perspective and draw attention to two interdependent ideas within the first articles of all the Earth Summit Declarations since 1992. First, development and environment are inseparable, as the failure of development policies and in the management of the human environment form one "interlocking crisis". It is argued that the economic and social development of nations cannot be understood separately from the capacity of the ecosystems to satisfy present needs and those of future generations, which implies developing a long-term perspective. Second, social justice and environmental justice are interdependent and integrated. Complex problems such as poverty, concentration of wealth, the gap between rich and poor countries, and the instability of democratic systems are closely related to environmental problems such as deforestation, desertification, acid rain, biodiversity loss and (more recently) climate change. The interdependence of these challenges was in fact behind the establishment in 1983 of the UN World Commission on Environment and Development. In each Earth Summit Declaration they have been inherently regarded as problems of socio-environmental justice in a global era.

For the UN, the three pillars of sustainability concept has become a primary communicational tool for disseminating these challenges, having achieved a high level of global prominence (Kates). However, after three decades of attempts to incorporate the sustainability triangle in prevailing sustainability practice, the relevance of this concept (along with the multidimensionality it purports to involve) has been called into question. This can be analysed from a variety of critical perspectives.

The primary critique of this particular sustainability trend targets its fundamentally economistic rationale. It is argued that the sustainability triangle is based on the simplistic inclusion of social and environmental variables within the dominant economic system (James). An integrated review of the Earth

Summit Declarations shows that the UN sustainability agenda has confined the social and environmental problems to an economic dimension, to such an extent that the latest declaration of Rio+20 launched "green economy" as the primary vehicle for sustainable development. In this way, social and environmental variables of the triangle emerge instrumentally, as a function of the economic dimension, rather than as elements calling for inspection from other perspectives. The economistic core of this perspective has been reinforced by a political unwillingness to seriously question the prevailing economic model. The foundation of this model is incremental capital and the promotion of constant economic growth as the central engine of development. In each of the Earth Summit Declarations, economic growth continues to be seen as part of the solution and not part of the problem of sustainability (Barlett), promoting the oxymoronic notion of "sustainable growth". The Rio Declaration of 1992 outlines that "States should cooperate to promote a supportive and open international economic system that would lead to economic growth and sustainable development in all countries, to better address the problems of environmental degradation" (Rio Declaration, Principle 12). The Rio+20 Declaration of 2012 states "we commit to work together to promote sustained and inclusive economic growth, social development and environmental protection and thereby to benefit all" (Rio+20 Declaration, Art. 6, 1). From an epistemological perspective, the three pillars of sustainability invite an even more profound critique. Its fundamental problem is that it does not distinguish between the ontological elements of human organisation – i.e. sustainability relations – and the structures that affect these relations (Becker). While social and ecological aspects are part of the relational ontology of human existence, economy is a structure, just like religion, education, science, technology, and cultural frameworks, in which this ontology develops. However, "Co-Designing in Love: Towards the Emergence and Conservation of Human Sustainable Communities" (Salazar) illustrates how, through its economistic approach, the prevailing sustainability agenda has systematically placed economy over other structures, and has even positioned it on par with the foundational socio-ecological aspect of the human dimension. Other important structures have not been given the same degree of attention. Religious-spiritual and cultural-aesthetic aspects are, without a doubt, a significant "missing pillar" of sustainability (Hawkes; Littig and Griessler; Nurse). In this sense, Burford *et al.* surmise that the various authors who have referred to the missing pillar of sustainability share "a concern with human values and how they are manifested in people's personal and professional lives" (3038). Although value systems are inherent to the three pillars of sustainability, there are structures derived from the humanities, the arts and religion that are vital to the emergence of aspects such as environmental ethics and ecological consciousness, and which are not included (or only to a secondary degree) in the Earth Summit Declarations. It is for this reason that the multidimensionality to which the figure of the three pillars of sustainability alludes loses its conceptual grounding.

Finally, from a more functional perspective, by not establishing a clear relational dynamic between the three variables, this model has been highly manipulated by corporations, governments and other institutions. According to this economistic model, even the minimal introduction of a social and environmental variable into the economic and financial framework of a company suffices to proclaim its sustainability. In this way, many institutions and corporations from various spheres have claimed to operate under the principles of sustainability, having incorporated the three E's into their decision-making processes, without having really discussed or understood how these dimensions are dynamically inter-connected (Farley and Smith). As predicted by O'Riordan, the result has been a random proliferation of actions for sustainability.

IV.2 Sustainability trend based on a socio-ecological epistemology

At the opposite end of the spectrum, the second sustainability trend is para-doxically also present in every Earth Summit Declaration and is clearly illus-trated by *Agenda 21*. Adopted by 178 governments participating in the Rio Earth Summit of 1992, *Agenda 21* emerges as an exhaustive plan that seeks to put the concept of sustainability as previously defined by the Brundtland Report of 1987 into action. *Agenda 21* represents a hallmark of sustainable development, as it sought to trigger a transition from public policy that was focused on a diagnosis of the crisis to one focused on agency. While the Stock-holm Declaration of 1972 emphasised the importance of assessing and com-municating the existence of a multidimensional crisis, the Rio Declaration of 1992 is focused on how to address this crisis. Several authors (Burford *et al.*; Pfahl; Spangenberg *et al.*) have suggested that, as the objectives and demands of *Agenda 21* became operationalised, a "fourth pillar of sustainability" emerged: institutional sustainability. *Agenda 21* is emphatic in its call to develop and strengthen (on local, national, regional and international scales) both organisations and institutional mechanisms (i.e. laws, policies, plans and administrative and political processes), in order to achieve sustainable develop-ment. As a result of *Agenda 21*, the United Nations Division for Sustainable Development incorporated the institutional component as a fourth pillar of sustainability in several reports and studies (Burford *et al.*). An evaluation of how institutional sustainability, as defined by *Agenda 21*, has been imple-mented exceeds the scope of this study. However, the rationale behind this form of understanding sustainability is pivotal.

First of all, *Agenda 21* has sought the development of a localised and partici-patory sustainability practice, involving actors and institutions on various scales. The following Earth Summit in 2002 reinforced this objective:

> Agenda 21 is a dynamic programme. It will be carried out by the various actors according to the different situations, capacities and priorities of coun-tries and regions in full respect of all the principles contained in the Rio Declaration on Environment and Development. It could evolve over time in the light of changing needs and circumstances. This process marks the beginning of a new global partnership for sustainable development.
>
> (United Nations, *Agenda 21*, Art 1.6)

We recognise that sustainable development requires a long-term perspective and broad-based participation in policy formulation, decision-making and implementation at all levels. As social partners, we will continue to work for stable partnerships with all major groups, respecting the independent, important roles of each of them.

(United Nations, *Report of the World*, Art. 26)

Such statements imply a fundamentally socio-ecological rationale, far from the economistic rationale represented by the sustainability triangle. More generally, here sustainability practice is understood and promoted as: (1) a contextual phenomenon, as its definition and operationalisation depend on particular socio-ecological dynamics; and (2) a political phenomenon, which leads to a focus on the organisation and reorganisation of deliberative processes of inclusive decision-making (Barton; Guimaraes; Spangenberg *et al.*). In other words, in this trend, sustainability practice is understood as a dynamic phenomenon that depends on particular socio-ecological relations, in which various actors play a central role as sustainability agents.

The socio-ecological rationale centres on a social and ecological ontology of the human dimension, which is illustrated in various ways in the declarations of 1972, 1992 and 2002. Principle 1 of the 1992 Rio Declaration establishes that "Human beings are at the centre of concerns for sustainable development. They are entitled to a healthy and productive life in harmony with nature". Article 2 of the 2002 Johannesburg Declaration states the commitment "to building a humane, equitable and caring global society, cognizant of the need for human dignity for all". Referring to Africa as not only the place where the 2002 Summit was held but also as "the cradle of humanity", Article 6 emphasises "our responsibility to one another, to the greater community of life and to our children".

This does not imply leaving the economic aspect aside. To the contrary, *Agenda 21* places much emphasis on current global economic dynamics, and on measures to be taken in this field. Based on a socio-ecological rationale, *Agenda 21* does not engage in defence of the economic growth that is characteristic of global neoliberalism. It makes a call to develop an economy that addresses the "unsustainable patterns of production and consumption", and to produce public policies and global, national and local strategies "to encourage changes in unsustainable consumption patterns". Economy thus becomes a function of current social and environmental problems (and not the opposite). Economic growth, which in its essence requires constant increases in consumption, began to be understood as part of the problem, not the solution:

Poverty and environmental degradation are closely interrelated. While poverty results in certain kinds of environmental stress, the major cause of the continued deterioration of the global environment is the unsustainable pattern of consumption and production, particularly in industrialized countries, which is a matter of grave concern, aggravating poverty and imbalances.

(*Agenda 21*, Art. 4.3)

Growing recognition of the importance of addressing consumption has also not yet been matched by an understanding of its implications. Some economists are questioning traditional concepts of economic growth and underlining the importance of pursuing economic objectives that take account of the full value of natural resource capital. More needs to be known about the role of consumption in relation to economic growth and population dynamics in order to formulate coherent international and national policies.

(*Agenda 21*, Art. 4.6)

However, this socio-ecological sustainability rationale, which reaches a crescendo in the Johannesburg Declaration of 2002, is undermined by the other political trend presented in the very same declarations. These two trends do not appear as merely two epistemologically and politically conflicting forces; rather the economistic and pro-growth trend has become politically dominant. The Rio+20 Declaration is a clear example of how the most fundamental socio-ecological principles of sustainable development have lost sway. These principles appear as a function of an economistic and pro-growth way of thinking based on the "three pillars of sustainability", as expressed in Article 6, as well as on the inclusion of "green economy" as the primary engine of sustainable development:

> We recognize that people are at the centre of sustainable development and, in this regard, we strive for a world that is just, equitable and inclusive, and we commit to work together to promote sustained and inclusive economic growth, social development and environmental protection and thereby to benefit all.
>
> (Rio+20 Declaration, Art. 6)

> We emphasize that [green economy] should contribute to eradicating poverty as well as sustained economic growth, enhancing social inclusion, improving human welfare and creating opportunities for employment and decent work for all, while maintaining the healthy functioning of the Earth's ecosystems.
>
> (Rio+20, Art. 56)

The document conveys the impression that all social and environmental problems could be solved through green economy, and are not considered as primary elements that shape the conceptualisation and practice of sustainability. Therefore, green economy, as promoted in the Rio+20 Declaration, tends to reduce sustainability practice to technological innovation (Articles 43, 48, 58, 74, 174) and economic growth, which are two fundamental pillars of global neoliberalism.

V The prevailing sustainability trend as a mask for global neoliberalism: local effects

The two conflicting sustainability trends here described lead to a differential conceptual framework and practical implications regarding sustainability on a local level. On one hand, the "three pillars of sustainability" imply a fundamentally universalistic model of action: this rationale is predicated upon an economistic and pro-growth paradigm defined a priori, which has been mainstreamed through inter-scalar public policies. The local scale appears as a depository, where these paradigms are operationalised under a structuralist ideal. In this way, it can be suggested that a sustainability agenda based on such a paradigm presupposes an essentially top-down approach, in which a hegemonic and homogenising conceptualisation prevails. Such an approach denies the ecology of organisations and local human spaces as fundamental elements of sustainability practice.

On the other hand, the paradigm based on a socio-ecological rationale, as illustrated through the principles of *Agenda 21*, implies definitions and practices of sustainability emerging from local organisations. *Agenda 21* operates on the idea that many complex environmental problems can be drawn to local scales. For the same reason, local organisations have an important role in defining, planning and implementing forms of institutionalisation that allow sustainability to be socially, culturally, ecologically and spiritually appropriate to their particular contexts. In other words, this perspective promotes sustainability practice as an inherently localised phenomenon, along the same lines as concepts such as "situated knowledge", "ecological literacy" and "ecosophies", as previously described. It can be argued that this trend adopts a bottom-up approach, in which sustainability practice is embedded in particular contexts. This does not mean that institutions on larger scales do not also have an important role to play. On the contrary, *Agenda 21* establishes key objectives on international institutions, which are necessary to facilitate the implementation and mainstreaming of multiple local sustainability agendas aligned with globally established principles.

This contradiction regarding the role of the local level in the global sustainability agenda leads to practical controversies and incongruences on this scale. On one hand, localities are asked to develop comprehensive local agendas based on local knowledge of particular actors and institutions. On the other hand, as local organisations become part of the global sustainability agenda, they are also "part and parcel" of public policies that promote and adopt processes of planning and development based on the paradigm of progress. In this way, the sustainability agenda emerging at a local level will have to address multiple issues. It may end up reproducing homogeneous and hegemonic visions of sustainable development, inhibiting the emergence of local sustainability visions based on cultural heterogeneity and a diversity of cosmovisions and paradigms. It may provide incentives for economistic visions that commodify and exploit social, cultural and ecological "resources". It may perpetuate social and environmental problems, which were the initial reasons for the creation of the UN Development and Environment program in 1983.

In practice, this would produce ineffective local sustainability practice. From an economistic perspective, the Rio+20 Declaration recognises that the past 20 years since Rio 1992 "have seen uneven progress" in different countries and localities regarding the objectives established at previous Earth Summits. As a response to this, the document emphasises the need "to accelerate progress in closing development gaps between developed and developing countries and to seize and create opportunities to achieve sustainable development through economic growth" (Rio+20 Declaration, B19). However, it is clear that ineffective sustainability at a local level, resulting from the ambiguity of global public policy, would provide dangerous scenarios for maintaining the status quo of the prevailing economic system. By informing ethical and political frameworks for local sustainability practice, sustainability can shun the risk of becoming a mask for hegemonic global neoliberalism.

VI Conclusion

In this chapter, two sustainability trends that are epistemologically and politically opposed have been identified and examined. The tension between them suggests the existence of a fundamental sustainability contradiction that is producing a high level of theoretical and practical ambiguity. The global sustainability agenda led by the UN has been marked by an economistic and pro-growth paradigm based on an epistemological attachment to progress. This is the perpetuation of an anti-sustainable system.

However, if sustainability as a practical phenomenon is our only possible future, it is essential to encourage a critical view of this concept, evolving from the core ideas of the environmentalist thinking of the 1960s. This will strengthen the idea that the challenge of sustainability has a fundamentally ethical and epistemological dimension. As such, socio-ecological ethics and epistemology of the human dimension must be the basis for the conceptual framework of sustainability as well as its entire practical agenda in a global era. If this occurs, the socio-ecological paradigm will be capable of transcending the hegemonic vision of sustainability and trigger the emergence of local sustainability agendas supported by global sustainability agency. If localisation is an unavoidable phenomenon, it is essential for the global sustainability agenda to facilitate these processes in socially and ecologically appropriate ways. This implies generating political will directed towards a more comprehensive and appropriate sustainability practice as well as producing an epistemological shift allowing sustainability practice to transcend the rhetorical realm and become effective.

Acknowledgements

This research was supported by: CONICYT Research Project Fondecyt Iniciación N° 11130519; Centro de Desarrollo Urbano Sustentable (CONICYT, FONDAP N°15110020); Centro de Estudios Interculturales e Indígenas (CONICYT, FONDAP N° 15110006).

Bibliography

Aberley, Doug. "Interpreting Bio-Regionalism: A Story from Many Voices". *Bioregionalism*, edited by Michael V. McGinnis, Routledge, 1999, pp. 13–42.

Agrawal, Arun. "Sustainable Governance of Common-Pool Resources: Context, Methods, and Politics". *Annual Review of Anthropology*, vol. 32, no. 1, 2003, pp. 243–262.

Barlett, Albert A. "Reflections on Sustainability, Population Growth, and the Environment-2006". *The Future of Sustainability*, edited by Marco Keiner, Springer, 2010, pp. 17–37.

Barton, Jonathan R. "Sustenabilidad Urbana como Planificación Estratégica". *Eure*, vol. 32, no. 96, 2006, pp. 22–45.

Bateson, Gregory. *Steps to an Ecology of Mind: Collected Essays in Anthropology, Psychiatry, Evolution, and Epistemology*. Intertext, 1972.

Becker, Christian U. *Sustainability Ethics and Sustainability Research*. Springer, 2012.

Berry, Wendell. "A Native Hill". *At Home on the Earth: Becoming Native to Our Place: A Multicultural Anthology*, edited by David Landis Barnhill, University of California Press, 1999.

Bosselmann, Klaus. *The Principle of Sustainability*. Ashgate, 2008.

Bruckmeier, Karl. "Sustainability between Necessity, Contingency and Impossibility". *Sustainability*, vol. 1, no. 4, 2009, pp. 1388–1411.

Burford, Gemma, Elona Hoover, Ismael Velasco, Svatava Janoušková, Alicia Jimenez, Georgia Piggot, Dimity Podger, and Marie K. Harder. "Bringing the 'Missing Pillar' into Sustainable Development Goals: Towards Intersubjective Values-Based Indicators". *Sustainability*, vol. 5, no. 7, 2013, pp. 3035–3059.

Capra, Fritjof. *The Web of Life: A New Synthesis of Mind and Matter*. HarperCollins, 1996.

Caradonna, Jeremy L. *Sustainability: A History*. Oxford University Press, 2014.

Carson, Rachel. *Silent Spring*. Houghton Mifflin, 1962.

Castro, Carlos J. "Sustainable Development: Mainstream and Critical Approaches". *Organization and Environment*, vol. 17, 2005, pp. 195–225.

Costanza, Robert, and Herman E. Daly. "Natural Capital and Sustainable Development". *Biology*, vol. 6, no. 1, 1992, pp. 37–46.

Das, Ashok K., and Lois M. Takahashi. "Evolving Institutional Arrangements, Scaling Up, and Sustainability Emerging Issues in Participatory Slum Upgrading in Ahmedabad, India". *Journal of Planning Education and Research*, vol. 29, no. 2, 2009, pp. 213–232.

Douthwaite, R. J. *Short Circuit: Strengthening Local Economies for Security in an Unstable World*. Green Books/Lilliput Press, 1996.

Elkington, John. *Cannibals with Forks: Triple Bottom Line of 21st Century Business*. Capstone, 1997.

Farley, Heather M., and Zachary A. Smith. *Sustainability: If It's Everything, Is It Nothing?* Routledge, 2014.

Foster, John. *After Sustainability*. Routledge, 2014.

Fung, Archon, and Erik Olin Wright, eds. *Deepening Democracy: Institutional Innovations in Empowered Participatory Governance*. Verso, 2003.

Geddes, Patrick. *Cities in Evolution: An Introduction to the Town Planning Movement and to the Study of Civics*. BiblioBazaar, 2010.

Guimaraes, Roberto P. *Tierra de Sombras: Desafíos de La Sustentabilidad y del Desarrollo Territorial y Local ante la Globalización Corporativa*. Naciones Unidas – Cepal: División de Desarrollo Sostenible y Asentamientos Humanos, 2003.

Haraway, Donna. "Situated Knowledges: The Science Question in Feminism and the Privilege of Partial Perspective". *Feminist Studies*, vol. 14, no. 3, 1988, pp. 575–599.

Hawkes, Jon. *The Fourth Pillar of Sustainability: Culture's Essential Role in Public Planning*. Common Ground Publishing/Cultural Development Network, 2001.

Hess, David J. *Localist Movements in a Global Economy: Sustainability, Justice, and Urban Development in the United States*. MIT Press, 2009.

Hopkins, Rob. *The Transition Handbook: From Oil Dependency to Local Resilience*. Green Books, 2008.

Husserl, Edmund. *Cartesian Meditations. An Introduction to Phenomenology*. Translated by Dorion Cairns. The Hague, 1960.

Ingold, Tim. *The Perception of the Environment: Essays on Livelihood, Dwelling and Skill*. Routledge, 2000.

James, Paul. *Circles of Sustainability*. Routledge, 2015.

Kates, Thomas W. "What Is Sustainable Development? Goals, Indicators, Values, and Practice". *Environment: Science and Policy for Sustainable Development*, vol. 47, no. 3, 2005, pp. 8–21.

Leopold, Aldo. *Una Ética para la Tierra*. Los Libros de la Catarata, 1999.

Leopold, Aldo. *A Sand County Almanac: With Essays on Conservation from Round River*. Ballantine Books, 1990.

Littig, Beate, and Erich Griessler. "Social Sustainability: A Catchword between Political Pragmatism and Social Theory". *International Journal of Sustainable Development*, vol. 8. nos. 1–2, 2005, pp. 65–79.

Massey, Doreen. *For Space*. SAGE, 2005.

Maturana, Humberto. "Reality: The Search for Objectivity, or the Quest for a Compelling Argument". *The Irish Journal of Psychology*. vol. 9, no. 1, 1988, pp. 25–82.

McHarg, Ian L. *Design with Nature*. Turtleback Books, 1995.

Meadows, Donella H., Dennis L. Meadows, Jørgen Randers, and William W. Behrens III. *The Limits to Growth: A Report for the Club of Rome's Project on the Predicament of Mankind*. Universe Books, 1974.

Merleau-Ponty, Maurice. *Phenomenology of Perception*. Routledge/Kegan Paul, 1962.

Mumford, Lewis. *The Pentagon of Power*. Harcourt Brace Jovanovich, 1964.

Naess, Arne. *Ecology, Community, and Lifestyle: Outline of an Ecosophy*. Translated by David Rothenberg. Cambridge University Press, 1990.

Naess, Arne. "The Shallow and the Deep, Long-range Ecology Movement. A Summary". *Inquiry*, vol. 16, no. 1, 1973, pp. 95–100.

Nurse, K. *Culture as the Fourth Pillar of Sustainable Development*. Commonwealth Secretariat, 2006.

O'Riordan, Tim. "The Politics of Sustainability". *Sustainable Environmental Management: Principles and Practice*, edited by R. Kerry Turner, Westview Press, 1988, pp. 31–49.

Orr, David W. *Ecological Literacy: Education and the Transition to a Postmodern World*. SUNY Press, 1992.

Pfahl, Stefanie. "Institutional Sustainability". *International Journal of Sustainable Development*, vol. 8, nos. 1–2, 2005, pp. 80–96.

Princen, Thomas. *Treading Softly: Paths to Ecological Order*. MIT Press, 2010.

Rozzi, Ricardo. "De las Ciencias Ecológicas a la Ética Ambiental". *Revista Chilena de Historia Natural*, vol. 80, no. 4, 2007, pp. 521–534.

Salazar, Gonzalo. "Localización del Diseño Ecológico: del Paisaje al Hacer-del-hogar". *Revista 180*, no. 31, 2013, pp. 44–49.

Salazar, Gonzalo. "Co-Designing in Love: Towards the Emergence and Conservation of Human Sustainable Communities". University of Dundee, Centre for the Study of Natural Design, 2011.

Sale, Kirkpatrick. *Human Scale*. Secker & Warburg, 1980.

Schumacher, E. F. *Small Is Beautiful*. Abacus, 1974.

Shuman, Michael. *Going Local: Creating Self-Reliant Communities in a Global Age*. Routledge, 2000.

Singer, Merrill. "Eco-Nomics: Are the Planet-Unfriendly Features of Capitalism Barriers to Sustainability?" *Sustainability*, vol. 2, no. 1, 2010, pp. 127–144.

Spangenberg, J. H., Stefanie Pfahl, and Kerstin Deller. "Towards Indicators for Institutional Sustainability: Lessons from an Analysis of Agenda 21". *Ecological Indicators*, vol. 2, nos. 1–2, 2002, pp. 61–77.

Thrift, Nigel. *Non-Representational Theory: Space, Politics, Affect*. Routledge, 2008.

UN-Habitat. *Planning Sustainable Cities. Global Report on Human Settlements 2009*. Earthscan, 2009.

United Nations, *The Future We Want*. Rio+20 United Nations Conference on Sustainable Development, Rio de Janeiro, 2012.

United Nations, *Report of the World Summit on Sustainable Development*. United Nations, Johannesburg, 2002.

United Nations, *Report of the United Nations Conference on Environment and Development*. United Nations, Rio de Janeiro, 1992.

United Nations, *Agenda 21*. United Nations Conference on Environment and Development, Rio de Janeiro, 1992.

United Nations (Brundtland Report), *Our Common Future*. World Commission on Environment and Development, 1987.

United Nations, *Report of the United Nations Conference on the Human Environment*. United Nations, Stockholm, 1972.

I.3   Slow living and sustainability

The Victorian legacy

Wendy Parkins

Recently, there has been a noticeable trend in humanities scholarship to respond to pressing twenty-first century concerns such as climate change, peak oil, and the consequences of globalisation. Those in the humanities have begun to argue that such contemporary concerns and events are not simply a matter for the scientific community but also for scholars who can address the implications for culture, politics, aesthetics and ethics, as well as – or alongside – the economic and environmental implications. Writing on the issue of sustainability, for example, Leerom Medovoi has recently argued that:

> Much of the appeal of 'sustainability' to both political progressives and humanities scholars is that it apparently calls for reflection on what is worth sustaining, and hence on the ethical and political stakes of the properly nourished life, one in particular that seeks [balance] between the needs of human beings and those of their natural and social environments.
>
> (130)

Reflecting on what is worth sustaining, moreover, often proceeds from a heightened sense of what the feminist philosopher Judith Butler has called "precarious life" (Butler). In a similar vein, Claire Colebrook, describing our tenuous sense of "attachment to a fragile planet", writes:

> We are at once thrown into a situation of urgent interconnectedness, aware that the smallest events contribute to global mutations, at the same time as we come up against a complex multiplicity of diverging forces and timelines that exceed any manageable point of view.
>
> (52)

While the debates and phenomena associated with climate change and the depletion of natural resources seem particularly galvanising now, in the early part of the twenty-first century, an earlier moment in modernity – the Victorian period – also led writers and thinkers to consider the consequences of rapid change on a global scale caused by human intervention. From observations of the widespread environmental despoliation in industrial areas, or the impact of exponential population growth in cities, to reports of the consequences of imperial or commercial adventures in distant locations (massacres, famines, wars), Victorians were forced to confront a previously unimagined scale of human endeavour and its consequences. Such a confrontation had a paradoxical effect: it emphasised the reach and agency of human intervention around the planet, while at the same time it enabled a new awareness of the vulnerability of each individual when faced with the global impact of the forces of modernity. In the Victorian age, we see the first responses, then, to a dawning comprehension that people were living in an economy of scarcity, of finitude, that the rapid transformation of society they experienced was premised on a myth of inexhaustible resources (e.g. coal) and unending growth that was ultimately unsustainable.

As early as the 1840s, an anxiety about environmental limits or the risk of resource depletion as a result of energy-intensive practices was expressed in Britain (MacDuffie 10–11). For example, the liberal philosopher John Stuart Mill bleakly envisaged an environment ravaged by the increasing demands of population and the industrialisation of agriculture, describing a "world with nothing left to the spontaneous activity of nature ... every flowery waste or natural pasture ploughed up, all quadrupeds or birds not domesticated for man's use exterminated as his rivals for food" (756). In 1865 in *The Coal Question*, the economist W. S. Jevons raised the alarming reality of the finite resource of coal on which the Victorian economy relied and, as MacDuffie puts it, it "helped make the question of resource exhaustion a part of the national conversation" (50). In the same decade, the pessimistic cultural critic John Ruskin lamented what he saw as the inevitable consequences of industrialisation and the Victorian dependence on fossil fuels and other finite natural resources:

> All England may, if it so chooses, become one manufacturing town; and Englishmen, sacrificing themselves to the good of general humanity, may live diminished lives in the midst of noise, of darkness, and of deadly exhalation. But the world cannot become a factory nor a mine. No amount of ingenuity will ever make iron digestible by the million, nor substitute hydrogen for wine.
> (110)

In their various ways, then, Mill and Ruskin tried to envisage where the processes of industrial and commercial expansion would end and all they could foresee was environmental catastrophe where life in any sense cannot be sustained. Their searing critique of the myth of progress – the belief that the general good will transcend suffering in the present and lead to a better future for all – rests on an implicit ecological vision, an understanding of the interdependence of organisms within an environment, that leads them both to reject the idea of limitless growth or endless development. Such an outcome, they contend, is neither possible nor desirable.

In the twenty-first century, the expression of such concerns is common in debates about sustainability but the Victorians did not yet use the word in this sense. As these introductory examples make clear, however, there was an emerging awareness of crisis from around the mid-nineteenth century onwards, derived from a growing sense that exponential increase in population, technological capabilities, and the rate of usage of natural resources were impinging on, or in conflict with, the limits of environmental conditions and, as MacDuffie notes, there is at times a more modern understanding of the term sustainability "lurking around the edges of discussion" in the work of Victorian authors (102–103).

There is, then, much that warrants our attention in an emerging discourse of sustainability in the nineteenth century and by examining some of these early ideas and debates concerning sustainability, we may enhance our own thinking about the challenges of everyday life in the twenty-first century. In what follows, I will consider in more detail two other Victorian writers whose ideals of sustainability were grounded in the centrality of culture and everyday life; sustainability was about the personal as much as the global. Whether our concerns are with the impact of climate change, global agriculture, or fossil-fuel economies, our response is often enacted at the level of the local and the everyday – how we cook and eat, what we grow or recycle, how we travel from home to work, how we spend our time. In turn, these everyday acts are conceptualised within the cultural beliefs and structures we inhabit – patterns of family and community, customs of food preparation and consumption, the aesthetics of domestic architecture and design. In the nineteenth century, William Morris and Edward Carpenter similarly began from the micro-level of everyday life in thinking about how the deleterious global consequences of industrial modernity could be challenged and, they hoped, reversed. They advocated an alternative approach to daily life and work in which the values of nature, beauty, and creativity were championed against the alienation, standardisation and increased social disparities they attributed to the industrial mode of production and the growth of Victorian consumer society. Celebrating the handmade and the artisanal, Morris and Carpenter variously began to formulate ideals of sustainability that they believed would enhance everyday life and restore the natural environment. It may now be easy to dismiss some of these ideas as naïve idealism but if we look more closely at some of the writings of Morris and Carpenter, we may discern a radical and complex response to the problem of sustainability in Victorian modernity that still resonates today.

William Morris (1834–1896)

William Morris is best known today as a designer and craftsman but he first acquired fame as a poet in the mid-nineteenth century. His long narrative poem, *The Earthly Paradise*, was a best-seller in the 1860s and 1870s. Around the same time, he began to diversify into what was a quite startling range of skills and fields: through his company Morris & Co, he designed furniture and patterns for fabric and wallpaper; he revived traditional crafts of printing and book-making, and traditional forms of weaving and dyeing techniques for cloth and carpets. He began to write and lecture on art, design and interior decoration in the 1870s and was one of the founders of the newly formed Society for the Protection of Ancient Buildings. In the 1880s, he formally committed to socialism and was an indefatigable campaigner for the cause until his death in 1896. He was a sought-after public speaker, he wrote political pamphlets and essays, as well as being the financier and editor of the socialist newspaper, *The Commonweal*. He participated in a number of socialist organisations, with the Hammersmith Socialist League meeting in Morris's home. At the same time, Morris continued to write and publish poetry and romances, and translations of classical and Nordic literature. His best-known literary work today is his utopian novel of 1890, *News from Nowhere*.

Certain threads run through all of Morris's interests, skills and passions: a love of nature; the value of work as a form of creative self-expression; the significance of beauty as a vital dimension of daily life; and the importance of fellowship and community. In both his art and his politics, the centrality of everyday life as the space where we have our being, perform our meaningful work, experience love and conviviality, express our creativity was always given priority. He never doubted that the political and the personal were interconnected and indissoluble, just as he never saw art and nature as opposing domains.

Morris is also considered one of the first "ecosocialist" thinkers. In 1881, Morris reminded his audience that "tis we ourselves, each one of us, who must keep watch and ward over the fairness of the earth", urging them not to leave the earth a "lesser treasure" than they had received ("Prospects of Architecture" 119, 120). Not surprisingly, Morris's attack on the environmental consequences of industrialisation was also bound up with his critique of capitalist production and the social relations it fostered. In one of his first public lectures, "The Lesser Arts" in 1877, for instance, a lecture best known for Morris's celebration of decorative arts that was crucial to the Arts and Crafts movement, he wrote:

> Is money to be gathered? Cut down the pleasant trees among the houses, pull down ancient and venerable buildings for the money that a few square yards of London dirt will fetch; blacken rivers, hide the sun and poison the air with smoke and worse, and it's nobody's business to see to it or mend it: that is all that modern commerce, the counting-house forgetful of the workshop, will do for us herein.
>
> (16)

Morris placed environmental destruction at the centre of his social critique, even before he had formally committed himself to the socialist movement, and this remained a consistent association throughout his political writings.

In a late essay called "Makeshift", for example, Morris linked the beauty of the natural environment with larger issues of social justice, arguing that urban workers were robbed of a natural right to experience the pleasures of the senses in open green spaces, a pleasure he believed to be as important – and sustaining – as daily nutrition. Morris always insisted that all members of society were impoverished by a depleted natural environment and so for him a radical transformation of society needed to be grounded in a rediscovery of "the simple joys of the lovely earth" ("Society of the Future" 193). In a lecture in 1887 called

"The Society of the Future", Morris delineated what he saw as the consequences of modern civilisation:

> It has covered the merry green fields with the hovels of slaves, and blighted the flowers and trees with poisonous gases, and turned the rivers into sewers; till over many parts of Britain the common people have forgotten what a field or a flower is like, and their idea of beauty is a gas-poisoned gin-palace or a tawdry theatre. And civilization thinks that is all right, and it doesn't heed it.
>
> (193)

In response, Morris outlined his desire for a future society united by a "wish to keep life simple, to forgo some of the power over nature won by past ages in order to be more human and less mechanical, and willing to sacrifice something to this end" (183).

As we will shortly see again with Edward Carpenter, Morris here suggests that to counter the devastation he associated with industrial modernity would require a relinquishment of some degree of mastery over nature and a simpler way of life. In other words, Morris's idea for how to live sustainably on the planet was fundamentally different from some expressions of sustainability in our own time which seek to accommodate sustainability within corporate culture or to offer market-based solutions to the destructiveness of industrial capitalism (Medovoi 137). As Medovoi has argued provocatively, corporate sustainability discourse is committed to the indefinite continuation of surplus value extraction that effectively disavows the degree of depletion and destruction that this continuity would require. Nothing could be further from Morris's acknowledgement of the limits of human dominance and the beginning of an ecological awareness of how humans could exist sustainably within the natural environment. In his utopian novel set in the future, *News from Nowhere*, for example, Clara describes an instrumentalist attitude to nature by people in the past (i.e. the nineteenth century) that has come to seem misguided:

> Was not their mistake once more bred of the life of slavery that they had been living? – a life which was always looking upon everything, … animate and inanimate – 'nature,' as people used to call it – as one thing, and mankind as another. It was natural to people thinking in this way, that that should try to make 'nature' their slave, since they thought 'nature' was something outside them.
>
> (219)

An early reviewer of *News from Nowhere* was critical that Morris had "exaggerated the dependence of human nature upon its environment" and described the novel – in a pun on the title of Morris's best-selling poem – as "not an earthly, but an earthy, Paradise". But Morris would not have seen this as a criticism: the better society he hoped for was one where a sense of the "earthiness" of the environment acknowledged our coexistence in an interdependent ecosystem, where humans did not tame nature but, in Val Plumwood's words "recognize[d our] dependency on the earth as sustaining other" (195).

Edward Carpenter (1844–1929)

Edward Carpenter resigned his Fellowship in Mathematics at Cambridge to become a lecturer in the new university extension scheme in northern England in the 1870s, a move that reflected his increasing desire for social activism. Having grown up in an affluent middle-class home and enjoyed the privileges of wealth and education, Carpenter's experience of living and teaching in the industrial north further radicalised his politics and from the 1880s for the rest of his life he, like Morris, was a committed socialist. Carpenter's interests were wide-ranging, however: he had a lifelong interest in spirituality beyond Christian orthodoxy, studying Hindu and Buddhist texts; he was a pioneer environmentalist and anti-pollution campaigner; he was a vegetarian and early animal-rights advocate; and, from the 1890s, he wrote essays challenging Victorian views of sexuality and campaigned for homosexual equality. His greatest writing success was probably a book called *Civilisation: Its Cause and Cure*, first published in 1889 but running to 18 editions over the next 40 years and appearing in multiple trans-lations around the world.

The chief reason why Carpenter may be associated with the beginnings of sustainability in the nineteenth century, however, was because of his move to a rural lifestyle from the 1880s which he saw as consistent with his socialism and his repudiation of industrial modernity. Carpenter purchased a small parcel of land at Millthorpe near Sheffield on which he established an orchard and market garden and built a cottage that he shared with his lover and his family. Carpenter advocated the virtues of manual labour, simple living, and self-sufficiency, writing and lecturing on these topics while also continuing to work at Millthorpe. From this point on, Carpenter's writing and political activism was grounded in the importance of everyday life – how we live our lives, the daily choices we make, how and what we consume, how we work, how we relate to others, the environment in which we live. These were all opportunities not only to express our values but to begin to change the world around us. In works such as *Towards Democracy*, a long prose poem first published in 1883 but revised and expanded in successive editions, and *Civilisation: Its Cause and Cure* Carpenter celebrated nature and the body, love and friendship, work and community, in ways which resonated with contemporaries who were also seeking alternatives to Victorian values. At Millthorpe, he received many visitors (including William Morris) who were keen to see how Carpenter lived and learn from him. Some went on to sim-ilarly return to the land and adopt lives of self-sufficiency.

Carpenter's essays were often addressed to readers in the social class he had renounced, calling on them to recognise the superficiality and lack of purpose in their lives. He believed that adopting a life of simplicity was a means not only to ensure greater social equality but would – paradoxically – enrich the lives of the privileged whose wealth protected them from experiencing life to the full. Consuming less not more, Carpenter believed, was the means to living the good life – good in both senses, as a virtue and a pleasurable experience. As Carpenter wrote in one essay:

> We will show in ourselves that the simple life is as good as any, that we are not ashamed of it – and we will so adorn it that the rich and idle shall enviously leave their sofas and gilded saloons and come and join hands with us in it.
>
> ("Co-operative Production" 24)

Carpenter's idealism is obvious in such statements but what is important here is that he associated simplicity of living with an enhanced everyday life, a life of pleasure, sensory experience and satisfaction through meaningful work (values also espoused in Morris's essays). Carpenter refused to see any contradiction between his socialism and his advocacy of personal change through personal action. His essays also called for more typical socialist goals such as the redistribution of wealth and nationalisation of land, along with a national system of cooperative production, but it was his articulation of an alternative way of life that attracted attention. And although he insisted he did not intend to be prescriptive on how to live the good life, his followers often interpreted his ideas as a kind of formula, adopting his practices such as wearing (and making) sandals, growing your own vegetables, or wearing simple woollen clothing. One man influenced by Carpenter's ideals, for instance, Henry Salt, resigned his teaching position at Eton, moved to a country cottage, and cut his academic gown into strips to tie up climbing vegetables in his garden.

Such earnest idealism is very easy to dismiss but Carpenter's advocacy of ethical consumption and simplicity was ahead of his time. In an essay in 1886 he urged:

> Keep at least one spot of earth clean; actually to try and produce clean and unadulterated food, to encourage honest work, to cultivate decent and healthful conditions for the workers and useful products for the public.
>
> ("Does It Pay?" 103)

Here, his association of environmental practices, safe food production, reduced consumption and ethical working conditions, have a direct connection to contemporary phenomena such as the fair-trade movement or the Slow Food organisation, based in Italy but with a global membership base, and its recent campaign "Buono, Pulito e Giusto" which seeks to link the consumption practices of the developed world with the sustainability of agricultural production and communities in the developing world.

Carpenter's idealism also presents a challenge to us as we think about sustainability in the twenty-first century. Carpenter had a theory about the relation between consciousness and the external environment or circumstances in which we find ourselves. "Desire, or inward change", he wrote, "comes first, action follows, and organization or outward structure is the result" (*Civilisation* 186). Carpenter remained committed to the agency of the individual to provide the starting point of a wider social transformation.

Both Morris and Carpenter, however, also stressed the value and importance of community. Their writings – and in Carpenter's case, his everyday practice – were premised on the idea that, as one recent scholar has put it, "the object and subject of sustainability efforts must be community" (Keough 67). Noel Keough has recently argued that a "sustaining ecological community" should be the goal of, and conduit for, achievable sustainability in our own context (and thus rejects the ecological modernisation framework of much literature on sustainable development). At Carpenter's Millthorpe, and in Morris's utopian society depicted in *News from Nowhere*, we can see an idea of what such a sustaining ecological community might look like.

Carpenter's most recent biographer, Sheila Rowbotham, has observed:

> Morris and Carpenter were utopians in the sense that they regarded politics as the means to an end; the end being a new way of living. Their conception of socialism was broad in sweep, carrying perceptions, relationships, daily life, the environment, along with it.
>
> (83)

In the writings and practice of Morris and Carpenter, I see forerunners of what I have called in previous research on contemporary culture, "slow living" (Parkins, "Out of Time"). Slow living is an approach to everyday life, arising in the late twentieth century, which advocated values of sustainability such as the preservation of local foods, environments and species, as well as cultures and communities (Parkins and Craig, *Slow Living*). Such values were linked to the adoption of a slower pace of life, which would not only allow more time for practices such as cooking, craft and forms of self-sufficiency, but also allowed slowness to take on a political valency, articulating a critique of the speed of modernity associated with globalisation, homogenisation of culture, over-consumption, and environmental depletion.

Describing our contemporary context, one recent writer on sustainability has described the problem with contemporary life to which "slow living" poses a response:

> Our practices bifurcate into mindless labour and distracting consumption, drawing us into an increasingly careless stance toward those things that sustain us. Practices [become] increasingly centred on objects that produce what we want without our attention, aid, or skill, and thus without our joy.
>
> (Davidson 112)

This critique of twenty-first century life bears a striking resemblance to the critical response to Victorian modernity found in the writings of Morris and Carpenter. The lack of care, attention and joy that modern life seemed to foster, they argued, was a damning indictment of the unsustainability of industrial modernity and the social relations it created. Listening to such Victorian voices reminds us that the solution to the sustainability crisis can never be purely technological or scientific but must be grounded in the daily practices of life and the feeling, sensing bodies through which we experience the world around us. As Davison concludes in his study of sustainability, in words that could have come directly from Carpenter, "Sustainability is nothing less, in late modernity, than the craft of a moral life" (177).

Bibliography

Butler, Judith. *Precarious Life: The Powers of Mourning and Violence*. Verso, 2004.

Carpenter, Edward. "Does It Pay?" *The Selected Works*. 1886. Prism Key Press, 2012.

Carpenter, Edward. *Civilisation: Its Cause and Cure*. 1889. Allen & Unwin, 1920.

Carpenter, Edward. *Towards Democracy* (1st pub. 1883). 3rd edn, T. Fisher Unwin, 1892.

Carpenter, Edward. "Co-operative Production: with reference to the experiment of Leclaire. A Lecture Given at the Hall of Science, Sheffield, 1883". John Heywood, 1883.

Clark, Nigel. "Volatile Worlds, Vulnerable Bodies: Confronting Abrupt Climate Change". *Theory, Culture and Society*, vol. 27, nos. 2–3, 2010, pp. 31–53.

Colebrook, Claire. "Framing the End of the Species: Images without Bodies". *Symplokē*, vol. 21, nos. 1–2, 2013, pp. 51–63.

Davison, Aidan. *Technology and the Contested Meanings of Sustainability*. SUNY Press, 2001.

Keough, Noel. "Sustaining Authentic Human Experience in Community". *New Formations*, no. 64, 2008, pp. 65–77.

MacDuffie, Allen. *Victorian Literature, Energy, and the Ecological Imagination*. Cambridge University Press, 2014.

Medovoi, Leerom. "A Contribution to the Critique of Political Ecology: Sustainability as Disavowal". *New Formations*, no. 69, 2010, pp. 129–143.

Mill, John Stuart. "The Principles of Political Economy with Some of Their Applications to Social Philosophy". 1848. *Collected Works of John Stuart Mill*, Vol. III, edited by John M. Robson, University of Toronto Press, 1965.

Morris, William. *News from Nowhere*, edited by Stephen Arata, Broadview, 2003.

Morris, William. "The Lesser Arts". 1877. *William Morris on Art, and Socialism*, edited by Norman Kelvin, Dover Publications, 1999, pp. 1–18.

Morris, William. "The Society of the Future". 1887. *Political Writings of William Morris*, edited by A. L. Morton, Lawrence and Wishart, 1990, pp. 188–203.

Morris, William. "The Prospects of Architecture in Civilization". 1881. *The Collected Works of William Morris*, Vol. XXII, edited by May Morris, Longmans Green and Company, 1910–1915, pp. 119–152.

Parkins, Wendy. "Out of Time: Fast Subjects and Slow Living". *Time and Society*, vol. 13, nos. 2–3, 2004, pp. 363–382.

Parkins, Wendy, and Geoffrey Craig. *Slow Living*. Berg, 2006.

Plumwood, Val. *Feminism and the Mastery of Nature*. Routledge, 2002.

Rowbotham, Sheila. *Edward Carpenter: A Life of Liberty and Love*. Verso, 2009.

Ruskin, John. *Unto This Last*. 1860. *The Works of John Ruskin*, Vol. 17, edited by E. T. Cook and Alexander Wedderburn, George Allen, 1905.

I.4   Innovation and consumption in the evolution of capitalist societies

Pier Paolo Saviotti

I Introduction

In the last 200 years economic development has produced an unprecedented wealth but also endangered sustainable life on earth. In this chapter, the evolution of consumption since the Industrial Revolution will be analysed in relation to the overall pattern of economic development. Consumption was limited to basic necessities until the twentieth century and only then started to change towards more differentiated and higher quality goods and services. The evolution of consumption will be compared to predictions that John Maynard Keynes made in 1930. Finally, emphasis will be laid on the need to change our present pattern of consumption towards a more sustainable one.

II Structural change and economic development

Cumulative economic growth started with the Industrial Revolution. Previously the output of the economic system did not seem to have changed systematically since medieval times but rather oscillated around a constant value. This lack of cumulative growth led some scholars to define the period preceding the Industrial Revolution as a Malthusian period (Galor and Weil). Malthus stressed the role of food production, and thus of land, as a constraint that would forever limit population growth. According to him any spurt of growth in food production would induce a growth of the population, leading to a shortage of food and a renewed reduction of population. As a consequence, both output and population could be expected to oscillate around a constant level. There could never be any cumulative growth. This upper limit to growth described by Malthus seemed to be inescapable until the Industrial Revolution. Malthus' emphasis on food production reflected the fact that until the Industrial Revolution most people were poor and could only satisfy basic needs such as food, clothing and housing.

There is now a general consensus that innovation was an extremely important factor involved in the Industrial Revolution and subsequent economic development. Examples of important innovations are the new machinery used in the textile industry, steam engines, new forms of energy and new materials (Landes; Mokyr; Hobsbawm). However, innovation alone could not have given rise to the

Industrial Revolution. New forms of organisation of labour, such as the factory system, new social classes formed by capitalists and labourers and new forms of finance for industry were required. Finally, innovation could not have contributed to economic development unless there was a demand for new innovative products and services. The formation of such a demand required both a technology that could produce goods and services at affordable prices and an adequate disposable income. Such a combination only became available at the beginning of the twentieth century.

At the time of the Industrial Revolution the environmental impact of human activities was not perceived as an important problem except at the level of the workplace. In fact, the global environmental impact was initially limited but increased with the size of human activities. It is now comparable to that of natural phenomena, which has suggested the name Anthropocene for our present epoch. The economic development that began with the Industrial Revolution is no longer sustainable and consumption spreading from industrialised to emerging countries is an important component of our environmental problem. The evolution of consumption is at the core of this chapter.

In the United Kingdom during the Industrial Revolution, when growth started to be cumulative, the limit represented by the land did not matter any longer. Population grew, but the output of basic needs could barely be satisfied. Although industrialisation gave rise to a spectacular increase in productive efficiency, working-class households could barely afford to purchase basic goods during most of the nineteenth century in the richest country in the world (Hobsbawm). Starting from the beginning of the twentieth century a growing disposable income allowed most consumers to purchase goods and services more sophisticated than the ones aimed at satisfying basic needs.

In spite of the considerable increase in efficiency since the beginning of the Industrial Revolution, most people in industrialised countries started to consume a wide variety of goods only in the twentieth century. In order to understand why it is necessary to introduce some basic notions of economic development and then come back to the role of consumption. Growth is a quantitative concept that tells us how much "bigger" the output of the economic system becomes in a given period. Knowing the rate of growth does not tell anything about the internal structure of the economic system or the changes it has undergone. Yet there are very good reasons to believe that quantitative growth cannot continue in the long run unless the structure of the economic system changes. Thus, as we will see, growth is driven by structural change.

The structure of an economic system is characterised by its components and their interactions. Examples of components can be industrial sectors, fields of human activity such as health care, or institutions such as the education system or the labour market. Examples of interactions are the flows of goods and knowledge between different industrial sectors, the flows of knowledge between research institutions and industrial sectors, the interactions between health care or tourism and sectors such as pharmaceuticals or aviation. Typically, such components and interactions define a structure that is relatively stable over the course of time. For example, since the Industrial Revolution the economic systems of industrialised countries have relied heavily on fossil fuels, coal first and then oil and gas. The transition to a sustainable economy will involve the disruption of the links between fossil fuels suppliers and users and their replacement by new links between suppliers and users of renewable inputs. This is going to be a major structural change, which can be understood by gaining an insight into the most important types of structural change which have occurred since the Industrial Revolution and by assessing how they have contributed to the overall process of economic growth. Particular focus will be laid on changes in consumption and demand.

Stylised facts are general tendencies detected in empirical data and selected by researchers for their potential importance in understanding economic development. They are considered particularly significant by those economists who do not expect to be able to deductively derive all observed economic phenomena from a basic set of laws and axioms, but who expect to start from observations in order to understand patterns of economic development. Three stylised facts, consisting of increases in productive efficiency, variety and intra-sector differentiation, have kept occurring throughout economic development since the Industrial Revolution. They are related to two basic properties of economic activities: efficiency and creativity.

An economic activity is considered here as the transformation of inputs into outputs. The efficiency of economic activities is determined by the ratio of outputs produced to inputs used. However, when inputs (I) and outputs (Q) are measured by value, as is normally the case, their ratio (Q/I) is not a pure measure of economic efficiency since it includes the effect of changes in values of I and Q. An increase in efficiency happens when a greater output is produced per unit of input, or conversely, when a smaller input is required to produce a unit of output. The ratio Q/I measures an increase in efficiency over a given period (t2-t1) only when the output produced is qualitatively unchanged. If during the same period there is an increase in both efficiency and the value of output, for example when there is an increase in the quality of the output justifying a rise in its price, the ratio Q/I will be a combined measure of efficiency and value. For example, if in a process aimed at producing a given type of shoes we reduce the quantity of all the inputs required per pair of shoes, we achieve a quantitative increase in efficiency because with the same amount of resources we can produce more shoes. However, if we change the type of shoes to a much higher quality, we are combining a change in efficiency with a change in value. Even in the unlikely case in which we needed a lower amount of the same inputs than the amount needed for the low-quality shoes, we would produce a qualitatively different type of output which would appeal to a different type of customer and fetch a higher price. Furthermore, it is quite unlikely that we can use the same inputs for the low and the high-quality shoes. Typically, we would expect the production of a higher quality output to require more sophisticated physical inputs and higher competencies, which would require higher wages. In other words, a pure measure of efficiency is possible only if the output is qualitatively unchanged and the change is only quantitative.

An example of increase in pure productive efficiency is given by the evolution of spinning in the early years of the Industrial Revolution. Spinning in the textile industry refers to the stage in which raw cotton is transformed into a thread. The efficiency of spinning increased significantly owing to the adoption of new textile machines, such as the spinning jenny or the mule. Assuming that the thread obtained through the new textile machines was identical to the one obtained through a spinning wheel, the only effect of the mechanisation of textiles was a reduction in the cost of their output.

Productive efficiency alone could not have led to constant growth since the onset of the Industrial Revolution, but it could only raise the output per unit of input in the same output types existing at the time. History shows that while productive efficiency increased in the existing types of output, two other trends contributed to observed economic development: new sectors in which the output was qualitatively different from pre-existing ones emerged; the quality and the differentiation of the output of pre-existing sectors increased. These two trends enhanced the variety of the economic system at the inter-sectoral and intra-sectoral level of aggregation respectively.

While increased efficiency can reduce the cost of a given type of output, it cannot produce qualitatively different types of goods and services. Only a different kind of property, called creativity, can generate completely new types of output. For example, when bicycles, cars, airplanes or computers were created, nothing similar existed. They were qualitatively different from anything that existed before and qualitatively different from one another. While efficiency gives rise to quantitative change, creativity gives rise to qualitative change.

The variety of an industrial system is given by the number of its distinguishable sectors. Each sector produces an output qualitatively different from all the others. An economic system includes organisations such as educational establishments or hospitals that are not industrial, although they interact with industrial sectors. What follows will focus on the variety of the industrial system, which measures the extent of its differentiation.

The differentiation of an economic system can occur at different levels of aggregation. The lowest possible level of aggregation is that of individuals. Firms and industrial sectors are higher levels of aggregation. Typically, firms correspond to the micro level, industrial sectors to an intermediate or meso-level (Dopfer *et al.*) and the whole economic system to the macro level. Variety can be measured at different levels of aggregation. Here the intra-sector and inter-sector levels will be examined.

The quality of the output of a given sector is defined by a combination of the services supplied by the same output. In principle, an output can be either a physical good or a disembodied service. Physical goods are not purchased for their material nature but for the type of services they supply (Lancaster; Saviotti and Metcalfe), which correspond to product characteristics. For example, cars or aircraft supply transport services such as speed, payload or range. These individual services can then be combined into an overall output quality based on the preferences of users or consumers. Physical goods differ from pure services because they supply their services by means of an "internal structure", in which the combination of different types of matter is shaped to supply the services desired. A physical product can then be represented by two sets of characteristics, one corresponding to the internal structure and the other to the services supplied to users or consumers. The design of a physical product of technology involves the transformation of the knowledge required to conceive of and produce the internal structure into marketable services.

Examples of internal structure are the body and mechanical parts of a car, the mechanisms of a watch or camera, and the hardware of a computer. The environmental impact of a technology is due in part to the production of its internal structure and in part to the social use of the technology itself. For example, the environmental impact of a car is determined by the materials and equipment for the production of its internal structure, but also by the inputs (fuel) and infrastructures (motorways) required for its use. On the whole, a reduction in the environmental impact of cars involves different designs (e.g. electrical cars) and a more efficient utilisation of cars. This could be achieved, for example, by car sharing. Sharing economy can involve all the products which are used sparingly but are currently purchased for the whole time. Another example of a very common product that could be used in a sharing mode is that of tools.

Similarly, a pure service (e.g. teaching) is supplied either by pure labour or by labour and capital goods, but without the intervention of a physical device in the final supply. Recently, the production and delivery of pure services have become increasingly capital-intensive, due to the growing use of ICT. Pure services can sometimes compete with services supplied by means of physical products. For example, railway companies and airlines can compete with cars. Emphasis will be laid on physical goods, but most of the conclusions can be extended to pure or disembodied services.

Car models differ according to characteristics, or services, supplied to their users, such as speed, acceleration, number of passengers, size of boot or fuel consumption. The value of these characteristics tends to increase from the cheapest to the most expensive model. Within each product class (e.g. cars, aircraft, computers), corresponding to a sector, there are many product models, each with a different quality level. The range of quality levels and the number of models define the intra-sector differentiation.

The three stylised facts described above will now be considered in long-run trajectories. Trajectories are general trends in certain variables occurring over long periods. Each of the three trajectories presents a constant increase in one property: efficiency for T1, output variety for T2, output quality and intra-sector differentiation for T3. The three trajectories are described in a greater detail in Table I.4.1.

Trajectory 2 represents the growing variety of transport technologies. All the technologies represented are qualitatively different from one another. Trajectory 3 represents the increasing product quality and intra-sector differentiation of automobile technology. The quality improves along with the improvement of services such as speed, acceleration or fuel consumption.

Table I.4.1 Trajectories of economic development

Trajectory 1: growing efficiency (T1)	*Trajectory 2: growing output variety (creativity) (T2)*	*Trajectory 3: growing output quality and intra-sector differentiation (creativity) (T3)*
The efficiency of productive processes increases during the course of economic development. Efficiency = the ratio of the inputs used to the output produced, when the type of output remains *constant*.	Here such variety is measured by the number of distinguishable sectors. Sectors are distinguishable when their outputs are *qualitatively different*, and thus not comparable and not substitutable. A sector is defined as the set of firms producing a common although highly differentiated output	The average quality of the output of existing sectors improves in the course of time and the range of outputs with different quality levels becomes wider. This means that, if during the period of observation, the type of output changes, what we will observe is a combination of growing productive efficiency and of quality change

Why did these trajectories occur? Were they independent or related? If they were related, how did they interact? Answers involve the time of their emergence. Increases in efficiency began very early in human history. Inventions such as the wheel, wind and water mills, the wheelbarrow (Mokyr) are very ancient. However, they could not lead to cumulative economic growth before the Industrial Revolution. In particular, before the Industrial Revolution, in what has been called the Malthusian period, agriculture was practiced by most of the population and most people barely managed to feed themselves. This continued during most of the nineteenth century in spite of the considerable increases of efficiency. The innovations introduced in the nineteenth century were oriented towards capital goods or the production of basic goods such as textiles. Here the focus is not on the reasons for the late start take off of efficiency growth in the Industrial Revolution, but on the delay between efficiency growth (T1) on the one hand, and variety (T2) and sectoral quality and intra-sector differentiation (T3) on the other. Increases in T2 and T3 only started during the twentieth century. What are the causes for this delay in spite of the considerable rise in efficiency since the beginning of the Industrial Revolution? There are essentially two reasons why efficiency growth (T1) preceded T2 and T3. First, throughout human history most people barely managed to satisfy their primary needs, such as food, clothing and housing. They lacked the resources to satisfy other human wants. If this situation had persisted indefinitely, the collapse of capitalism predicted by Marx would have likely occurred. During the nineteenth century, the high rate of population growth absorbed all the rise in productive efficiency generated by the Industrial Revolution. In the twentieth century a number of factors, including growing levels of education, a slowdown in the rate of population growth and redistributive economic policies, contributed to create the disposable income required to go beyond basic needs. That human needs and wants were hierarchically organised was understood by economists and social theorists such as Carl Menger and Abraham Maslow by the end of the nineteenth century.

II.1 Consumption, demand and economic development

A hierarchical theory of wants has been discussed by leading economists at the end of the nineteenth century such as Léon Walras, William Stanley Jevons, Alfred Marshall and has received its most explicit and detailed treatment in the work of Carl Menger. This theory implies that wants can be ranked in order of absolute importance, with most basic wants at the bottom of the list (lower wants) and with the most sophisticated (higher wants) at the top. Although such hierarchical theory is obviously correct for what concerns the relationship between primary needs and higher wants, it is much less useful for the ranking of higher wants. However, here it suffices to state that in order for higher wants to be satisfied, a threshold represented by the minimum efficiency required to cover primary needs for the whole population had to be overcome.

Throughout the nineteenth century, consumption by working-class households was limited to primary needs because the high rate of population growth absorbed all the increases in productive efficiency that had occurred. The rate of population growth increased significantly from 1750 to 1850 for reasons that are not yet entirely clear, but include the opportunity to marry younger and raise a family, due in part to the increasing wealth created by the Industrial Revolution (Hobsbawm 42–45). As the rate of population growth started to decline in the twentieth century, a higher income per capita created the disposable income with which consumers could purchase higher goods and services. Furthermore, even without the effect of population growth the production of higher quality and more differentiated goods and services could not have created a high enough disposable income in the initial phases of the Industrial Revolution. Such a disposable income only became available when the growth of income per capita was sufficiently accelerated by the production of higher quality and more differentiated goods and services (T3), which required higher competencies supplied by education. Thus, in a co-evolutionary process education created the required competencies but also enhanced purchasing power by raising wages and creating further employment. Of course, the co-evolutionary process included other processes and institutions, such as the emergence of marketing or the growing recourse to loans in order to finance purchases. The transition from a pattern of consumption centred around primary needs to one in which manmade wants began to dominate demand was defined as a shift from what was necessary to what could be imagined (Saviotti and Pyka, "From Necessities to Imaginary Worlds").

Could economic development not have gone on producing growing quantities of food, clothing and housing for everyone? If no new goods and services had emerged, the demand for food, clothing and housing could not have grown as fast as efficiency, in principle allowing for the production of all the output demanded through a declining share of the labour force or shorter working times. This bottleneck, which would have been a huge obstacle to the long-run continuation of economic growth, could be avoided through emerging new goods and services, as their production would absorb the labour force potentially made redundant by the imbalance between demand and efficiency (Pasinetti, *Structural Change*; Pasinetti, *Structural Economic Dynamics*). Thus, economic development could not have continued in the long term unless the creation of new goods services determined a growing output variety.

Although growing efficiency alone is not sufficient to ensure the long-term continuation of economic growth, it is necessary to it. The creation of new goods and services requires resources to be invested in search activities. Such resources can only derive from a surplus generated by the growing productive efficiency in pre-existing activities. However, as already seen, if such a surplus had not been reinvested, it would have led to growing unemployment and a bottleneck in economic growth. Thus, growing efficiency and growing variety are complementary mechanisms of economic development. Both endogenous growth models and Marxist analysis failed to appreciate this complementarity. Endogenous growth models failed to clearly distinguish between efficiency and variety, although they introduced some form of variety (Romer) or product quality (Aghion and Howitt). Marx himself did not distinguish between efficiency and variety. As a consequence, he did not take into account the possibility that growing variety could provide an escape route from the bottleneck that an economic system driven exclusively by growing efficiency would encounter.

An alternative mechanism that contributed to partly compensate the effect of increasing productive efficiency was the reduction of working hours, which during the twentieth century dropped from approximately 3,200 per year to approximately 1,700 in ten industrialised countries (Huberman). In principle, this reduction would allow to preserve a constant employment by having each labourer work shorter times. This mechanism was in fact used, but it was combined with the employment created by the production of new goods and services. In fact, the reduction of working hours slowed down in the second half of the twentieth century, precisely when the rate of growth of intra-sector output quality and differentiation was increasing. Thus, reductions in working hours neither completely replace the emergence of new goods and services nor their growing quality and internal differentiation. In summary, reduced hours and the overall differentiation created by T2 and T3 were combined to generate the observed path of economic development. It is to be noticed that the observed reduction in working times included both a shorter working week and an increasing length of paid holidays. Growing leisure times have become an important component of all industrialised and post-industrial societies. In addition to improving people's welfare, they have contributed to provide additional employment in the form of the services required during leisure periods and holidays. Of course, all these changes would have been difficult to conceive of without increasing wages and achieving a more equitable income distribution.

II.2 Keynes' predictions

It is interesting to compare the observed pattern of economic development with the predictions made by Keynes in 1930. He thought that if productivity kept growing at the same pace as it had done in the previous century, it would have solved what he called the "economic problem" and would have drastically reduced the fraction of individual lifetime spent on working activities. He distinguished between "absolute needs", which we have regardless of the situation of other human beings, and "relative needs", which we have to make us feel superior to our fellow citizens. He considered the latter an unimportant component of consumption.

Keynes seems to have imagined that the future economic system would be a stable state in which there would be consumption satiation, no need for capital accumulation and very small savings. Owing to the increasing productive efficiency, the need to work would be drastically reduced to the point where individuals would be able to work no longer than 15 hours per week. This prediction was based on the hypothesis that consumption would be satiated. In fact, Keynes assumed that the "economic problem", which he considered equivalent to the satisfaction of absolute needs, coinciding with primary needs, would be definitely solved. Although a reduction of working hours has occurred, there has been a differentiation of consumption beyond what he called the "economic problem".

So, Keynes was wrong on a number of counts:

1   He saw the satisfaction of consumption as a part of a final steady state in which there would be no more innovation and capital accumulation would cease. This is contradicted by the extremely dynamic and restless character of capitalism (Metcalfe, *Evolutionary Economics*; Metcalfe *et al.* "Adaptive Economic Growth").
2   As a consequence of 1) he failed to predict the vigorous development of consumption by the emergence of new goods and services (e.g. consumer durables, health care, etc.).
3   An underestimation of new sources of consumption led him to overestimate the reduction of the fraction of life spent working. Although, as pointed out before, the working times per year diminished, they did so at a slowing pace in the second half of the twentieth century, precisely when the variety and quality of consumption items increased very quickly.
4   Keynes predicted that in the future final state everyone would have ample leisure time during which they would indulge in the cultural activities preserving the bourgeoisie. However, he thought that these activities would not have an economic character. Yet, in addition to purely productive activities, leisure activities can give rise to income and employment, as already remarked. Clear examples are tourist operators, the media providing entertainment, and sports.

Thus, economic development involved both a fast increase of the variety and quality of the goods and services consumers purchased and a reduction of working hours, although not as much as Keynes predicted. His failure to foresee the increase in consumption was rooted in a far too static conception of capitalism that led him to hypothesise a final steady state of the economic system in which consumption is satisfied. As previously pointed out, such a steady state would have been unstable and would have led to a development bottleneck determined by the imbalance between the rate of growth of productive efficiency and that of demand. In turn, this bottleneck could have been overcome by an increased production and consumption of a growing variety and quality of goods and services. Thus, the observed development path included both an increase of consumption and a reduction of working times, although the latter was inferior to the one predicted by Keynes.

Keynes considered the consumption corresponding to relative needs unimportant, compared to that corresponding to absolute needs. In fact, the former, called "imaginary worlds" by Saviotti and Pyka ("From Necessities to Imaginary Worlds"), was much more dynamic than the one related to primary needs. This development path, characterised by the continuous emergence of new economic activities, is rooted in a Schumpeterian vision. In *Capitalism, Socialism and Democracy* Schumpeter argued that capitalism, being essentially a method of economic change, never is and never can be stationary (82). Thus, we cannot expect history to reach a final state. The extent of economic change that occurred during the twentieth century was not limited to the emergence of sectors other than those corresponding to basic needs. Even within the sectors supplying basic needs the increase of quality and internal differentiation was extremely high. However, if higher quality and variety were functional requirements in a selected number of cases, in most cases their function was rather to let other people know who you are. This type of consumption, which is often called Veblen consumption (1899), was not as insignificant as Keynes had assumed and recently has become widespread even in the middle or lower classes (Saviotti and Pyka, "Innovation, Structural Change"). The huge rise in variety and quality of food, clothing and housing that happened in particular after the Second World War as well as the growing importance of brands show that most new forms of consumption are not of a strictly functional nature.

A further limitation of Keynes' prediction that working hours would drop to 15 hours per week was due to the changing nature of work. Up to his time poor people worked long hours in jobs that were stultifying and unhealthy, while rich people had plenty of leisure. Recently, the situation seems to have been reversed: working times have been reduced in jobs involving physically demanding tasks while executives and top managers spend longer and longer hours in their offices (Freeman). These changes result from the gap between work as drudgery and work as identity: top managers work long hours because they feel happy to do so while repetitive work is gradually being automated.

The value of Keynes' work is unquestionable. His predictions were not completely incorrect. The economic problem, intended as the complete satisfaction of the basic needs expressed by the majority of the population in industrialised countries, was overcome and a considerable reduction of working hours did occur. He failed to predict the considerable expansion of consumption, both in quality and quantity, that the citizens of the same countries would experience. If we were to repeat Keynes' exercise and foresee the future of consumption, we could hardly expect consumption to stabilise or retreat either in quantity or quality. Although a point might be reached where the variety of individual consumption could not increase any further, aggregate consumption variety could still rise if individual consumption became more and more differentiated. There is no reason why it should be impossible to pursue the differentiation of the industrial system without increasing our environmental impact. This objective can be attained by green economy or by circular economy. The former can create new products which combine environmental sustainability with growth potential. The latter considers all human activities as parts of an ecosystem in which the outputs of an activity can become the inputs of another.

The topics investigated are closely connected to the environmental impact of human activities. The twentieth-century consumption style is certainly unsustainable and should be modified. The simplest way would be to go back to a much simpler form of consumption closer to the satisfaction of basic needs. Of course, this would make us all much poorer. At present, nobody seems to contemplate this solution. Although we may have to discard particularly polluting forms of consumption, such as unrecyclable plastic bags, in most cases the solution is likely to involve reducing the environmental impact of the demanded goods and preserving the services they supply to consumers, instead of returning to previous epochs. The environmental impact of some existing technologies per unit of use has decreased in recent years. Inspired by this reduction, the environmental Kuznets curves (Stern *et al.*) hypothesise an inverted U shape in the relationship between environmental impact and per capita income. However, the growing international circulation of the same technologies has more than compensated for this reduction, thus leading to a growing global impact.

III Conclusions

Economic development after the Industrial Revolution has generated a considerable increase in the variety and quality of goods and services consumed by a large share of the population in industrialised countries. This increase in consumption did not start immediately after the beginning of the Industrial Revolution but had to wait until the twentieth century, owing to the very fast increase in population during the nineteenth century and to the lack of the disposable income required. An expanded consumption only became possible in the twentieth century through a co-evolutionary process linking education, innovation, a balanced distribution in the shares of salaries and profits, and the formation of an adequate disposable income enabling the purchase of new and high-quality goods and services.

In this chapter, the observed pattern of development has been compared to Keynes' predictions in 1930 about the future of consumption. The economic problem, which he equated to the satisfaction of basic needs, was solved, working hours could be reduced to 15 hours per week and most people could enjoy ample leisure time. His predictions contrasted with the observed economic development, because working hours dropped less than he had anticipated and consumption went well beyond basic needs by incorporating many new and high-quality goods and services.

The need to make our economic system sustainable will require a considerable transformation of the existing pattern of consumption, in order to reduce the environmental impact of production processes as well as change the social organisation of consumption. New forms of consumption, such as car sharing, are being developed. A widespread adoption of good practices requires profound changes in social habits and paradigms as well as a synergy of different fields of knowledge.

Bibliography

Aghion, Philippe, and Peter Howitt. "A Model of Growth through Creative Destruction". *Econometrica*, vol. 60, no. 2, 1992, pp. 323–351.

Dopfer, Kurt, John Foster, and Jason Potts. "Micro-Meso-Macro". *Journal of Evolutionary Economics*, vol. 14, no. 3, 2004, pp. 263–279.

Freedman, Robert, Editor. *Marx on Economics*. Penguin Books, 1962.

Freeman, R. B. "Why Do We Work More than Keynes Expected?" *Revisiting Keynes. Economic Possibilities for Our Grandchildren*, edited by Lorenzo Pecchi and Gustavo Piga, MIT Press, 2008, pp. 135–142.

Galor, Oded, and David N. Weil. "Population, Technology, and Growth: From Malthusian Stagnation to the Demographic Transition and Beyond". *The American Economic Review*, vol. 90, no. 4, 2000, pp. 806–828.

Hobsbawm, Eric John Ernest. *Industry and Empire*. Penguin Books, 1968.

Huberman, Michel. "Working Hours of the World Unite? New International Evidence of Worktime, 1870–1913". *Journal of Economic History*, vol. 64, no. 4, 2004, pp. 964–1001.

Jevons, William Stanley. *The Theory of Political Economy*. 4th edn, Macmillan, 1924.

Keynes, John Maynard, "Economic Possibilities for Our Grandchildren (1930)". *Revisiting Keynes. Economic Possibilities for Our Grandchildren*, edited by Lorenzo Pecchi and Gustavo Piga, Cambridge MIT Press, 2008, pp. 17–26.

Lancaster, Kelvin J., "A New Approach to Consumer Theory". *Journal of Political Economy*, vol. 74, no. 2, 1966, pp. 132–157.

Landes, David S., *The Unbound Prometheus. Technological Change and Industrial Development in Western Europe from 1750 to the Present*. Cambridge University Press, 1969.

Malthus, Thomas Robert. *An Essay on the Principle of Population*. Joseph Johnson, 1798.

Marshall, Alfred. *Principles of Economics*. 8th edn, Macmillan, 1949.

Marx Carl. *Capital: A Critique of Political Economy*. Lawrence & Wishart, 1887.

Maslow, Abraham H. "A Theory of Human Motivation". *Psychological Review*, vol. 50, no. 4, 1943, pp. 370–396.

Menger, Carl. *Principles of Economics*. 1871. Translated by James Dingwall and Berthold Frank Hoselitz, with an Introduction by Friedrich A. Hayek. New York University Press, 1981.

Metcalfe, John Stanley, John Foster, and Ronnie Ramlogan. "Adaptive Economic Growth". *Cambridge Journal of Economics*, vol. 30, no. 1, 2006, pp. 7–32. doi:10.1093/cje/bei055.

Metcalfe, John Stanley. *Evolutionary Economics and Creative Destruction*. Routledge, 1998.

Mokyr, Joel. *The Lever of Riches: Technological Creativity and Economic Progress*. Oxford University Press, 1990.

Pasinetti, Luigi L. *Structural Change and Economic Growth*. Cambridge University Press, 1981.

Pasinetti, Luigi L. *Structural Economic Dynamics*. Cambridge University Press, 1993.

Romer, Paul M. "Endogenous Technological Change". *Journal of Political Economy*, vol. 98, no. 5, Part 2, 1990, S71–S102.

Saviotti, Pier-Paolo, and John Stanley Metcalfe. "A Theoretical Approach to the Construction of Technological Output Indicators". *Research Policy*, vol. 13, no. 3, 1984, pp. 141–151.

Saviotti, Pier-Paolo, and Andreas Pyka. "Innovation, Structural Change and Demand Evolution: Does Demand Saturate?" *Journal of Evolutionary Economics*, vol. 26, 2015, pp. 1–22. doi 10.1007/s00191-015-0428-2.

Saviotti, Pier-Paolo, and Andreas Pyka. "From Necessities to Imaginary Worlds: Structural Change, Product Quality and Economic Development". *Technological Forecasting and Social Change*, vol. 80, 2013, pp. 1499–1512.

Schumpeter, Joseph. *Capitalism, Socialism and Democracy*. 1942. 5th edn, George Allen and Unwin, 1976.

Stern, David I., Michael S. Common, and Edward B. Barbier. "Economic Growth and Environmental Degradation: The Environmental Kuznets Curve and Sustainable Development". *World Development*, vol. 24, no. 7, 1996, pp. 1151–1160.

Veblen, Thornstein. *The Theory of the Leisure Class: An Economic Study in the Evolution of Institutions*. Macmillan, 1899.

Walras, Léon. *Éléments d'Economie Politique Pure, ou, Théorie de la richesse sociale*. 1896. 3rd edn, F. Rouge, 1988.

I.5   In a prescient mode

(Un)sustainable societies in the post/apocalyptic genre

Paola Spinozzi

I Apocalypse in utopia as a literary genre

The topos of the apocalypse is ancient and universal. Prophecies about a catastrophe causing the demise of civilisation resonate throughout the literature of Judaism and Christianity from 1 BC to the Middle Ages. The *Book of Revelation*, the last in the New Testament, is associated with two dates: AD 68–70, following the persecutions under Emperor Nero and preceding the destruction of Jerusalem in AD 70, and AD 90–95, the period of the persecutions in the last years of Emperor Domitian, ruling in AD 81–96. The apocalypse as destruction and renewal is at the core of millenarianism: vibrantly evoked, the end of the world is a divine flagellum, after which the Messiah announced in the Scriptures will come to establish his millennium, a new world of peace and wellbeing. The apocalyptic end is destructive as it is cathartic: it annihilates corruption, poverty, illness and brings a new order.

While marking an end, the downfall is inscribed within a cycle of eternal return, according to which disastrous events will recur again and again infinitely, leading humankind to always redefine its place and purpose in history. The concept of cyclical history presupposes that progression is also a reiteration. Circularity involves the awareness that highly advanced stages of development have been reached periodically and catastrophic events have occurred repeatedly. The notion of eternal recurrence implies that any collapse in the history of a civilisation also coincides with a starting point: the end comes as inevitably as does a new beginning. The apocalypse heralds palingenesis.

Utopia as a literary genre is a catalyst for terminal visions. Whereas the intention of apocalyptic writers *stricto sensu* is to anticipate the approaching end, early modern to contemporary utopian authors elaborate hypotheses for the benefit of future generations. Floods, droughts, fires, earthquakes, total eclipses and famines were understood as forms of divine retribution in pre-modern times, after which human intervention began to play an increasingly significant role. The emphasis on anthropogenic events gained momentum in the nineteenth century: since the inception of the Industrial Revolution the counter-effects of technological progress and the degradation of the environment have been denounced in tales of nuclear explosions, collisions of planets, climate change, and pandemics. In the contemporary age, authors of post-apocalyptic novels speculate on how humanity will adapt to different causes of risk and will cope with specific typologies of disasters. The ways in which humankind is portrayed responding to apocalyptic portents and cataclysms reveal the evolution of post-apocalyptic narratives over the last two centuries.

In the contemporary age risk has become, and is perceived to be, constant and all-pervasive: disaster strikes and will continue to strike. Preparedness and resilience have been adopted to explain how humanity adapts, first by coping with disaster and then by rebuilding the habitat. The United Nations Office for Risk Reduction have defined resilience as:

> The ability of a system, community or society exposed to hazards to resist, absorb, accommodate to and recover from the effects of a hazard in a timely and efficient manner, including through the preservation and restoration of its essential basic structures and functions.
>
> (UNISDR, "Resilience" 24)

How a community "resiles" or "springs back" depends on the availability of resources and the ability of organising itself before and during a shock. Resilience intersects adaptation, namely the ways in which a socio-economic system reacts to shocks in the short term, and adaptability, that is its capacity to adapt to new growth paths or to changing environment in the long term.

These concepts will be adopted as tools for interpreting literary representations of the apocalypse and the post-apocalypse. Apocalyptic writers focus on the end of the world, describing a natural disaster, a nuclear holocaust or the degeneration of society, while post-apocalyptic authors display the aftermath of catastrophe and the construction of a new world. Apocalyptic dystopias delve into the dynamics of the catastrophic event, showing how the protagonists are challenged by extreme circumstances. Post-apocalyptic novels raise questions as to whether risk and disaster can generate agency by strengthening capacity building in present-day societies.

II *Fin-de-siècle* apocalypse

Imagery in the *Book of Revelation* expresses different kinds of angst. Fear and pessimism feed apocalyptic narratives, inspiring eschatological visions. In Thomas More's *Libellus vere aureus* (1516) the protagonist Raphael Hythlodaeus bears the name of the angel who reveals the apocalypse. Destruction and reconstruction are entwined in utopian thought: catastrophe will strike and a new beginning will require sustained human volition.

Owing to its ductile imagery, the apocalypse has continued to resonate at the end of every century. In the 1880s, while Great Britain was facing the aftermath of the Industrial Revolution, fear of the future conjured up terminal scenarios. The paradigm of destruction and reconstruction defines William Delisle Hay's *Three Hundred Years Hence; or, A Voice from Posterity*, Richard Jefferies' *After London; or, Wild England*, William Henry Hudson's *A Crystal Age* and William

Morris' *News from Nowhere; or, An Epoch of Rest Being Some Chapters from a Utopian Romance*. While Britain became an industrialised nation and evolutionary theories questioned the existence of God, nightmarish visions of the future reverberated in fin-de-siècle dystopian novels.

Three Hundred Years Hence; Or, a Voice from Posterity is presented as a transcript of "Professor Meister's conversations", a series of university lectures on the history of mankind from 1880 to 2180, dealing with overpopulation, insufficient food supply and natural selection:

> Their climate was rigorous, their food of inferior quality and often scanty to the famine point, their lives were subject to various unwholesome influences, disease was rife, and medical and sanitary science in its darkest age among them.
>
> (19)

> How people were sallow and sickly in appearance, always ailing, shortlived, and depressed in spirit; how childhood was specially injured, and its growth and development impaired; how vegetation would not thrive amid the sulphurous fumes, thus abundantly testifying to the unwholesomeness of the respirable air.
>
> (100)

During the Terrane Exodus, described as a universal emigration from the land into the Cities of the Sea, the water surface was divided into 60 States and the dry land became common property for rent. The world map in 2180 comprises submarine towns of white koralla, such as the harbour of Krakenburgh, within huge aerated domes on the Ocean bed; supermarine cities like Londinova and towns like Aquamarina; hypogeic states with cities of metal buildings like Argenta, the capital of the Interior State of Sonoria. Wild animal life is extinct and nature has been equalised and balanced by bioengineering. The Great American Republic and its ideas of Freedom, Brotherhood, and Progress were adopted as the model for the Federate Nations of the Twentieth Century of Peace. Professor Meister shows what he defines as the scientific classification of humans back in the nineteenth century. On the basis of divergent physical types and corresponding mental faculties, "we look back upon the Yellow Race with pitying contempt, for to us they can but seem mere anthropoid animals, not to be regarded as belonging to the race that is summed and glorified in United Man" (W. D. Hay 248–249), while the Negroes are the "black abortions of Humanity" (253). By advocating the scientific validity of racial theories, Professor Meister explains that racial extermination on a global scale led to the fusion of the civilised races into the United States of Humanity. Hay combines Malthusianism, Herbert Spencer's popularisation of evolutionary theories and eugenics to produce graphs, statistics and axioms through which he can validate a new world founded on extreme racialisation.

Environmental phenomena changing the course of human history define the apocalypse in the 1880s. As I. F. Clarke notices in *The Pattern of Expectation (1644–2001)*:

the new theme of the return to nature entered the tale of the future – first, and partially, in the obliteration of industrial civilization, which Richard Jefferies so lovingly described in *After London* in 1885; second, and far more effectively, in the forest world of *A Crystal Age* of 1887, in which H. H. Hudson created the prototype for the modern arcadia of the future.

(146)

The return to a state of nature caused by a global catastrophe is the defining trait of the modern post-apocalyptic dystopia.

Jefferies recounts how, after an anthropogenic disaster, cities are overrun by vegetation and humans react by building up a neo-medieval society. The description of the apocalypse exhibits meticulous details and simultaneously draws upon Biblical language to convey the decadence that caused the end.

> Thus the low-lying parts of the mighty city of London became swamps, and the higher grounds were clad with bushes. The very largest of the buildings fell in, and there was nothing visible but trees and hawthorns on the upper lands, and willows, flags, reeds, and rushes on the lower.... It is a vast stagnant swamp, which no man dare enter, since death would be his inevitable fate. There exhales from this oozy mass so fatal a vapour that no animal can endure it. The black water bears a greenish-brown floating scum, which for ever bubbles up from the putrid mud of the bottom. When the wind collects the miasma, ... it becomes visible as a low cloud which hangs over the place.... It is dead. The flags and reeds are coated with slime and noisome to the touch; there is one place where even these do not grow, and where there is nothing but an oily liquid, green and rank.... For all the rottenness of a thousand years and of many hundred millions of human beings is there festering under the stagnant water, which has sunk down into and penetrated the earth, and floated up to the surface the contents of the buried cloacae.

(37–38)

While describing how cities collapsed and society reverted to barbarism, Jefferies achieves the high oratorical power of a preacher. The protagonist Felix Aquila is both a historian and a palaeographer who collects partial, incoherent sources, trying to decipher them. His need to preserve and understand cultural memory sets him apart from the others, who have reverted to a neomedieval lifestyle based on physical prowess:

> Besides the parchments ... there were three books, much worn and decayed, which had been preserved.... One was an abridged history of Rome, the other a similar account of English history, the third a primer of science or knowledge.... Exposed for years in decaying houses, rain and mildew had spotted and stained their pages; the covers had rotted away these hundred years, ... many of the pages were quite gone, and others torn by careless handling.... Felix had, as it were, reconstructed much of the knowledge

which was the common (and therefore unvalued) possession of all when they were printed.

(46–47)

Aquila's determination to overcome cultural amnesia drives his quest through the wilderness, which symbolises a ritual passage from stagnation to agency. Jefferies' post-apocalyptic world shows how a violent return to the state of nature can be gradually overcome by recovering historical and cultural memory, on which a new phase of civilisation will be built. Albeit cautious about human potentialities, Jefferies believes in the faculty of choice and self-determination.

Hudson, instead, offers a disquieting view of a perfectly sustainable society in which individual drives have been overcome by an all-pervasive care of the communal houses and people have a strong rapport with animals and the environment. While trying to adapt to a post-apocalyptic neo-Arcadia where sexuality has been strongly contained, the narrator Smith experiences a gap between his past and present existence. Life in the efficient but emotionless beehive structures that have replaced the cities produce a loss of historical and cultural roots and then a sense of estrangement and futility:

I was conscious now, as I had not been before, of the past and the present, and these two existed in my mind, yet separated by a great gulf of time – a blank and a nothingness which yet oppressed me with its horrible vastness. How aimless and solitary, how awful my position seemed! It was like that of one beneath whose feet the world suddenly crumbles to dust and ashes, and is scattered throughout the illimitable void, while he survives, blown to some far planet whose strange aspect, however beautiful, fills him with an undefinable terror ... my agitation, the strugglings of my soul to recover that lost life, were like the vain wingbeats of some woodland birds blown away a thousand miles over the sea, into which it must at last sink down and perish.

(190–191)

The conflict between memory and oblivion produces a short circuit plunging Smith into a state of mental stress which culminates in entropy and autism. Both memory and forgetfulness can be either disturbing or therapeutic: on the one hand recollections of traumas function as warnings, thus fulfilling a therapeutic function; on the other hand, amnesia soothes grief, yet also erases the awareness of events.

The Biblical imagery resonates both in *After London* and *A Crystal Age*: while Jefferies indulges in terrifying images, Hudson vehemently evokes all human ingenuities and institutions. As Clarke notices, "the licence to play lord of the universe, granted by the tale of the future, encouraged Hudson to purge his feelings of anger and resentment in an apocalyptic vision of the world renewed" (*The Pattern of Expectation* 148). *A Crystal Age* ends with the soliloquy of Smith who, after having realised he cannot fit within the neo-Arcadian community, voices his despair by stigmatising the world as a dysfunctional concoction of material objects and abstract ideas:

> a sort of mighty Savonarola bonfire, in which most of the things once valued have been consumed to ashes – politics, religions, systems of philosophy, isms and ologies of all descriptions; schools, churches, prisons, poorhouses; stimulants and tobacco; kings and parliament; cannon with its hostile roar, and pianos that thundered peacefully; history, the press, vice, political economy, money, and a million things more – all consumed like so much worthless hay and stubble. This being so, why am I not overwhelmed at the thought of it? In that feverish, full age – so full, and yet, my God, how empty! – in the wilderness of every man's soul, was not a voice crying out, prophesying the end?
>
> (Hudson 293–294)

Whether Jefferies' and Hudson's responses to the Industrial Revolution question or validate the construction of a sustainable pre-industrial society remains unclear. In contrast with their ambivalent views of the post-apocalypse, *News from Nowhere; or, An Epoch of Rest. Being Some Chapters from a Utopian Romance* unfolds William Morris' vision of messianic hope resulting in palingenesis. The Biblical apocalypse and the Old Norse myth of Ragnarök merge in the history of Victorian Britain, conjuring up the description of a civil war waged by the labourers. The tragic events that will ultimately erase class divisions and lead to the establishment of a communitarian society match the process of destruction, end and new beginning evoked both in the *Book of Revelation* and in the *Poetic Edda* and *Prose Edda*. In Chapter 17, explaining "How the Change Came", it is told that the great revolution would be followed by an intermediate stage of State Socialism finally overcome by the establishment of a society based on equality and fellowship. William Guest, the traveller from industrialised London, carefully listens to Old Hammond, the custodian of memory in the socialist England of the future: on 13 November 1887 the meeting summoned by the Social Democratic Federation and the Irish National League in Trafalgar Square was violently disrupted by the police. Three people were killed, over 100 injured and two arrested. That very evening Morris lectured on "The Society of the Future" at a meeting organised by the Socialist League at Kelmscott House in Hammersmith.

The London demonstration, known as Bloody Sunday, is reported in *News from Nowhere* with the realistic precision of a historical chronicle and the magniloquence of a prophetic revelation blending the apocalypse and Ragnarök:

> The closely packed crowd would not or could not budge, except under the influence of the height of terror, which was soon to be supplied to them … and to most men … it seemed as if the end of the world had come, and to-day seemed strangely different from yesterday … a hoarse threatening

roar went up from [the crowd]; and after that there was comparative silence for a little, till the officer had got back into the ranks.... 'Throw yourselves down! they are going to fire!' But no one scarcely could throw himself down, so tight as the crowd were packed. I heard a sharp order given, and wondered where I should be the next minute; and then – It was as if the earth had opened, and hell had come up bodily amid us. It is no use trying to describe the scene that followed. Deep lanes were mowed amidst the thick crowd; the dead and dying covered the ground, and the shrieks and wails and cries of horror filled all the air, till it seemed as if there were nothing else in the world but murder and death. Those of our armed men who were still unhurt cheered wildly and opened a scattering fire on the soldiers.... How I got out of the Square I scarcely know: I went, not feeling the ground under me, what with rage and terror and despair'.

... 'How fearful! And I suppose that this massacre put an end to the whole revolution for that time?'

'No, no,' cried old Hammond; 'it began it!'.

(154–156)

It is an extraordinary event that changes the course of history and gives rise to a new civilisation. Infused with Christian and Old Norse narratives of annihilation and rebirth, the description transcends the actual historical fact and acquires a universal significance. In the myth of Ragnarök the fate of the gods is followed by peace and the awakening of nature; in *News from Nowhere* the tragic events of the revolution fuel intense political activism directed to establish a socialist society:

But now that the times called for immediate action, came forward the men capable of setting it on foot; and a new network of workmen's associations grew up very speedily, whose avowed single object was the tiding over of the ship of the community into a simple condition of Communism; and as they practically undertook also the management of the ordinary labour-war, they soon became the mouthpiece and intermediary of the whole of the working classes; and the manufacturing profit-grinders now found themselves powerless before this combination.

(159–160)

Morris's ability to merge facts and prophecies, realism and visionary imagination transforms historical events into paradigmatic models of human actions taking shape within incessant cycles of death and rebirth:

the world was being brought to its second birth; how could that take place without a tragedy.... The spirit of the new days, of our days, was to be delight in the life of the world; intense and almost overweening love of the very skin and surface of the earth on which man dwells.

(175)

Morris's new world stems from the class revolution foreseen by Marx and the Socialists and thrives on the visionary intensity of apocalyptic and Old Norse symbolism.

Late nineteenth-century renditions of the apocalypse attempt to predict how technological development will cause environmental disasters. Writers oscillate between a dialectical sense of history and nostalgia: reminiscences of ideal epochs in the past interfere with radical visions of the future, caught between revivalism and innovation.

III The twentieth century as apocalypse

Late nineteenth-century speculative writers questioned optimistic positivism by voicing growing distrust in the myth of progress and emphasising the connection between millenarian and messianic elements. Pandemics, cataclysms, astronomical phenomena, or extra-terrestrial invasions are the predominant natural and supernatural causes of destruction in pre-1900 dystopias. The First and Second World War introduced a paradigm shift, showing that the lethal power of armaments could lead to total annihilation. After the atomic bombings of Hiroshima and Nagasaki in 1945, dystopias have persistently delivered warnings about global power clashes exacerbated by weapons of mass destruction.

Aldous Huxley's *Brave New World* described post-apocalyptic human beings engineered through reproductive technology. In the year AD 2540 life is managed by the World Controllers: children, born in the laboratory and educated through hypnopaedic processes, are encouraged to develop loose morals, follow strong impulses and enjoy material wealth. Culture, critical thinking and the desire for knowledge have been suppressed, natural processes have been reduced to a minimum and diversity is punished as an expression of anti-social deviance. Huxley's unsettling depiction of a regime imposing homogenisation epitomised the pessimistic view, escalating between the wars, that the apocalypse would not be followed by palingenesis.

Humankind doomed by global destruction was relentlessly portrayed by speculative writers in the 1950s. *The Day of the Triffids*, *The Kraken Wakes* and *The Chrysalids* by John Wyndham share the belief that the downfall of civilisation is the retribution for the errors of humankind as well as the requirement for salvation. In Ray Bradbury's *Fahrenheit 451* the ban on books is as destructive as the imminent nuclear war. Guy Montag, a fireman in charge of erasing every trace of culture, joins a small group of subversive people who preserve knowledge through oral transmission.

The nuclear era has deeply affected the collective imaginary and shaped new dystopias. For decades, the production of nuclear weapons and the following unstable power balance in the Western and Eastern Blocs have fuelled speculations and analyses leading to the theorisation of a Third World conflict. Civil and military authorities in many countries have enacted alert and defence measures against a nuclear attack, while scientists have conducted researches on the effects of a nuclear conflict on nature and human health. Conjectures on World War III have been stimulated by debates on the size of national nuclear arsenals and on policies preventing more nations from acquiring weapons. Dystopian writers have portrayed worlds in which people exposed to the risk of an atomic holocaust are overcome by anguish, or cling to their faith in the government and science, or resort to simple selective ignorance of the problem.

This is Tomorrow, launched on 9 August 1956 at the Whitechapel Art Gallery in London, explored popular culture in various media. The beginning of British Pop Art made an impression on 26-year-old J. G. Ballard. In the 80 projects produced over 15 years he saw an experimental quality he would incorporate in his goal as a writer. As he explained in *Miracles of Life: Shanghai to Shepperton, An Autobiography*: "The overall effect of *This is Tomorrow* was a revelation to me, and a vote of confidence, in effect, in my choice of science fiction" (188). Surrealism and estrangement in contemporary society inspired *Escapement* and *Prima Belladonna*, the first two science fiction stories he published only a few months later in the December 1956 issues of *New Worlds* and *Science Fantasy* respectively.

The literature of anticipation proved receptive to innovative ideas in the 1960s. Inspired by Nouvelle Vague directors Jean-Louis Godard and François Truffaut, British authors redefined science fiction by distancing themselves from the escapist plots favoured by Anglo-American writers. While serving as editor of the British magazine *New Worlds* from May 1964 to March 1971 and then again from 1976 to 1996, Michael Moorcock published J. G. Ballard, Brian Aldiss, John Brunner, and Christopher Priest. Whereas science fiction in the USA had become a mainstream genre tailored to scientific and technological progress as well as imperialistic politics, in Great Britain it questioned middle-class conventions and produced counterculture. Instead of portraying interplanetary journeys, British New Wave writers perused the human psyche and portrayed the Earth as a foreign place and an interzone. Intersecting popular and experimental literature, they incorporated and revitalised the apocalypse and near future, claiming that technology, science and mass communication affect our interior space, the only one truly far-off.

While invoking the apocalypse, Ballard and Aldiss deny the possibility of catharsis and regeneration, intimating that all things human are finite and the end of civilisation is inescapable. The protagonists of their dystopias share an extreme drive towards destruction and self-annihilation. As David Ketterer explains in *New Worlds for Old: The Apocalyptic Imagination, Science Fiction, and American Literature*, it is a mental apocalypse (93). Aldiss incorporated a parodic element by introducing the "cosy catastrophe", in which "the hero should have a pretty good time (a girl, free suites at the Savoy, automobiles for the taking) while everyone else is dying off" (Rogers 294). Aldiss coined this definition to criticise *The Day of the Triffids* and *The Kraken Wakes*, Wyndham's most successful novels, while he praised *The Chrysalids* as his masterpiece.

Ballard is defined as "Le Chirurgien de l'Apocalypse" by Robert Louit in his Preface to *Le livre d'or de la science-fiction no. 5074. J. G. Ballard*, a collection of short stories published in French in 1980. The reference to surgical precision captures the metamorphosis of the apocalypse in Ballard's novels, pervaded by the tragic belief that human beings yearn for annihilation and death. Regeneration is never pursued and the apocalypse, far from generating fear, is what they want to embrace. The terminus has become a goal, providing the only fulfilment possible.

Repetition-with-variation defines Ballard's "Elemental Apocalypse Quartet" of the early- and mid-1960s. In *Postmodernist Fiction* Brian McHale observes that:

> in each, Earth is subjected to a global disaster, whether a plague of sleeping-sickness, rising sea-level, a manmade drought, or the bizarre crystallisation of living matter. In each, a researcher, called Powers or Kerans or Ransom or Sanders (the last three near-anagrams), becomes obsessed with the strange new conditions of existence, and is drawn deeper and deeper into them, to his own annihilation. In each, the researcher forms a liaison with a mysterious woman, and suffers persecution at the hands of a demonic male figure, in some sense his double; and so on.
>
> (69–70)

"The Quartet" focuses on characters forced to develop decision-making skills in high-risk environments. In *The Drowned World* Ballard shows humankind struggling to adapt to a world in which temperature rapidly increases, the level of the sea rises, and resources are scarce. The effects of climate change have transformed London into a new and radically different tropical setting in which biologist Dr Robert Kerans and a few other survivors are fighting against the environment. As uncertainty erodes the notion of space and time, conjectures about what will happen sound like nonsensical statements: "If we return to the jungle we'll dress for dinner" (Ballard, *The Drowned World* 46). While the living conditions become unbearable and the habitable space shrinks, individuals must perform daily chores such as rationing, on which life expectantly has become entirely dependent:

> Fuel raised more serious problems. The reserve tanks of diesel oil at the Ritz held little more than 500 gallons, sufficient to operate the cooling system for at most a couple of months. By closing down the bedroom and dressing room and moving into the lounge, and by raising the ambient temperature to ninety degrees, he would with luck double its life, but once the supplies were exhausted the chances of supplementing them were negligible.
>
> (52)

In the middle of the jungle, which has become the new reality, the city is gone, its name is forgotten and the sense of estrangement generated by the lagoon festering with animals is enhanced by the association with the aquatic landscape in Charles Kingsley's 1863 fairy tale *The Water Babies*. The only way to cope with the disappearance of the urban fabric is to bring it back through childhood memories:

Part of [this city] used to be called London, not that it matters. Curiously enough, though, I was born here…. We left here when I was six, but I can just remember being taken to meet [my father] one day. A few hundred yards away there was a planetarium, I saw a performance once…. The big dome is still there, about twenty feet below water. It looks like an enormous shell, fucus growing all over it, straight out of 'The Water Babies'. Curiously, looking down at the dome seemed to bring my childhood much nearer … at my age all you have are the memories of memories. After we left here our existence became completely nomadic, and in a sense this city is the only home I've ever known.

(82–83)

As survival requires the ability to negotiate insecurity and displacement, global migration becomes the only certainty. Even without an itinerary, movement in itself becomes the primary goal in risk situations heightened by spatial and temporal singularities.

The sense of rupture reaches its climax in Anna Kavan's *Ice*. In a world weakened by global war and succumbing to ice, the protagonist embarks upon an all-absorbing quest: the young woman he is determined to find may be an actual person or just the projection of his disturbed psyche. Hyperrealism and surrealism blend into a nonlinear narrative following his trip as well as his absorption into an apocalypse shaped by hallucination and stream of consciousness.

I began to wonder if I would ever find her in the general disorder. It did not look as if any organized life could have been going on here since whatever disaster had obliterated the villages and wrecked the farms. As far as I could see, no attempt had been made to restore normality. No rebuilding or work on the land had been done, no animals were in the fields.

The road badly needed repairs, the ditches were choked with weeds under the neglected hedges, the whole region appeared to have been left derelict and deserted.

A handful of small white stones hit the windscreen, making me jump. It was so long since I had experienced winter in the north that I failed to recognize the phenomenon. The hail soon turned to snow, diminishing visibility and making driving more difficult. It was bitterly cold, and I became aware of a connexion between this fact and my increasing uneasiness….

An unearthly whiteness began to bloom on the hedges. I passed a gap and glanced through. For a moment, my lights picked out like searchlights the girl's naked body, slight as a child's, ivory white against the dead white of the snow, her hair bright as spun glass. She did not look in my direction.

Motionless, she kept her eyes fixed on the walls moving slowly towards her, a glassy, glittering circle of solid ice, of which she was the centre.

Dazzling flashes came from the ice-cliffs far over her head; below, the outermost fringes of ice had already reached her, immobilized her, set hard as concrete over her feet and ankles. I watched the ice climb higher, covering knees and thighs, saw her mouth open, a black hole in the white face, heard her thin, agonized scream. I felt no pity for her. On the contrary, I derived an indescribable pleasure from seeing her suffer.

(11–12)

Kavan instils a sense of estrangement by recounting normal, even banal events through a distorting mirror, in which known objects, viewed from an unfamiliar perspective, become uncanny. The visit to the young woman and her husband turns from realistic to surreal to deeply disquieting when readers realise that the world has become an endless wilderness and the human being vividly described by the narrator is an oneiric apparition. The apocalypse has fallen outside and inside.

In *Helliconia Spring*, *Helliconia Summer* and *Helliconia Winter* Brian Aldiss traces the evolution of humans facing extreme environmental conditions on planet Helliconia 6,000 years in the future. Helliconian astronomy, geology, biology and anthropology are observed from the Earth Observation Station Avernus, inhabited by 6,000 terrestrials. While the monitoring devices transmit Helliconian images and data to Earth for scientific purposes, the inhabitants, alienated by the hyper-technological environment, stop taking care of the space-ship equipment and infrastructures, develop aberrant perversions which decimate the population, become savages and start a civil war, leaving Avernus an almost empty shell. Life on Helliconia and Avernus shows parallelism with life on Earth, devastated by nuclear war and then reborn thanks to the repair processes activated by Gaia, the Earth-mother force. While continuing to observe the events on Helliconia, the new human beings have come to disregard technology and chosen to nurture a deep connection with nature and cosmic energy. The gift of empathy, a dominant survival trait, makes them respect nature and the others, including Helliconian people, whom they decide to help.

Helliconia, Avernus and Earth are bound by enantiodromia, defined by Carl Jung as the principle whereby the superabundance of any force inevitably produces its opposite: "things constantly turn in to their opposites; knowledge becomes by turns a blessing and a curse, as does religion; captivity and freedom interchange roles; phagors become by turns conquerors and slaves" (Aldiss, "Brian writes"). For Aldiss, this is the reason why humans cause destruction: by seeing themselves as radically different from nature, they suppress the irrational components of their psyche and lose the ability to communicate and share. Neither technology, nor science nor medicine will shun the end; only empathy can.

Louise Lawrence's *Children of the Dust* focuses on three generations of related individuals exposed to a nuclear apocalypse, coping with its effects on nature and human life, and adapting to new environmental conditions in a post-apocalyptic world. The first part focuses on the effects of the nuclear blast on a family, their arduous survival in the living room, Sarah's nihilistic thoughts, the long agony of their dog, pain and sickness from radiation, and the dead land-scape. Catherine, the only survivor and mother of post-apocalyptic mutants, is a new Eve on whom rebirth depends. In the second part, Ophelia is born and lives for 16 years in a government bunker under military control, while the few scientists still alive try to restore the world as it was. Catherine's daughter Lilith is born with white eyes and hair covering her body. The encounter between the new generation of humans, mutated and adapted, and the bunker survivors, conservative and predatory, reveals that science and technology cannot rebuild the pre-nuclear world and Homo Sapiens has become unable to live in the outer world. In the third part, set five decades after the nuclear war, the few humans in the shelter, now uninhabitable, must move out. Ophelia's son, Simon, meets Laura, a mutant girl who resembles Lilith. Simon and Laura descend from the same father and it is him who persuades his people to collaborate with her community by sharing technological knowledge and allowing them to fully develop their skills. Their physical features and psychic abilities such as telepathy and a strong sense of empathy make them perfectly suitable to the post-nuclear world:

> Dodos and dinosaurs and *Homo sapiens* had not been wasted. The human race no failed evolutionary experiment, their nuclear war no ultimate disaster. It had happened because it was meant to happen and nothing was lost. The ideas, the thoughts, the achievements, Timperley Abbey and the standing stones on the stark horizon, Boyle's law and Einstein's theory of relativity … she would inherit it all, Laura with her arms around him, warm and touching and covered with white fur. She and her kind would reap the whirlwind, the mind of man and life on earth. Simon stroked her hand. He did not begrudge her, did not begrudge any of them the knowledge he possessed. They were better than he was … *Homo superior*, the children of the dust.
>
> (Lawrence 174)

Both *Helliconia Winter* and *Children of the Dust* describe post-nuclear societies inhabited by different species. On Helliconia the humans, being the most intelligent, rule over the Phagors and the Nondads, even though these species are biologically more suitable to adapt to the environmental conditions of the planet. Aldiss and Lawrence share the belief that empathy, co-operation, and respect can overcome violence and discrimination. They also share ambivalent attitudes towards technological progress, which causes the extinction of the inhabitants on Avernus and in the anti-atomic bunker, because it impairs their ability to cope with new environmental conditions.

Since the bombing of Hiroshima and Nagasaki in August 1945, nuclear weapons and their potential effects have been a leitmotiv in popular culture. Pictures of atomic devices and mushroom clouds symbolise the Cold War period. Stanley Kubrick's *Dr. Strangelove or: How I Learned to Stop Worrying and Love the Bomb* deals with the nuclear threat in a satirical setting, while Sidney Lumet's *Fail Safe* describes an accidental thermonuclear first strike caused by a glitch in US alert systems. *99 Luftballons* by the German band Nena tells how a nuclear war is caused by an error in the alert system and *2 Minutes to Midnight* by Iron Maiden talks about the Doomsday Clock. Metal and punk artists such as Black Sabbath, Metallica or Megadeth have written songs about, or alluding to, the nuclear war theme. Sigur Rós' video for the song *untitled #1 (a.k.a. Vaka)* shows the surreal, disturbing scenario of a post-apocalyptic world polluted by the nuclear fallout, in which a group of children wear gas masks while playing among black ashes. The nuclear war theme features in many video games such as *StarCraft* (first released in 1998) and *Call of Duty* (first released in 2003).

IV Risk and resilience in the twenty-first century

Risk in the contemporary age has become an overarching phenomenon through which human activities can be understood in relation to the threat and irreversible damage they cause to the environment. Risks are the extreme consequences of human behaviour, resulting in the annihilation of nature, failure of civil society and political decay (Fukuyama). However, self-defence strategies generated by the sense of an ending are at the core of risk, "a systematic way of dealing with hazards and insecurities induced and introduced by modernisation itself" (Beck 21). Contemporary societies, constantly projecting themselves forward, are risk societies, as the act of growing more and more "preoccupied with the future (and also with safety) ... generates the notion of risk" (Giddens and Pierson 209). The sense of doom haunting the human psyche has also functioned as one of the most effective means to preserve the human species. Over the last few decades the emergency approach, according to which disasters should be dealt with after their occurrence, has been replaced by a focus on risk reduction and more recently by an integrated view of risk management, which focuses on anticipating whether a catastrophic event will occur as well as coping with its effects.

Corrective and prospective protocols for risk and disaster management are based on diverse theoretical approaches. According to traditional quantitative methods, risk can be measured and it is the lack or distortion of information that prevents people from being adequately prepared to face a disaster. According to contemporary constructivist views, risk is a social construction deeply connected to local systems of meaning. Thus, although risk refers to time-dependent events, temporal characteristics cannot be easily identified, owing to the uncertainty of the future: the risk of disaster is thus contingent, presenting different perspectives to different observers (Luhmann) and acquiring a local significance, as it involves specific cultural assets in different societies (Atkinson).

Social, economic, political, cultural, ideological and infrastructural factors susceptible to damage have all contributed to exacerbate the concept of vulnerability. Increasing population density in urban centres, the growth of large cities, environmental degradation, economic uncertainty, the exponential rise of consumption and climate change cause the proliferation of risks in everyday lives. Such a wide range of factors requires the concept of differential vulnerability, since not all groups are equally exposed to hazards and risks, nor do they have the same resources for coping with emergency. Social, cultural and economic factors are involved in the analysis of risk scenarios (Hoffman and Oliver-Smith).

Nowadays hazards, crises and disasters inform dystopian narratives of socio-economic recession, cultural displacement and rebuilding. Contemporary post-apocalyptic dystopias focus on people coping with risk society, trying to adapt to the impending end which transforms survival into the ultimate challenge. The circular dynamic of crisis, apocalypse and palingenesis matches the three phases of risk management: the pre-disaster scenario characterised by prevention, pre-paredness, and alert; the disaster and emergency scenario involving immediate responses; the post-disaster scenario including reconstruction, rehabilitation and redevelopment. The ways in which the various phases are tackled in con-temporary post-apocalyptic plots raise questions about readers' responses to the possibility or prospect of facing those very events.

The post-apocalyptic genre has proliferated since the beginning of the twenty-first century: Philip Reeve's *The Hungry Cities Chronicle*, comprising *Mortal Engines*, *Predator's Gold*, *Infernal Devices*, and *A Darkling Plain*, exemplifies steampunk for young adults. Blending nostalgia for the pre-industrial world and attraction for technological progress, steampunk authors imagine a post-apocalyptic world, in which steam power has maintained its primary importance, and technological progress has produced anachronistic inventions, resembling the ones people in the nineteenth century could have imagined. *Mortal Engines* begins with the description of the Hunting Ground:

It was a dark, blustery afternoon in spring, and the city of London was chasing a small mining town across the dried-out bed of the old North Sea. In happier times, London would never have bothered with such feeble prey. The great Traction City had once spent its days hunting far bigger towns than this, ranging north as far as the edges of the Ice Waste and south to the shores of the Mediterranean. But lately prey of any kind had started to grow scarce, and some of the larger cities had begun to look hungrily at London. For ten years now it had been hiding from them, skulking in a damp, mountainous, western district which the Guild of Historians said had once been the island of Britain. For ten years it had eaten nothing but tiny farming towns and static settlements in those wet hills. Now, at last, the Lord Mayor had decided that the time was right to take his city back over the landbridge into the Great Hunting Ground. It was barely halfway across when the look-outs on the high watch-towers spied the mining town, gnawing at the salt-flats twenty miles ahead. To the people of London it seemed like a sign from the gods, and even the Lord Mayor (who didn't believe in gods or signs) thought it was a good beginning to the journey east, and issued the order to give chase. The mining town saw the danger and turned tail, but already the huge caterpillar tracks under London were starting to roll faster and faster. Soon the city was lumbering in hot pursuit, a moving mountain of metal which rose in seven tiers like the layers of a wedding cake, the lower levels wreathed in engine-smoke, the villas of the rich gleaming white on the higher decks, and above it all the cross on top of St Paul's Cathedral glinting gold, two thousand feet above the ruined earth.

(1)

The description of predatory London exemplifies Municipal Darwinism, which Reeve skilfully develops from evolutionary theories. National consensus in the Victorian age was based on the assumption that the world expansion of Great Britain would occur owing to specific national features: inborn predisposition to government, administrative autonomy and dominion. British imperialism was sustained by the idea of a race born to predominate. Newspapers and periodicals published between 1880 and 1914 highlighted a particular kind of nationalism exploiting social Darwinism, according to which conflict among the nations was but a biological need, a remedy able to transform weak individuals into men, and nations into world empires. The idea that war could rejuvenate the nation was emphasised over and over again in late-Victorian propaganda. Pseudo-scientific social Darwinism was based on the idea that mankind did not aim at peace, and survival was only for the morally and physically superior. Such view would undermine theories of natural law and humanitarian ideals as well as the Marxist concept of class struggle, fuelling the national-popular idea of an eternal antagonism among the populations and the newly emerging ideology of the conflict among the races.

Reeve's steampunk interpretation of social Darwinism shows how young adult fiction can portray our world, endangered by man-made events. The Japanese Katsuhiro Ōtomo has been able to render the steampunk world through powerful visual imagery in the 2004 animated action film *Steamboy*.

In steampunk fiction, the issue of reconstruction is emphasised through the DIY (Do It Yourself) lifestyle of the tinkerer and producer of artefacts. Such premise is predicated on an ideology that affects society at multiple levels. The practice of reusing/recycling is a form of ecological intervention showing that individuals can take responsibility towards the environment. However, if restricted to the private sphere and lacking feasible collective goals, the steampunk frame of mind generates escapism and aestheticizing postures.

David Mitchell's *Cloud Atlas* is a book of books self-consciously displaying a pyramidal structure, in which the post-apocalyptic narratives "An Orison of Sonmi~451, Part 1", "Sloosha's Crossin' an' Ev'rythin' After", "An Orison of Sonmi~451, Part 2" meet at the vertex. Described from a very far future, our world appears to be flawed by ageing and decaying people who would end their lives in houses for the elderly:

> no fixed-term life spans, no euthanasium. Dollars circulated as little sheets of paper and the only fabricants were sickly livestock. However, corpocracy was emerging and social strata was demarked, based on dollars and, curiously, the quantity of melanin in one's skin.
>
> (234–235)

"Sloosha's Crossin' an' Ev'rythin' After", the sixth story occupying the central position in the novel, revolves around Zachry, belonging to the peaceful farmers of the valley folk often raided by the Kona tribe, and Meronym, who has come to stay with the villager and represents technologically sophisticated people known as the Prescients. Her parents and their generation, she explains to Zachry in an English idiom of the future, hoped that the great, famous cities – Melbun, Orkland, Jo'burg, Buenas Yerbs, Mumbay, Sing'pore – had survived. Five decades after landing at Prescience, the Ship was relaunched, but the expedition revealed that the world was still in ruins.

> *They finded the cities where the old maps promised, dead-rubble cities, jungle-choked cities, plague-rotted cities, but never a sign o' them livin' cities o' their yearnin's. We Prescients din't b'lief our weak flame o' Civ'lize was now the brightest in the Hole World, an' further an' further we sailed year by year, but we din't find no flame brighter. So lornsome we felt. Such a presh burden for two thousand pairs o' hands!*
>
> (271)

Their post-apocalypse stimulates Mitchell's analysis as to whether it is better to return to nature and be savage or live in a technologically advanced habitat and be civilised. The hegemony of technology over humankind has proved fatal, but the return to a prelapsarian dimension poses many constraints. The savages satisfy their hunger, anger, and sexual urges: their will prevails. The civilised always ponder as to whether they should follow their instincts or comply with social rules: their will is subservient to rationality. However, the dichotomy is misleading, as humans are both wild and cultivated. Mitchell re-presents the well-known contrast between faith in progress and desire for an idyllic retreat, when humans are innocent and untainted by technology. Better to be regulated by artificial intelligence or go back to a rural lifestyle? While offering a vivid description of both scenarios, he leaves the antithesis as is.

> *More what?* I asked. *Old Uns'd got ev'rythin'. Oh, more gear, more food, faster speeds, longer lifes, easier lifes, more* power, *yay. Now the Hole World is big, but it weren't big 'nuff for that hunger what made Old Uns rip out the skies an' boil up the seas an' poison soil with crazed atoms an' donkey 'bout with rotted seeds so new plagues was borned an' babbits was freak-birthed. Fin'ly, bit'ly, then quicksharp, states busted into bar'bric tribes an' the Civ'lize Days ended, 'cept for a few folds 'n' pockets here 'n' there, where its last embers glimmer.*
>
> (286)

It is certain that capitalism and consumerism, rather than placate, have reached a climax followed by regression to a primitive state. Human hunger cyclically gives birth to civilisation and destroys it.

The latest post-apocalyptic novels show a penchant for parody. In Will Self's *The Book of Dave* a disturbed taxi driver called Dave Rudman collects and prints on metal all his invectives against women and elucubrations about custody rights for fathers. His book also includes The Knowledge, referring to the detailed information about London acquired through his job. Dug out centuries later, the book becomes the sacred text for an aggressive, misogynistic religious cult spreading in southern England and London after a catastrophic flooding. By imagining that a minor phenomenon in the past has caused a majestic chain reaction in the future, where the term AD labels years After Dave, the author deliberately erodes the reader's suspension of disbelief and shows that narrating the (post)apocalypse always requires artifice. Self introduces a twist in the post-apocalyptic genre by suggesting that a pseudo-ideal society can all too easily be erected on an initial misinterpretation and promulgate a preposterous value system. The utopian drive towards an ideal world is corroded by the sense of shallowness pervading the contemporary age.

In *The Red Men* Matthew De Abaitua displays a penetrating touch of reverse engineering when his narrator suggests that the revelation is still paramount, but we will have to ask ourselves what to do if the apocalypse continues to elude us. In *If Then*, set in the small English town of Lewes after a financial apocalypse, an algorithm called the Process now calculates the utopian optimum for people, provides them with jobs, monitors production as well as thoughts and behaviours. Even more disconcertingly, it selects who can stay in Lewes or must be evicted, and is recreating war on the outskirts of the town. As love and war unfold, dystopian and post-apocalyptic fiction, post-humanism and cyberpunk boldly intermingle.

Post-apocalyptic novels in the contemporary age avoid exact descriptions of what caused the apocalypse and revolve around displacement, an adaptation skill in contexts of uncertainty, vulnerability and struggle for survival. Post-disaster scenarios presuppose yet another risk scenario, in which pre-existing social and cultural patterns are reconfigured, often regenerating symbols and systems of meaning that had been consolidated in previous societies.

Preparedness requires planning and scientific calculation (Scoppetta), involves institutions and authorities, encompasses response, management, recovery, rehabilitation and reconstruction. The Sendai Framework for Disaster Risk Reduction 2015–2030 (UNISDAR) states that:

> Policies and practices for disaster risk management should be based on an understanding of disaster risk in all its dimensions of vulnerability, capacity, exposure of persons and assets, hazard characteristics and the environment. Such knowledge can be leveraged for the purpose of pre-disaster risk assessment, for prevention and mitigation and for the development and implementation of appropriate preparedness and effective response to disasters.
>
> (9)

Reimagining the urban fabric, society and cultural heritage allows authors to explore whether it is wiser to build up cultural values, ways of thinking, and social rules radically different from the ones in the pre-apocalyptic world, or it is more productive to work starting from the debris, questioning and reshaping old parameters, preserving and regenerating. Tabula rasa of the past as opposed to palimpsest of beliefs, traditions, and heritage: apocalypse and palingenesis in contemporary dystopias call for comparisons with current definitions of disaster and rebuilding. In particular, post-apocalyptic and palingenetic narratives draw attention to ways of negotiating differential vulnerability, existentially and pragmatically, within the private and the public sphere.

Bibliography

Aldiss, Brian Wilson. "Brian writes about the creation of the Helliconia Trilogy". *Brian Aldiss. Science Fiction Grandmaster*, http://brianaldiss.co.uk/writing/novels/novels-h-l/helliconia/.html. Accessed 11 August 2016.

Aldiss, Brian Wilson. *Helliconia Winter*. Atheneum, 1985.

Aldiss, Brian Wilson. *Helliconia Summer*. Triad Granada, 1983.

Aldiss, Brian Wilson. *Helliconia Spring*. Triad Granada, 1982.

Atkinson, Christopher L. *Toward Resilient Communities. Examining the Impacts of Local Governments in Disasters*. Routledge Research in Public Administration and Public Policy, 2014.

Ballard, James Graham. *Miracles of Life: Shanghai to Shepperton, An Autobiography*. Fourth Estate, 2008.

Ballard, James Graham. *The Drowned World* and *The Wind from Nowhere*. Doubleday 1962.

Beck, Ulrich. *Risikogesellschaft. Auf dem Weg in eine andere Moderne*. Suhrkamp, 1986. *Risk Society. Towards a New Modernity*. Sage Publications, 1992.

Bradbury, Ray D. *Fahrenheit 451*. Ballantine Books, 1953.

Brians, Paul. *Nuclear Holocausts: Atomic War in Fiction*. Kent State University Press, 1987.

Clarke, I. F. *The Pattern of Expectation (1644–2001)*. Book Club Associates, 1979.

De Abaitua, Matthew. *If Then*. Angry Robot, 2015.

De Abaitua, Matthew. *The Red Men*. Snowbooks, 2007.

Fukuyama, Francis. *The End of History and the Last Man*. Free Press, 1992.

Giddens, Anthony, and Christopher Pierson. "The Politics of Risk Society". *Conversations with Anthony Giddens: Making Sense of Modernity*. Stanford University Press, 1998, pp. 204–217.

Hay, William Delisle. *Three Hundred Years Hence; or, A Voice from Posterity*. Newman and Co., 1881.

Heffernan, Teresa. *Post-apocalyptic Culture: Modernism, Postmodernism, and the Twentieth-Century Novel*. University of Toronto Press, 2008.

Hoffman, Susanna, and Anthony Oliver-Smith. *Catastrophe and Culture. The Anthropology of Disaster*. School of American Research Press, 2002.

Hudson, William Henry. *A Crystal Age*. T. Fisher Unwin, 1887.

Huxley, Aldous L. *Brave New World*. Chatto & Windus, 1932.

Hyogo Framework for Action, *Building the Resilience of Nations and Communities to Disasters*, 2005.

Jefferies, Richard. *After London; Or, Wild England*. Cassell & Company, 1885.

Jordison, Sam. "Literary Apocalypse Now, and Then". *Guardian*, 4 December 2007, www.theguardian.com/books/booksblog/2007/dec/04/literaryapocalypsenowandth. Accessed 11 August 2016.

Kavan Anna, *Ice*. 1967. With a Foreword by Christopher Priest. Peter Owen Publishers, 2011.

Kermode, Frank. *The Sense of an Ending. Studies in the Theory of Fiction*. Oxford University Press, 1967.

Ketterer, David. *New Worlds for Old: The Apocalyptic Imagination, Science Fiction, and American Literature*. Indiana University Press, 1974.

Kingsley, Charles, *The Water Babies*. Macmillan, 1863.

Kubrick, Stanley. *Dr. Strangelove or: How I Learned to Stop Worrying and Love the Bomb*. Columbia Pictures, 1964.

Lawrence Louise. *Children of the Dust*. Red Fox, 1985.

Lichtig, Toby. "Apocalypse Literature Now, and Then". *Guardian*, 20 January 2010, www.theguardian.com/books/booksblog/2010/jan/20/apocalypse-literature-now. Accessed 11 August 2016.

Louit, Robert. "Le chirurgien de l'apocalypse". *Le livre d'or de la science-fiction no. 5074. J. G. Ballard*. Presses Pocket, 1980, pp. 7–40.

Luhmann, Niklas. *Soziologie des Risikos*. De Gruyter, 1991.

Lumet, Sidney. *Fail Safe*. Columbia Pictures, 1964.

McHale, Brian. *Postmodernist Fiction*. 1987. Routledge, 2004.

Mitchel, David. *Cloud Atlas*. Sceptre, 2004.

More, Thomas. *Libellus vere aureus, nec minus salutaris quam festivus, de optimo rei publicae statu deque nova insula Utopia*. Dirk Martens, 1516.

Morris, William. "The Society of the Future". 1887. *Political Writings of William Morris*, edited by A. L. Morton, Lawrence and Wishart, 1990, pp. 188–203.

Morris, William. *News from Nowhere; or, An Epoch of Rest. Being Some Chapters from a Utopian Romance*. Roberts, 1891.

Ōtomo, Katsuhiro. *Steamboy*. Sunrise 2004.

Reeve, Philip. *Mortal Engines*. 2001. Scholastic Inc., 2012.

Rogers, Jane. "Jane Rogers's Top 10 Cosy Catastrophes". *Guardian*, 5 July 2012, www.theguardian.com/books/2012/jul/05/jane-rogers-top-10-cosy-catastrophes. Accessed 11 August 2016.

Scoppetta, Nicholas. "Disaster Planning and Preparedness: A Human Story". *Social Research*, vol. 75, no. 3, 2008, pp. 807–814.

Self, Will. *The Book of Dave*. Viking Press, 2006.

United Nations International Strategy for Disaster Reduction – UNISDR. *Sendai Framework for Disaster Risk Reduction 2015–2030*, www.unisdr.org/we/coordinate/sendai-framework. Accessed 11 August 2016.

United Nations International Strategy for Disaster Reduction – UNISDR. "Resilience". *2009 UNISDR Terminology on Disaster Risk Reduction*. United Nations, 2009, www.unisdr.org/files/7817_UNISDRTerminologyEnglish.pdf. Accessed 19 March 2017.

Wagar, W. Warren. *Terminal Visions: The Literature of Last Things*. Indiana University Press, 1982.

Wyndham, John. *The Chrysalids*. Michael Joseph, 1955.

Wyndham, John. *The Kraken Wakes*. Michael Joseph, 1953.

Wyndham, John. *The Day of the Triffids*. Michael Joseph, 1951.

Part II

Policies and institutions for wellbeing

Prologue

Massimiliano Mazzanti

The development of sustainable wellbeing requires that market functioning should be corrected by new public policies and enhanced awareness of key challenges ahead of us, among others the decarbonisation of our societies by 2050, the achievement of a zero-waste circular economy, the improvement of the overall quality of seas and oceans, and the reversal of biodiversity losses. The reduction of inequality is an additional development goal pertaining to the social and environmental spheres. A different notion of development based on a sound perception of growth as a driver, not an objective, should be part of alternative economic and political discourses. The true social aim is to enhance wellbeing, which is diversely dependant on the consumption of private and public tangible goods, on the use of intangible goods such as social and relational capital, on the key role of interconnected factors such as human capital, education and employment. Their quality and quantity are an ultimate goal of our societies, given its monetary, educational and relational contents.

Technological development itself is not a societal goal. As a factor which serves to achieve higher efficiency in the use of labour and natural resources, it should be strongly perceived as a complement to investments in new skills and competences towards the achievement of sustainable wellbeing. Education and training are the real sources of social equity, more than income redistribution, since capabilities are the basis of equal opportunities within and across generations. Societies need reskilling processes, as institutions and their governance strictly depend on the skill contents of social organisations. It is necessary to reskill our (interdisciplinary) competences in the face of environmental-social challenges; a radical reskilling process which should characterise and reach all the relevant spheres of society: consumers, citizens, entrepreneurs, scientists, and policymakers.

Part II presents complementary perspectives on how policies can help achieve economic and environmental sustainability. Policies are necessary to correct the several market failures in realms such as the environment, health, and knowledge creation. Theoretical and empirical analyses must scrutinise whether policies are well designed. Economic efficiency, effectiveness, political feasibility and social acceptance are all together features that characterise a well-designed long-run oriented policy intervention. The challenge is tough. Bringing together different views and paradigms will improve both the design and implementation of public policies.

"Contextualising Sustainability: Socio-Economic Dynamics, Technology and Policy" by Massimiliano Mazzanti and Marianna Gilli aims at defining a comprehensive framework by extending the political economy approach to sustainability and human development. The aim is to broaden the perspective on the role and effects of environmental policy by highlighting its connections with institutional and market dynamics, where learning and adaptation through technological and behavioural changes are crucial. Quantitative analyses are commented on to highlight the necessity to enhance the generation and use of high-quality datasets for better understanding whether societies are on sustainability tracks. The dynamic perspective of the chapter is coherent with the historical analysis of innovation patterns presented by Pier-Paolo Saviotti in Part I.

The human development approach complements Paul James's arguments, also in Part I, and Valeria Andreoni's discourse in "Social Equity and Ecological Sustainability through the Lens of Degrowth". The concept of degrowth further extends the sustainability arena. Based on the idea that an infinite economic growth is impossible in a finite and limited environment, degrowth economy proposes solutions based on voluntary reduction of consumption, sustainable and ethical productions, voluntary work, reciprocity activity and overall redefinition of values and social relationships. In this chapter, an overview of the main pillars of degrowth is provided together with examples of the bottom-up initiatives oriented to promote more sustainable and human-focused economic practices. The degrowth type of criticism adopted by Andreoni finds a complementary reading in the chapter by Gonzalo Salazar in Part I. Andreoni and Salazar convey messages rooted in the socio-economic realm and touching on cultural issues that contribute to activating a worldwide sustainability transition.

The first two chapters provide a very rich and diversified picture of the issues associated with economics and political economy, framed as social sciences. This explicitly points to the various interactions (ecological and socio) economics and political economy can fruitfully establish with other disciplines within and outside the social sciences. Mazzanti, Gilli, and Andreoni share the message that intangibles are especially relevant: technological development, cultural development, institutional changes are complementary assets for sustainability transitions.

A similar broad perspective, with a focus on meso-micro issues, is supported by Gjalt Huppes and Ruben Huele. In "Institutionalist Climate Governance for Pleasant Cities and the Good Life" they discuss the pros and cons of decentralising policy governance. Climate policy is at a governance crossroads, with far-reaching consequences, also for cities. Cities are currently involved in schemes for centrally planned emission reduction, with goals, targets, and means decided centrally, which they merely help implement. This form of Planning and Control optimises climate decisions in a top-down fashion. Instead, institutionalist governance enables cities to act autonomously, with goals, incentives, and some options being decided centrally, and means being decided locally. Autonomous cities take their responsibility proactively, creating both legitimacy and a good life for their citizens in the process.

This theoretical discussion is of extreme importance for the implementation of policies, given the increasing role that urban areas will play in the future as economic centres and producers of environmental externalities. Current central planning tendencies can be reversed by novel institutions equipped to pursue global climate stability and local resilience. Local infrastructure of cities is one key element, focusing on low speed transport, like walking and cycling, in compact integrated cities, close to nature. Research and education can also flourish in the incentivised surroundings. The chapter by Huppes and Huele enriches the economics-oriented discussion and establishes a dialogue with urban planners, policymakers, and industrial ecologists. It is a strong example of inter-disciplinary thinking and open-minded social science approach.

Along a path from macro to micro, "Assessing Public Awareness about Bio-diversity in Europe" analyses the impact of the *Natura 2000* network. Anna Kalinowska addresses the loss of biodiversity, which is a paramount global policy challenge. Biodiversity encompasses all spheres of human life and activity and yet its loss has accelerated to an unprecedented level, both in Europe and worldwide. It has been estimated that the current global extinction rate is 1,000 to 10,000 times higher than the natural background extinction rate. It is a major problem not only for ethical reasons but also because it bears serious consequences on economic and social stability, a cause for deep concern that seems to be underestimated in public perception. Kalinowska examines public familiarity with the idea of biodiversity and perceived seriousness of biodiversity loss across the EU countries. Socio-demographic analyses reveal that the level of education and self-employment strongly influence the knowledge of biodiversity issues and the *Natura 2000* network. It is clear that biodiversity conservation cannot be achieved without the engagement of society as a whole. The active involvement of stakeholders, key policy sectors and civil society are thus fundamental. One message for interdisciplinary research is that education and engagement are as crucial as the economic willingness to pay for supporting policies that aim at increasing the quantity and quality of environmental public goods.

Cultural development is one of the intangible assets that lead to the achievement (or failure) of sustainability aims. Within the cultural sphere, language and linguistic differences are aspects often analysed for their impact on growth and wellbeing. For example, economic studies have examined the importance of language skills and geography behind international knowledge diffusion, as drivers of growth. The role of language in sustainability is relatively overlooked.

"Is the Current Global Role of English Sustainable?" by Richard Chapman evidences that most conceptualisations of English in geopolitical, linguistic and social terms view it as a language enjoying a fundamental, if not dominant, role in the world. This is seen as true in all the "important" walks of life: economics, diplomacy, education, science and the arts. There is largely a consensus about the importance of English as *the* global language, and the contrasts in the debate mainly regard the question of whether or not this is a beneficial state of affairs and how long this hegemony will continue. This debate affects sustainability as well: English might easily be described as the language of sustainability, the international code that renders debate and co-operation possible and facilitates global action. English is defined as the language of emergency rescue services, NGOs and much of the published work in environmental studies and anti-terrorist agencies. We could add that sustainability is itself theorised mostly in English and the present publication is an example of this. While describing the current state of English in more detail, Chapman defines its present roles in order to be able to critique claims as to its future. Does English have a sustainable future, and will this future be conducive to sustainability in a wider and integrated socio-economic cultural sense?

The five chapters illustrate possible ways of achieving an innovative and equitable sustainable society by analysing the features of specific policies and the governance level of application. It is very clear that policies should be coherent with the overall system of values and social infra-structures: whether policies can enhance wellbeing depends on diverse tangible and intangible goods and on the production of other tangible and intangible social goods.

II.1 Contextualising sustainability

Socio-economic dynamics, technology and policies[1]

Massimiliano Mazzanti and Marianna Gilli

I Conceptual background

Gross domestic product (GDP) indicators are limited in their capacity to completely reflect socio-economic performances and need either alternative or
complementary indicators (van den Bergh and Antal; Jones and Klenow).[2]
While alternative indicators to GDP might be useful, this chapter focuses on
complementary ones, namely theories and empirical indicators that attempt to
integrate GDP with other measures of welfare. Human development, stemming
from the work of the Nobel Prize winner Amartya Sen, is one such indicator.
In 2009 the French government established the "Stiglitz-Sen-Fitoussi" Commission to develop an updated conceptual framework. Other indicators beyond
GDP have emerged out of environmental and ecological economics (Perman *et
al.*; Turner and Pearce), such as sustainability-oriented genuine saving
accounts (Costantini and Monni, "Sustainability"; Costantini and Monni,
"Environment") and hybrid environmental-economic accounts (European
Environment Agency, *Resource-Efficient Green Economy*; Marin *et al.*; Marin
and Mazzanti; Mazzanti and Montini; Costantini *et al.*).

Genuine savings highlight the necessity to continually invest in economic and
natural resources to ensure sustainability. Hybrid accounts (e.g. integrated and
coherent GDP, employment, emission data) are useful for elaborating upon
economic-environmental efficiency-oriented indicators such CO_2/GDP and waste
production/GDP in order to promote data-informed policymaking and socially
efficient firm/sector resource management.[3] This synergistic view of economic
and environmental factors and objectives (e.g. reducing emissions while creating
more jobs) is coherent with the relatively recent notion of a green economy
(European Environment Agency, *Resource-Efficient Green Economy*; Barbier
and Markandya).

The objective of this chapter is to offer both a conceptual and empirical
overview of the factors that might positively correlate with positive human
development (HD) and sustainable development (SD) performance, particularly innovation investments, knowledge transfers and environmental policy.
Our analysis falls within the beyond GDP and sustainability-oriented conceptual realms.

In summary, social and environmental indicators are necessary for complementing GDP analysis for two main reasons. First, GDP, although an important and iconic indicator in our society, is only an instrument for reaching the goal of higher social welfare. Defined in either strictly economic utilitarian terms (Jones and Klenow) or broader conceptions of welfare, prices for market goods as reflected by the GDP represent only a fraction of value. Social value is also attributed to goods that are public and extra-market in nature (e.g. environmental externalities such as pollution and greenhouse gases, innovation and knowledge spillovers, health, education, etc.), to which we may assign a price (e.g. a carbon tax defines a price for pollution). Relevant socio-economic dimensions go beyond what GDP can capture in terms of efforts to increase welfare that are strictly linked to poverty reduction, pollution reduction and increasing the quantity and quality levels of both private and public goods. Economics, political economy, public economics and economic development issues thus extend well beyond profits, finance and mere growth. Those are important, but only as components of a larger picture in which proper economic "management" incorporates all relevant social resources (economics as oikos-nomos). By relevant, we mean resources that have a social value due to absolute or relative scarcity and because economic agents (consumers, taxpayers and firms) are willing to pay for increasing levels of consumption.

II Economic development and sustainability

In the model of long-run economic development, human capital accumulation and innovation (R&D as input, patents as formal outcome, tangible and intangible knowledge diffusion, etc.) drive income (GDP) per capita. GDP is thus the basis for measuring enhanced HD and sustainability performance (higher GDP levels might be used to fund strategic public goods, health, education and environment quality).[4] It is worth noting that economic performance in the short run does not automatically lead to positive long-term dynamics. The issue is "how we spend" and "on what we spend" our current income as a society.[5] Again, to establish a sustainable society (using broad conceptions of sustainability), the notion of "investments" should be placed at the centre.

Investment (compared to consumption) is the only factor that jointly increases current economic demand (what we need today to escape stagnation) and generates new foundations for capital forms. A sustainable society is an "investing society" that is able to (more than) compensate for its natural erosion of capital (natural, man-made and human)[6] and transfer to future generations at least the same stock of total/natural/human/technological capital that it received.[7] The potentially different strategies which countries implement in terms of investments and policies assign importance to analysis of convergence in relation to GDP and overall welfare. Much of the current economic turmoil and imbalances stem from improper public policies and lack of income distribution that have generated poor convergence, notwithstanding the important achievements by emerging and developing countries over the past 20 years. Convergence is an empirical fact and is a pre-condition for a sustainable economic and political global community. The lack of convergence risks creating political turmoil, excessive migration and economic imbalances within integrated areas.

A sustainable society is thus a society that properly invests and introduces "well designed" policies (Costantini and Mazzanti, "The Dynamics"; Costantini and Mazzanti, "On the Green") that correct the various market failures and encompass environmental, knowledge and health components, among others. In those realms, markets "fail" when policies and/or state intervention are necessary to increase social welfare. In addition to socio-environmental goals, environmental policies may provide support to economic performances if "well designed policies" (Baumol and Oates; Porter and van der Linde; Porter; Ambec *et al.*) are able to spur adequate (environmental) innovation[8] dynamics. Nevertheless, this is a general statement that should be verified case by case at the macro-, meso- and microeconomic levels (Borghesi *et al.*, "Carbon Abatement"; Mancinelli *et al.*). Among others, Albrizio *et al.* have recently showed that a tightening of environmental policy is associated with a short-term increase in industry level productivity growth in the most technologically advanced countries.

III Environmental policies and ecological taxes

Among environmental policies, ecological tax reforms (ETR) comprise a factor that could characterise a country's approach to enhancing welfare. Ecological tax reform serves as an umbrella under which market-based instruments can be optimally designed and coherently implemented. In *The European Environment – State and Outlook 2005* published by the European Environment Agency it is explained that:

> Tax reform can contribute to a more sustainable, healthy environment. A gradual shift of the tax base away from taxing 'good resources' such as investment and labour and towards 'bad resources' such as pollution and inefficient use of energy would also help to internalise external costs into service and product prices. This would in turn create more realistic market price signals.
>
> (22)

ETR follows the work of Arthur Cecil Pigou, a mentor of John Maynard Keynes, who formally demonstrated about a century ago that environmental taxation enhanced social welfare. The only tax in microeconomics that increases net social welfare is a green (Pigouvian) tax because its economic costs are lower than the social benefits (reduced pollution) it generates. In addition, a tax – broadly a market-based instrument – provides a price incentive. Firms are continuously incentivised to reduce pollution in order to decrease tax expenditures. Firms are thus "forced" to innovate and be more efficient. Such green innovations may produce win-win scenarios in which economic and environmental performances are jointly increased through innovation. The "social" role of innovation is then explained as a major driver of economic growth in the long run, enhancing the possibility to transition to a greener, more sustainable society.[9]

For a comprehensive conceptual and empirical examination of ETR's political economy arguments, we rely on Ekins and Speck. Ecological/environmental tax reform (ETR) is an essential element in long-term sustainable growth/development. During strong (social) crises, ETR may serve to call into question the entire (social) "model of development". ETR modifies relative prices by targeting resource scarcity and externalities and relaxes burdens on factors such as labour and/or using revenues to support public investments in R&D, technology, shaping green consumption options, energy saving choices/investments and measures for direct environmental protection. ETR could be used to reduce labour taxes, eventually targeting the weak part of the labour market: young people, women, the long-term unemployed, etc. ETR can fund education and training, finance direct public expenditures for increasing demand and green features (technology and infrastructure) and subsidise prices and interest rates in the market to rebalance relatively static and inter-temporal prices in favour of greener options. The increase in revenues might be substantial over the long term. The 2014 *Study on Environmental Fiscal Reform Potential in 12 Member States* (European Commission) calculates €101 billion from environmental taxes in 2025, given current figures of €35 billion. Especially in the EU, which still faces less than full employment production levels and very low inflation, ETR could easily gain momentum (European Environment Agency, *Environmental Taxation*).

Building upon the aforementioned conceptual background, with roots in the realms of political economy, environmental economics and public economics, this chapter attempts to serve as a vehicle for engaging with other disciplines along the wellbeing-sustainability "fil rouge". It explains the broad potential of economic discourse for future inter/multi/trans-disciplinary projects. It links conceptual analysis with updated macro-empirical figures, thus presenting an applied economics way of thinking; the empirical analyses take a global or more restricted European/OECD perspective, depending on data availability for key indicators. Finally, it offers some policy recommendations that are useful for policymakers who operate in economic or more hybrid departments, as well as for researchers seeking to develop projects that span disciplines.

The following sections present descriptive and quantitative evidence. First, recent data are used to describe some macro trends about (green) innovations, environmental policies and human development trends and to develop a new framework. Second, sustainability arguments are addressed by presenting a global picture about greenhouse gas increases and their socio-economic drivers, namely innovation/efficiency, economic composition and growth. Convergence is finally addressed by looking at European HD trends. The final comments summarise the main issues and point to future inter/multi/trans-disciplinary projects.

IV Innovation, policies and human development: macro trends

We present updated macro trends based on available data on the key sustainability-oriented socio-economic indicators presented and discussed above. On the innovation and knowledge side, trends about inventions (patents), innovation adoption of environmental innovations and R&D are presented. In a typical "innovation function" framework, R&D is the input, while patents and adoption are the outputs. Environmental policies and ETR trends are presented both worldwide and at the EU level, where ETR has been a major issue since the publication of the European Commission white paper back in the 1990s under the presidency of Jacques Delors (Commission of the European Communities). It remains an innovative political agenda oriented towards growth, job creation and sustainability. We support the hypothesis that stronger innovation investments and more stringent environmental policies might increase future wellbeing (Brunel and Levinson).

IV.1 Innovation and invention: towards a green economy

Figure II.1.1 shows the trend for patents filed under the PCT (Patent Cooperation Treaty) from 1977 to 2013. For one group of countries, namely Germany, France, the United Kingdom, Italy, Japan and the United States, the trend has arisen over the period. In particular, the filing of patents has dramatically increased in the US since the end of the 1980s, reflecting the significant funding

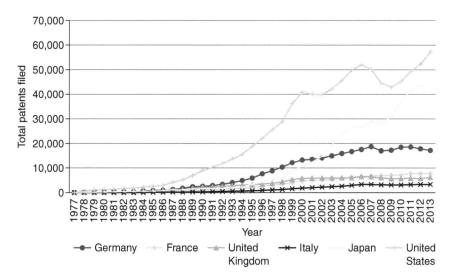

Figure II.1.1 Total patents filed under the PCT, 1977–2013, based on the inventor's country of residence.

Source: authors' elaboration of data from the OECD Patent Statistics in the public domain, https://stats.oecd.org/Index.aspx?DataSetCode=PATS_REGION.

for R&D and inventions. By the 1990s, the quantity of patents had increased by over four times what it was in 1985 (2,559.47). The trend escalated until the 2000s, when it encountered some difficulties, probably due to the economic and financial instability caused by the Internet bubble of 2001 and the more recent crisis of 2009. In addition, Japan experienced fast growth in the number of patents, especially from 1998–1999, when 10,895 patents were filed. Before the economic crises (less than ten years later), around 30,000 patents were filed annually. The effect of the crisis on Japan's inventory flow has been minor compared to the United States, with 1,000 fewer patents filed compared to the previous year. Germany is a leader in patenting activity in Europe. Since the second half of the 1980s, the growth in the number of patents filed has increased exponentially in comparison with other EU countries. In 2007, the number of patents filed under the PCT by Germany was approximately six times higher than the number of patents filed by Italy, the country with the lowest growth in patenting activity (18,743.48 for Germany; 3,361 for Italy). As with Japan, Germany, France, the United Kingdom and Italy experienced less of an effect of the economic recession compared to the US. Overall, the trend is more volatile in the US than in the EU, possibly a product of two different capitalism models: the Anglo-Saxon market base and the "continental" one (comprising Germany, France and Italy, among others) in which the state plays a greater role as an economic player and regulator. The role of the state as innovation funder is in any case strong in the US. It is finally worth noting that the EU has not reached the 3 per cent target for GDP expenditure set by the Lisbon agenda for growth and jobs in 2000.

Figure II.1.2 presents the R&D figures for some major countries (GERD is the total R&D spending). The global role of the US is clear. Germany is the leader of the EU, though it does not meet the 3 per cent R&D/GDP Lisbon agenda target. Figure II.1.3 focuses on the EU and depicts the variation of R&D expenditures in the EU in 2005–2013. It is worth noting that eastern European countries are the most dynamic, likely due to their wider access to European-financed programmes for technological improvements. Northern countries such as Norway, Sweden, Germany and Austria are leaders in the development of new productive processes and products, so the positive growth in their R&D expenditures is not surprising (Gilli *et al.*, "Sustainability and Competitiveness"). Notice that France, which can also be included in the group of the innovation leaders, did not notably increase its investment in R&D during the period. Among southern European countries, the only noticeable increase in R&D was in Portugal. Those are significant changes. The R&D investment is still at 2 per cent of GDP in 2014, versus a target of 3 per cent. Northern European countries present higher R&D investments as a share of GDP.

Moving back to the global picture in Figure II.1.4, it is worth noting that in the majority of BRIC countries the growth of patents filed under the PCT was close to 0 until the early 2000s, with the exception of Russia, where patenting activity has been growing since the mid-1980s, and Brazil, where patenting activity slightly increased during the 1990s.

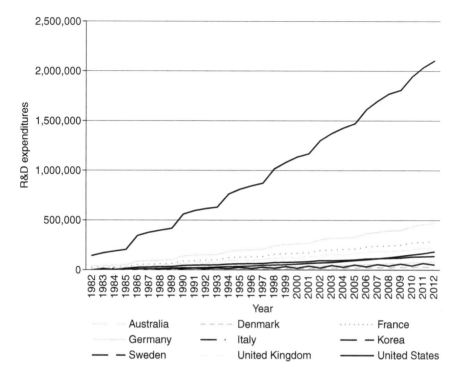

Figure II.1.2 R&D expenditures in selected OECD countries.

Source: authors' elaboration of data from the OECD database in the public domain, www.oecd-ilibrary.org/content/indicator/d8b068b4-en.

Since the late 1990s India and China have experienced a very fast increase in patenting activity. Since 1999, the number of patents filed in China increased from 579 to 22,184 in 2013, a growth of 3,731 per cent. In India, the number of patents increased from 203 in 1999 to 1,970 in 2013 (844 per cent growth). These numbers are driven, among other factors, by the quick economic growth that most Asian countries have experienced over the last two decades.

Looking at the specific segment of *green patents* in Figure II.1.5, four global players emerge (OECD, 2011, *Invention and Transfer*): the USA and Japan (typical patent developers), Germany (the EU leader) and South Korea (in an emerging role). South Korea serves as a stunning case study for growth and HD performance. It has a strong role in invention as well, affirming the link between growth and innovation. It is worth noting that South Korea devoted the bulk of its 2009–2012 fiscal recovery packages to green investments (30 out of 38 billion), part of the global strategy set at the 2009 G20 summit in London to tackle the economic and financial crisis through public investments.[10]

Green patents by GDP further highlight the astonishing South Korean performance, which overtakes the US in 2005 and then continues to move further up.

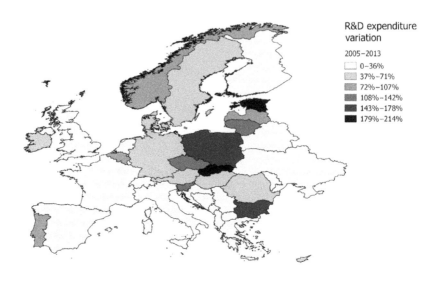

Figure II.1.3 Total R&D expenditure variation in Europe (2005–2013).

Source: authors' elaboration of data from the EUROSTAT database in the public domain, http://ec.europa.eu/eurostat/web/products-datasets/-/t2020_20&lang=en.

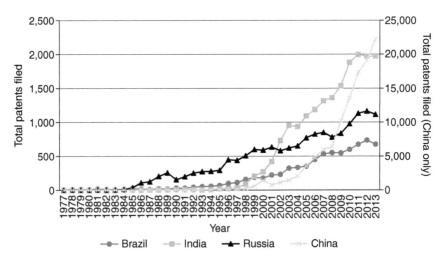

Figure II.1.4 Total patents filed under the PCT, 1977–2013, by inventor's country of residence (China on the right axis).

Source: authors' elaboration of data from the OECD Patent Statistics in the public domain, www.oecd-ilibrary.org/science-and-technology/data/oecd-patent-statistics/patents-by-main-technology-and-by-international-patent-classification-ipc_data-00508-en.

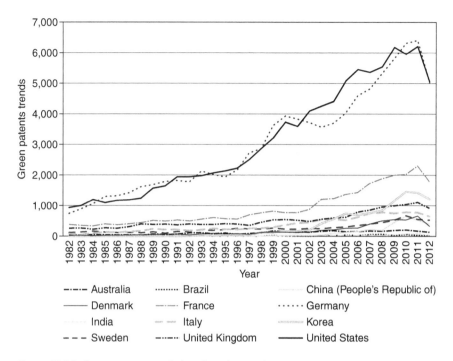

Figure II.1.5 Green patents trends in selected countries.

Source: authors' elaboration of data from the OECD Patent Statistics in the public domain, https://stats.oecd.org/Index.aspx?DataSetCode=PATS_REGION.

The green invention performances by Germany, the UK, France and Italy are notable too in relative GDP terms (Figures available on request). Germany, by far the EU leader for the entire period, interestingly, and perhaps worryingly, has recently had a declining pattern. The gap with respect to the US has widened. One explanation is due to the different economic policy directions the two countries have taken after the recession, with an expansion of the public budget in the USA and the opposite in Germany. This might be one reason, given the importance of public spending in innovative capacity (Mazzucato). In addition to the problems deriving from insufficient demand in the short run, "austerity" measures might generate lower innovation performances. In fact, innovation and invention depend largely on basic funding (R&D) and market expectations (GDP growth).

IV.2 Environmental policy stringency and ecological taxes

Figure II.1.6 shows the trend of the OECD Environmental Policy Stringency indicator, a country-specific and comparable measure of the stringency of environmental policies.[11] Among the considered countries, most of the variation

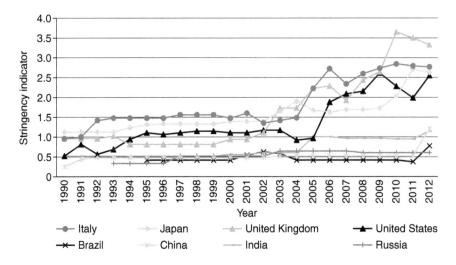

Figure II.1.6 OECD environmental policy stringency indicator (EPS), 1990–2012.

Source: authors' elaboration of data from the OECD Environmental Policy Stringency Index (EPS), https://stats.oecd.org/Index.aspx?DataSetCode=EPS.

occurs in Italy, Japan, the United Kingdom and the United States, while China, Brazil and Russia present a stable trend throughout the period. India shows a slight increase in stringency after 2005, when the index increased from approximately 0.6 to 1. With regards to the first group of countries (Italy, Japan, the United Kingdom and the US), the variation in the stringency is quite hectic, in some cases increasing and decreasing from year to year. Despite the yearly variation, the trend has been generally increasing since the second half of the 2000s, with a peak of 3.65 in 2010 by the United Kingdom. Overall, then, there is evidence that stringency positively correlates with income levels. As with many other public goods, the environment is economically speaking a "luxury good", with its "consumption" increasing more than proportionally with income. Two issues are thus at play. First, emerging and developing countries can imitate and follow strategies adopted by more advanced countries. This has happened in the EU with the compulsory EU policy implementation for recent members (primarily eastern EU countries). This might happen today if we look at the increasingly important role of China in global environmental policy, especially climate change policy (Mazzanti and Rizzo). China, again due to increasing income levels and high pollution, has started implementing policies that are more stringent. Second, in the "Trump era", with the US playing possibly an even weaker role in climate policy, China can emerge as the new big player, alongside the EU. For example, China has set a new strategy regarding emission trading systems for carbon emissions (Borghesi *et al.*, *The European Emission*).

Ecological tax reforms are important elements of environmental policy implementation, especially in the EU (see European Environment Agency,

Environmental Taxation, for a wide analysis of trends). Figure II.1.7 shows the yearly change in total environmental taxation from 2005 to 2013 (environmental taxation revenues are presented in current prices). We thus focus mainly on comparative figures (across countries and ET categories). Figures II.1.7 and II.1.8 depict the change in total environmental taxation and per capita environmental taxation, respectively.

Looking at Figure II.1.7, it is worth noting that environmental taxation[12] in northern as well as southern European countries has changed less than in eastern European countries. Once again, Eastern European countries have been among the last to join the Union and probably did not have a well-developed environmental fiscal system prior to the introduction of the minimum excise tax rates by the Energy Taxation Directive (2003/96EC). Consequently, they might have needed to introduce environmental taxation to comply with the European Union legislation. The average environmental taxation in eastern European countries is well below the average taxation elsewhere: €19,640,000, compared to €128,324,700 in southern EU countries and €180,131,600 in northern ones (figures and data are available on request).

In line with the results depicted in Figure II.1.7, Figure II.1.8 shows that per capita environmental fiscal pressure has increased the most in eastern European countries, especially Estonia, Latvia, Romania and Bulgaria, testifying to the importance of focusing on total and relative changes as well. Income and population dynamics are heterogeneous across EU countries.

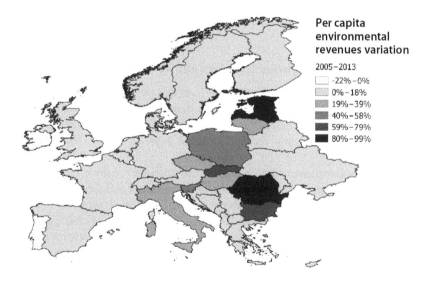

Figure II.1.7 Variation in total environmental taxes in Europe (2005–2013).

Source: authors' elaboration of data from the EUROSTAT database in the public domain, http://ec. europa.eu/eurostat/en/web/products-datasets/-/T2020_RT320.

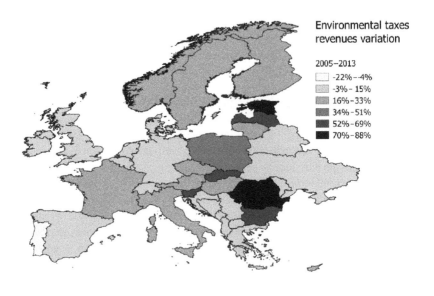

Figure II.1.8 Variation in per capita environmental taxes in Europe (2005–2013).

Source: authors' elaboration of data from the EUROSTAT database in the public domain, http://ec. europa.eu/eurostat/en/web/products-datasets/-/T2020_RT320.

V Innovation, knowledge transfers and sustainability

Innovation and economy composition effects drive environmental and economic performances (Quatraro; European Environment Agency, *Resource-Efficient Green Economy*; Gilli *et al.*, "Sustainability and Competitiveness"). Societies evolve by introducing new technologies that improve efficiencies in the use of resources and introduce new goods and services. Innovation and knowledge development complement the changes in the "content" of the economy (moving from industry to services, changing the mix of industry and services, etc.). High-income countries have naturally a very high share of services, while developing and emerging countries are today's global manufacturers. Within high-income countries, Italy and Germany are outliers in terms of industry relevance, but the industry share today does not exceed 20 per cent of GDP. Thus, two important issues emerge.

The first is linked to the more innovative contents of manufacturing with respect to services. The increased share of services might generate reductions in income per capita. For this reason, the EU has launched a "manufacturing strategy" so that industry makes up 20 per cent of the GDP by 2020. It is worth noting that in the short term these strategies increase emissions. On the other hand, the innovation contents of manufacturing might then reduce emissions.

This testifies to the fact that environmental sustainability in some ways depends on the capacity of "innovation" (technological and behavioural) to more than compensate for the scale effect of growth.

Second, stronger globalisation and economic integration lead to lower "national" thinking, even in terms of environmental policies and accounting. If it is true that technological improvements drive emissions reductions, part of the reductions we observe could be dependent on the delocalisation of production (Levinson, "Technology"). This means that both the sustainability of production and consumption would be at stake. High-income regions activate production in other areas of the world: the derived environmental emissions are shared in terms of responsibility. National emissions figures only capture the "production perspective", not the consumption side (e.g. imports from China that embody emissions and are activated by the demand of the US and EU) (European Environment Agency, *Resource-Efficient Green Economy*; Marin *et al.*; Gilli *et al.*, "Sustainable Development"). The toy global industry is a stunning case study. The UNIDO Industrial Development report is a key reference for understanding the global relevance of innovation and structural change as sources of economic growth, development and sustainability (for a more analytical extract see Gilli *et al.*, "Sustainable Development").

Figure II.1.9 shows an interesting analysis of how environmental performance can be disentangled, looking closer at innovation (efficiency), composition, and

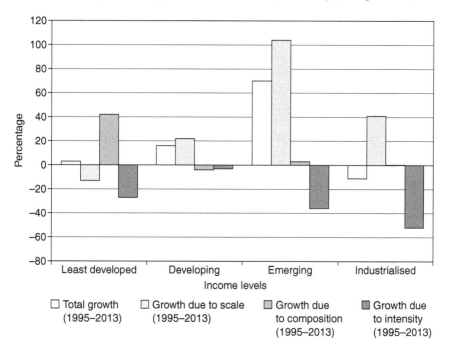

Figure II.1.9 Production-based CO_2 emissions across income levels.

Source: authors' elaboration of data from the Eora MRIO Database, http://worldmrio.com/.

growth effects. We decompose environmental pressures per capita of the manufacturing sector (either direct or "consumption-based") into various components: level of value added per capita (scale), share of production or consumption of a specific manufacturing sector over total production or consumption (composition); environmental pressures per unit of production or consumption (intensity). The analysis is based on EORA data (Lenzen *et al.*, "Mapping the structure", "Building EORA").

Striking differences appear globally across the four different groups of countries. First, we note that industrialised countries are the only group associated with a negative trend for CO_2 emission in the analysed period (1995–2013), while the other three groups registered a significant increase in emissions. Among the three components, the wealth effect always has a positive impact on total emission with the exception of the "least developed" countries, where it is negative. On the other hand, the composition effect has a similar and negligible impact on the four income groups, while the efficiency/technical effect (intensity of emissions in GDP) shows some important heterogeneity. In particular, technical improvement has reduced total emissions in all income groups with the exception of the "least developed" countries in which the emission associated with this effect grew in the analysed period.

The picture is extremely interesting, given the critical economic and environmental situation of the least developed countries that have suffered economic stagnation, even within a positive period of growth for developing and emerging countries, taking into account the 2008–2009 downturn. Developing countries instead grew but did not sustain efficient economic activities. With composition still negligible, the compensation effect of efficiency factors remains marginal. In the post-Kyoto era, notwithstanding the diffusion of Clean Development Mechanisms projects worldwide (Costantini and Sforna), low income and developing countries have not taken advantage of either policy-induced effects or technological diffusion. The new global carbon fund (CGF) should take this evidence into serious consideration when financing mitigation and adaptation projects. The GCF derives from the conference of the Parties in Cancun. It is a key pillar of climate change policy after the COP21 in Paris and should fund adaptation and mitigation strategies at an expense of around $100 billion a year by 2020. The sources of funding might be international donors and the usual government taxation, including long-term debt accumulation, or more specific and innovative solutions such as carbon taxes and emission trading auction revenues.

Emerging countries show an expected growth-led emission path, with some signs of efficiency compensation (innovation offsets), a signal that internal innovation mechanisms and international transfers of technologies have affected overall emissions trends.

Within a more stable composition of the economy, industrialised countries have succeeded in compensating scale effects with higher efficiency. The striking differences in emission intensity call for knowledge transfer, not just in terms of north–south (namely, richer northern areas selling technologies to poorer countries), but south–south and south–north feedback as well. In fact, knowledge should be transferred through various channels in which the emphasis is on the co-creation of knowledge and mutual feedback from the involved parties. It is worth noting that the effectiveness of knowledge transfer depends on the capacity of recipients to absorb knowledge, a product of the aggregate level of R&D investments, but also on the coherence between developed technologies and the cultural-institutional systems that host and adopt it. In addition, technological change is always effectively implemented if integrated with education and training investments.

Within this global integrated discourse, two innovation issues are pivotal: the role of frugal and grassroots innovations and non-codified knowledge. The key element linking them is the necessary expansion of the meaning of innovation to understand the real innovation phenomenon. Innovation is much more than (patented) technology, consisting of both (1) tangible and intangible knowledge flows and (2) complementary technological advances, organisational change and training in terms of human capital formation, skill redevelopment, etc. In fact, better social, environmental and economic performance, namely greater wellbeing and enhanced capabilities to create welfare and income, arise from the complementary use of diversified forms of innovations. Small innovations are as important as large-scale innovations: the effect is crucial to understanding its value for society and people. The central factor of interest is the context for adopting the innovations and the synergies that are present, which of course are also highly dependent on the context, such as geography, the sector, etc. Questions such as "which innovation(s) and how and where they are to be adopted" must be conceptually and empirically addressed. Education and training play a special role in the process. Though often overlooked, the formation of (new) skills and competences always complements techno-organisational innovations, given the primary relevance of human capital across development levels. Human capital is the necessary factor for achieving a sustainable society where technology increasingly transforms into social values and capabilities through enhanced access to resources.

VI Convergence in human development

This section provides an example of a convergence analysis using HD indicators. Other indicators and areas can be used to replicate the exercise. Convergence analysis is a simple and effective tool that shows how economic theories are quantitatively tested. It is worth recalling the policy and political flavour of convergence/divergence analyses. The ultimate goals of our societies are (1) to increase wellbeing in terms of (human) development (as a whole and in its specific components) and sustainability, defined in terms of "investing society",[13] and (2) to increase those indicators with an eye to the equitable allocation of resources worldwide and in sub regions. It is clear within economics that "equity" is not just a social aim (Picketty). Inequality can in fact hamper economic growth, which is based on a proper balance between wages and profits

(Kaldor; Pasinetti). Excess profits might mean a lack of demand, which leads to a typical capitalistic crisis. Sustainable economic growth thus depends on a balanced increase of investments, demand and profits.

The motivations behind investigating convergence in this realm relate to the striking differences in the trends of Human Development Index (HDI) and other variables among the different geographical areas considered, namely Northern Europe, Southern Europe and Eastern Europe. Convergence analysis will evaluate HDI in countries with variable speeds of economic and social development. In other words, we assess whether the human development gap between more and less developed countries is closing.

As outlined in Barro and Sala-i-Martin, convergence analysis is an important tool to assess various aspects of economy. They originally applied it to the rate of growth of an economy (i.e. the growth of GDP) and identified two concepts of convergence, β-convergence and σ-convergence. According to their definition, β-convergence exists when "poor economies tend to grow faster than the rich ones" (Barro and Sala-i-Martin 3), while σ-convergence exists when "the dispersion of the real per capita GDP levels tends to decrease over time" (Sala-i-Martin 3). Thus, according to the authors, β-convergence relates to the mobility of the different economies within the given distribution of world income, while σ-convergence concerns whether the cross-country distribution of world income shrinks over time.

The two definitions are certainly related; in fact, if a poor economy grows faster than a rich one (β-convergence), their GDP levels will become more similar, therefore causing a reduction in cross-country variability of income (σ-convergence). It is tautological that if the levels of two economies become similar over time (i.e. the dispersion of GDP decreases in time and there is σ-convergence), the poor ones grow faster than the rich ones (β-convergence).

The methodology proposed by Barro and Sala-i-Martin for income growth rate can be applied to the HDI growth rate. Variables that may influence the process of convergence are included as drivers, such as GDP, environmental taxes, R&D expenditure and capital and labour. The results are depicted in Figure II.1.10.

Less developed European countries are actually catching up with the most developed ones; moreover, environmental taxation, especially for transportation, appears to have a positive and effective role in favouring this process. R&D expenditures are also of vital importance, bearing relevance across the various convergence analyses). Income, capital and labour do not have a significant role in fostering β-convergence. These comments relate to econometric estimations, which are available upon request.

To investigate σ-convergence, we computed for each year the standard deviation of HDI across countries. The results plotted in Figure II.1.11 are that for each year the value of the y-axis corresponds to the standard deviation, that is HDI variability within the sample of countries. As expected, the variability of HDI is reduced over time, meaning that economic and social development have levelled off throughout the European Union.

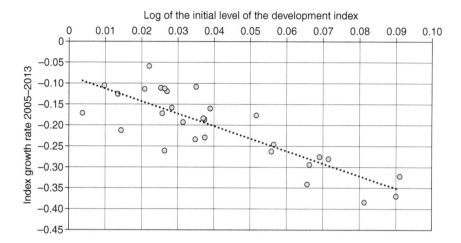

Figure II.1.10 Beta convergence in the development index in the EU.

Source: authors' elaboration of data from the United Nations Development Programme, http://hdr.undp.org/en/data.

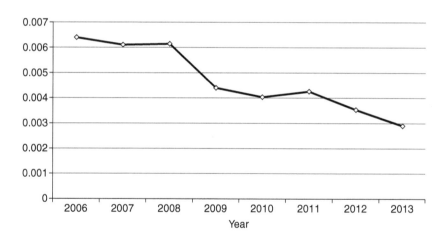

Figure II.1.11 Sigma convergence in the development index in the EU.

Source: authors' elaboration of data from the United Nations Development Programme, http://hdr.undp.org/en/data.

In summation, different areas of the European Union vary in terms of economic and social development, with the northern countries being more advanced than the southern and eastern ones. Moreover, while southern European countries seem less able to overcome the obstacles and the challenges posed by the economic crisis, Eastern Europe has responded more readily and has been able to sidestep the trends of 2008–2010. The convergence analysis shows that (1) less developed countries are catching up with more developed ones; (2) both environmental taxation and R&D expenditures are relevant in this process and can therefore be used as means to favour social and economic cohesion. Finally, (3) the level of development among European countries is actually levelling off, as shown by the presence of σ-convergence.

This is possibly unexpected news, considering the high political turmoil the EU has experienced due to the economic downturn and the (unsolved) consequences of ineffective economic and social policies. Overall, the EU integration process, with all of its drawbacks, seems to have generated a convergence in development. Policy integration is a main driver behind that trend. Replications of the analysis can be provided on a global level and for specific regional subsystems (Marin, "Closing the Gap?").

VII Economic and policy implications

GDP is an important but limited indicator of welfare and economic development. The higher the value societies assign to the environment, income equality and access to resources, health and other public goods, the more urgent is the need to use and exploit complementary theories and ways to measure socio-economic welfare performances. In essence, complementary, or satellite, indicators are necessary to enrich and measure core GDP performance in a hybrid fashion. Human development indexes and hybrid economic-environmental accounts move towards this direction. The environment and other public goods can be integrated in the economic sphere by monetising or merging. Non-market goods such as the environment might be monetised by analysing the social willingness to pay (e.g. for enhancing the environment, or increasing public goods). Hybrid accounts such as "CO_2 emissions on GDP" are useful.

One of the main conclusions to be drawn from discussions on economic development and sustainability is the need for a balanced pattern based on consumption and investments. Economic agents and societies tend to under-invest and over-consume available income due to inter-temporal myopia and short-term looking incentives. Sustainability can be defined as a society's willingness to invest in the future; these investments are necessary to sustain current economic demands towards full employment and compensate for the intrinsic erosion of capital stocks. Investments in human capital, intangibles such as education and innovation, man-made capital and renewable natural capital are needed to transition to a more equitable, greener and therefore sustainable society.

The second noteworthy takeaway is that investments in innovation and knowledge capital are essential for creating the preconditions for a more sustainable and equitable society. Innovation is the main driver of long-term economic growth; it can increase the efficiency of natural resource use (eco-innovations), open new markets for new and existing firms and, if adopted in synergies with human capital, enhance the skills, capabilities and wages of the workforce. In a word, innovation and, broadly, knowledge are at the basis of balanced and sustainable growth of profits and wages. To be most effective, technological development should be adopted as a complement to behavioural changes and human capital investments.

The third consideration refers to the role of policies, namely environmental and innovation policies. Public policies are the key to human development and sustainability for three main reasons. First, they correct the effects of private markets that are not capable of producing public goods, as market prices do not capture non-market values. Therefore, the quality and quantity of environmental goods remain under-delivered. Second, market-based environmental regulations, such as ecological taxes and emissions trading, are potentially able to incentivise continuous innovation in more efficient production. Cost-minimising households and firms seek to reduce their tax payment. The current situation is undergoing an evolution determined by political resistance to, but also gradual acceptance of, ecological taxes. Emissions trading is also being diffused worldwide following the establishment of the EU carbon market by the 2003 EU ETS Directive. The main economic appeal of economic instruments is that, in addition to providing incentives for using the environment better, they generate revenues that can fund other objectives (labour and innovation) and specific projects. Carbon taxes at a national or global level, congestion and pollution charges at an urban level are among the most relevant examples.

Empirical evidence shows that green innovation and environmental policies might positively correlate with economic development. Innovation and invention patterns indicate that high-income countries still possess a competitive advantage, but the gaps have decreased and new players are coming from the emerging and developing world. Innovation capacity is one of the main drivers behind sustainable development. Besides China, South Korea is probably the most striking example. The commitment of South Korea to green investment is an example of how a technologically oriented green strategy is synergic to growth and development. Larger gaps are present with respect to the stringency of environmental policy and adoption of ETR. It is necessary for emerging and developing countries to move towards environmental policies. These policies may nevertheless be transported through international trade, since greater economic integration requires playing by similar rules of the game. It is worth noting that trade exchanges facilitate pollution but also knowledge flows that are necessary to closing the existing efficiency gaps. Multilateral and bilateral north/north, north/south and south/south exchanges will transfer green knowledge, eco-innovations and sustainable practices on a global scale. The sustainability of production and consumption is a new framework to which we must adapt; it is a product of the increasing global integration of markets and knowledge. Countries are no longer self-sufficient nor are they islands unto themselves; along with sectors, firms and consumers, they are highly interconnected and mutually integrated. We all share a global responsibility for sustainability. It is not only an ethical statement but also an economic fact. Economies are integrated, knowledge quickly moves worldwide, problems and solutions are global.

In this global picture composed of many regional areas, convergence is key. It is desirable for economic reasons, since the ultimate goal of human societies is to enhance welfare and create convergence mechanisms across the globe. It is politically valuable, since countries and regions that diverge – or do not catch up – are at risk of cultural and economic isolation. Convergence is not manna from heaven; it stems from well-designed macroeconomic policies, sound objectives, recognition of regional specialisation, and investments in technology and human capital across income levels. To extend convergence beyond mere economic growth, societies need to widen the policies they adopt, including environmental policies. The world in which we live is full of global and regional convergences and divergences that researchers and policymakers need to continuously assess. Convergence of welfare and sustainable development is a concrete possibility. We have the knowledge and the financial resources to do this more frequently. The construction of bridges in the scientific world helps to improve the sources for applied research and our understanding of the facts we observe.

Notes

1 We would like to particularly thank Roberto Zoboli, Stefan Speck, Giovanni Marin and Nicola Cantore. This chapter follows key research trends we have shared over recent years, and research is always the intended or unintended output of a shared effort. It touches on topics investigated under the umbrella of the European Topic Centre on Waste Material and the Green Economy, 2014–2018 (funded by the EEA) and the H2020 project Green.eu (www.inno4sd.net) of which SEEDS-University of Ferrara is a partner unit. We are grateful to Saptorshee Chakraborty for his support to empirical analysis.

2 "While welfare is highly correlated with GDP per capita, deviations are often large" (Jones and Klenow 2426). "Leisure, inequality, mortality, morbidity, crime and the natural environment are just some of the major factors affecting living standards within a country that are incorporated imperfectly, if at all, in GDP" (Jones and Klenow 2426).

The chapter intentionally avoids addressing the degrowth issue, which is broader but nevertheless originates out of the limits of the GDP approach. We consider it a more general umbrella, while the chapter (empirically) addresses some of the specific issues. We refer the reader to van den Bergh, "Environment Versus Growth", and D'Alisa *et al.*, who offer diversified ideas and comments on the degrowth agenda.

3 The European NAMEA (National Accounting Matrix of Environmental Accounts) is a key example. WIOD (World Input Output Dataset) is another rich integrated dataset (www.wiod.org/home). Hybrid data permits the investigation of the role of innovation and structural change in a dynamic setting.

4 This theoretical reasoning links development and sustainability through the idea developed by Nobel laureate Simon Kuznets, who studies the existence of a bell-shaped curve linking inequality and GDP per capita. The same hypothesis has been tested for environmental quality and GDP per capita dynamic relationships (Carson; Mazzanti *et al.*, *Environmental Efficiency*).

5 The UNDP report clearly shows that countries at the same level of income generate different HD performances, depending on how they invest in health, education, poverty reduction, etc. Using different conceptual lines compared to HD indicators, Jones and Klenow show that "average western European living standards appear much closer to those in the USA when we take into account Europe's longer life

expectancy, additional leisure time, and lower inequality" (2427). Mortality seems to be one of the most important factors in explaining welfare differences across countries. In fact, "Most developing countries – including much of sub-Saharan Africa, Latin America, Southern Asia and China – are substantially poorer than incomes suggest because of a combination of shorter lives and extreme inequality" (2427–2428). Adding environmental considerations would probably worsen the scenario for those countries (and for the US as well).

6 The human capital share of the total stock is increasing across income levels. Human capital is the key factor for HD and sustainable development. The importance of its relationship with technological development is often overlooked; technology by itself cannot generate social improvements and therefore should not be its own goal. Enhanced human capital levels are crucial to support higher wages and new inventions; it is now empirically evident that the transition towards a greener society moves along a strict link between new skills and new technologies (Vona *et al.*).

7 Whether a society focuses and sets constraints on total or specific forms of capital is a political economy decision that differentiates weak and strong sustainability models (Neumayer). The policy consequences are substantial: weak sustainability (economics-oriented sustainability) implies full substitution between different forms of capital and emphasises an increase in total stock. In a sustainable society, an increase in the entire amount of capital should ensure increased capabilities to generate income and welfare.

8 For definitions of eco- and environmental innovations see Kemp; Barbieri *et al.*; Mazzanti *et al.*, "Firm Surveys".

9 This adds an "efficiency view" to the sustainability definition. Theoretically speaking, it refers to the IPAT (Impact Population Affluence Technology) identity (Marin and Mazzanti), common in ecological economics, industrial ecology and contiguous disciplines. It highlights the role of technology/efficiency as a major force that compensates for the detrimental effects of growth on the environment.

10 The report *Towards a Global Green Recovery* published on 2 April 2009 is available at www.lse.ac.uk/GranthamInstitute/publication/towards-a-global-green-recovery-recommendations-for-immediate-g20-action/.

11 Stringency is defined as the degree to which environmental policies put an explicit or implicit price on polluting or environmentally harmful behaviour. The index ranges from 0 (not stringent) to 6 (highest degree of stringency). The index covers 28 OECD and six BRICS countries for the period 1990–2012. The index is based on the degree of stringency of 14 environmental policy instruments, primarily related to climate and air pollution (https://stats.oecd.org/Index.aspx?DataSetCode=EPS).

12 We refer to total environmental taxation. Its three components are: transport taxation, energy taxation, air pollution taxes and resource taxes. The latter are the "real" environmental taxes if we follow an economic theoretical definition. Those are the smallest in share in the EU and most countries.

13 Though the HDI and sustainability frameworks are interrelated and could be placed under the "beyond GDP" discussion, it is also useful to keep them separate. HDI enriches GDP information by adding the key public goods produced by GDP (taxes). Sustainability adds components (the environment) and focuses on the role of investments. A crucial link is the necessity to invest in socially critical public goods to enhance growth, sustainability and wellbeing.

Bibliography

Albrizio, Silvia, Tomasz Kozluk, and Vera Zipperer. "Environmental Policies and Productivity Growth: Evidence across Industries and Firms". *Journal of Environmental Economics and Management*, vol. 81, 2017, pp. 209–226.

Ambec, Stefan, Mark A. Cohen, Stewart Elgie, and Paul Lanoie. "The Porter Hypothesis at 20: Can Environmental Regulation Enhance Innovation and Competitiveness?" *Review of Environmental Economics and Policy*, vol. 7, no. 1, 2013, pp. 2–22.

Barbier, Edward B., and Anil Markandya. *A New Blueprint for a Green Economy*. Routledge, 2012.

Barbieri, Nicolò, Claudia Ghisetti, Marianna Gilli, Giovanni Marin, and Francesco Nicolli. "A Survey of the Literature on Environmental Innovation Based on Main Path Analysis". *Journal of Economic Surveys*, vol. 30, no. 3, 2016, pp. 596–623.

Barro, Robert J., and Xavier Sala-i-Martin. *Economic Growth*. 2nd edn, MIT Press, 2003.

Bauer, Nico, Alex Bowen, Steffen Brunner, Ottmar Edenhofer, Christian Flachsland, Michael Jakob, and Nicholas Stern. "Towards a Global Green Recovery: Recommendations for Immediate G20 Action", April 2009, LSE, London. Policy Paper.

Baumol, William J., and Wallace E. Oates. *The Theory of Environmental Policy*. 2nd edn, Cambridge University Press, 1987.

Borghesi, Simone, Massimiliano Montini, and Alessandra Barreca. *The European Emission Trading System and Its Followers: Comparative Analysis and Linking Perspectives*. Springer, 2016.

Borghesi, Simone, Francesco Crespi, Alessio D'Amato, Massimiliano Mazzanti, and Francesco Silvestri. "Carbon Abatement, Sector Heterogeneity and Policy Responses: Evidence on Induced Eco Innovations in the EU". *Environmental Science and Policy*, vol. 54, 2015, pp. 377–388.

Botta, Enrico, and Tomas Kozluk. "Measuring Environmental Policy Stringency in OECD Countries: A Composite Index Approach". *OECD Economics Department Working Papers* no. 1177, 4 December 2014. OECD Publishing, doi.org/10.1787/5jxrjnc45gvg-en.

Brunel, Claire, and Arik Levinson. "Measuring the Stringency of Environmental Regulations". *Review of Environmental Economics and Policy*, vol. 10, no. 1, 2016, pp. 47–67.

Carson, Richard T. "The Environmental Kuznets Curve: Seeking Empirical Regularity and Theoretical Structure". *Review of Environmental Economics and Policy*, vol. 4, no. 1, 2010, pp. 3–23.

Commission of the European Communities. "Growth, Competitiveness, and Employment. The Challenges and Ways Forward into the 21st Century". Commission of the European Communities, 1993.

Costantini, Valeria, and Giorgia Sforna. "Do Bilateral Trade Relationships Influence the Distribution of CDM Projects?" *Climate Policy*, vol. 14, no. 5, 2014, pp. 559–580.

Costantini, Valeria, and Massimiliano Mazzanti, editors. *The Dynamics of Economic and Environmental Systems: Innovation, Policy and Competitiveness*. Springer, 2013.

Costantini, Valeria, and Massimiliano Mazzanti. "On the Green and Innovative Side of Trade Competitiveness? The Impact of Environmental Policy and Innovation on EU Exports". *Research Policy*, vol. 41, no. 1, 2012, pp. 132–153.

Costantini, Valeria, Massimiliano Mazzanti, and Anna Montini, editors. *Hybrid Economic Environmental Accounts*. Routledge, 2012.

Costantini, Valeria, and Salvatore Monni. "Sustainability and Human Development". *Economia Politica*, vol. XXV, no. 1, 2008, pp. 11–32.

Costantini, Valeria, and Salvatore Monni. "Environment, Human Development and Economic Growth". *Ecological Economics*, vol. 64, no. 4, 2008, pp. 867–880.

D'Alisa, Giacomo, Federico Demaria, and Giorgos Kallis. *Degrowth: A Vocabulary for a New Era*. Routledge, 2014.

Ekins, Paul, and Stefan Speck. *Environmental Tax Reforms*. Oxford University Press, 2011.

European Commission, DG Environment. *Study on Environmental Fiscal Reform Potential in 12 Member States*. European Commission, 2014.

European Environment Agency. *Environmental Taxation and EU Environmental Policies*. European Environment Agency, 2016.

European Environment Agency. *Resource-Efficient Green Economy and EU Policies*. European Environment Agency, 2014.

European Environment Agency. *The European Environment – State and Outlook 2005 – Synthesis Report*. Office for Official Publications of the European Communities, 2005.

Gilli, Marianna, Massimiliano Mazzanti, and Francesco Nicolli. "Sustainability and Competitiveness in Evolutionary Perspectives: Environmental Innovations, Structural Change and Economic Dynamics in the EU". *Journal of Socio-Economics*, vol. 45, 2013, pp. 204–215.

Gilli, Marianna, Giovanni Marin, Massimiliano Mazzanti, and Francesco Nicolli. "Sustainable Development and Industrial Development: Manufacturing Environmental Performance, Technology and Consumption/Production Perspectives". *Journal of Environmental Economics and Policy*, vol. 6, no. 2, 2017, pp. 183–203, http://dx.doi.org/10.1080/21606 544.2016.1249413.

Jones, Charles, and Peter K. Klenow. "Beyond GDP? Welfare across Countries and Time". *American Economic Review*, vol. 106, no. 9, 2016, pp. 2426–2457.

Kaldor, Nicholas. "A Model of Economic Growth". *Economic Journal*, vol. 67, no. 268, 1957, pp. 591–624.

Kemp, Rene. "Eco-innovation: Definition, Measurement and Open Research Issues". *Economia Politica*, vol. 3, 2010, pp. 397–420.

Lenzen, Manfred, Daniel Moran, Keiichiro Kanemoto, and Arne Geschke. "Building EORA: A Global Multi-regional Input-Output Database at High Country and Sector Resolution". *Economic Systems Research*, vol. 25, no. 1, 2013, pp. 20–49.

Lenzen, Manfred, Keiichiro Kanemoto, Daniel Moran, and Arne Geschke. "Mapping the Structure of the World Economy". *Environmental Science and Technology*, vol. 46, no. 15, 2012, pp. 8374–8381.

Levinson, Arik. "Technology, International Trade, and Pollution from US Manufacturing". *American Economic Review*, vol. 99, no. 5, 2009, pp. 2177–2192.

Marin, Giovanni. "Closing the Gap? Dynamic Analyses of Emission Efficiency and Sector Productivity in Europe". *The Dynamics of Environmental and Economic Systems. Innovation, Environmental Policy and Competitiveness*, edited by Valeria Costantini and Massimiliano Mazzanti, Springer, 2013, pp. 159–177.

Marin, Giovanni, and Massimiliano Mazzanti. "The Evolution of Environmental and Labour Productivity Dynamics". *Journal of Evolutionary Economics*, vol. 23, 2013, pp. 357–99.

Marin, Giovanni, Massimiliano Mazzanti, and Anna Montini. "Linking NAMEA and Input Output for 'Consumption vs. Production Perspective' Analyses: Evidence on Emission Efficiency and Aggregation Biases Using the Italian and Spanish Environmental Accounts". *Ecological Economics*, vol. 74, 2012, pp. 71–84.

Mazzanti, Massimiliano, and Ugo Rizzo. "Diversely Moving towards a Green Economy: Techno-Organisational Decarbonisation Trajectories and Environmental Policy in EU Sectors". *Technological Forecasting and Social Change*, vol. 115, 2016, pp. 111–116, doi:10.1016/j.techfore.2016.09.026.

Mazzanti, Massimiliano, and Anna Montini, editors. *Environmental Efficiency, Innovation and Economic Performances*. Routledge, 2010.

Mazzanti, Massimiliano, Davide Antonioli, Claudia Ghisetti, and Francesco Nicolli. "Firm Surveys Relating Environmental Policies, Environmental Performance and Innovation. Design Challenges and Insights from Empirical Application". OECD Environment Working Paper no. 103, 12 April 2016. OECD Publishing, doi: 10.1787/5jm0v405l97l-en.

Mazzanti, Massimiliano, Giovanni Marin, Susanna Mancinelli, and Francesco Nicolli. "Carbon Dioxide Reducing Environmental Innovations, Sector Upstream/Downstream Integration and Policy: Evidence from the EU". *Empirica*, vol. 42, no. 4, 2015, pp. 709–735.

Mazzucato, Mariana. *The Entrepreneurial State*. Anthem Press, 2013.

Musolesi, Antonio, Massimiliano Mazzanti, and Robert Zoboli. "A Panel Data Heterogeneous Bayesian Estimation of Environmental Kuznets Curves for CO_2 Emissions". *Applied Economics*, vol. 42, no. 18, 2010, pp. 2275–2287.

Neumayer, Eric. *Weak versus Strong Sustainability: Exploring the Limits of Two Opposing Paradigms*. Edward Elgar Publishing, 1999.

Organization for Economic Cooperation and Development. *Patents by Main Technology and by International Patent Classification (IPC), OECD Patent Statistics (database)*. OECD Publishing, 2017, doi.org/10.1787/data-00508-en.

Organization for Economic Cooperation and Development. *Gross Domestic Spending on R&D (Indicator)*. OECD Publishing, 2017, doi:10.1787/d8b068b4-en.

Organization for Economic Cooperation and Development. *Invention and Transfer of Environmental Technologies*. OECD Publishing, 2011.

Pasinetti, Luigi L. "Rate of Profit and Income Distribution in Relation to the Rate of Economic Growth". *The Review of Economic Studies*, vol. 29, no. 4, 1962, pp. 267–279.

Perman, Roger, Yue Ma, Michael Common, David Maddison, and James McGilvray. *Natural Resource and Environmental Economics*. 4th edn, Pearson, 2011.

Picketty, Thomas. *Capital in the 21st Century*. Belknapp Press, 2013.

Porter, Michael E., and Claas van der Linde. "Green and Competitive: Ending the Stalemate". *Harvard Business Review*, no. 73, vol. 5, 1995, pp. 120–134.

Porter, Michael. "America's Green Strategy". *Scientific American*, vol. 264, no. 4, 1991, p. 168.

Quatraro, Francesco. *The Economics of Structural Change in Knowledge*. Routledge, 2013.

Turner, R. Kerry and David W. Pearce. *Environmental Economics: An Elementary Introduction*. Johns Hopkins University Press, 1993.

United Nations Industrial Development Organization. *Industrial Development Report 2016. The Role of Technology and Innovation in Inclusive and Sustainable Industrial Development*. UNIDO, 2015.

Van den Bergh, Jeroen. "Environment Versus Growth – A Criticism of 'Degrowth' and a Plea for 'A-Growth'". *Ecological Economics*, vol. 70, no. 5, 2011, pp. 881–890.

Van den Bergh, Jeroen, and Miklós Antal. "Evaluating Alternatives to GDP as Measures of Social Welfare/Progress". WWWforEurope Working Paper no. 56, March 2014. WWWforEurope, www.foreurope.eu/fileadmin/documents/pdf/Workingpapers/WWW-forEurope_WPS_no056_MS211.pdf.

Vona, Francesco, Giovanni Marin, Davide Consoli, and David Popp. "Green Skills". NBER Working Paper no. 21116, April 2015. The National Bureau of Economic Research, www.nber.org/papers/w21116.

II.2 Social equity and ecological sustainability through the lens of degrowth

Valeria Andreoni

I The pillars of degrowth

World societies are facing complex, interrelated crises. Environmental and climate change disasters, the economic and financial crash, the increasing social discontent voiced in developed and developing countries are problems that human societies must face in the present and near future (Garcia-Olivares and Solé). The increasing scale of the global economy and the consequent demands on the Earth's source and sink functions show that the paradigm of sustainable economic growth dominating policies since the Second World War has failed to generate a system where technological progress and resources efficiency should be able to ensure consumption increase, environmental protection and social equity (Andreoni). The present crisis has thus opened up opportunities to test new paradigms, policies and lifestyles aimed at shaping global societies according to the capacity of our environment. It is in this context that the concept of degrowth has been proposed (Andreoni and Galmarini, "How to Increase Wellbeing").

Degrowth has been defined as "an equitable downscaling of production and consumption activities that increases human wellbeing and enhances ecological conditions at the local and global level, in the short and long term" (Schneider *et al.* 511). Following the principles of thermodynamics, it presupposes that economy is a subsystem of the environment, limited and thermodynamically closed. Clearly, an infinite economic growth in a finite system is impossible.

According to this approach, degrowth economy proposes a quantitative reduction of the impacts on the environment along with changes in our culture, attitudes and system of values. In particular, the concept of degrowth is based on the idea that in order to fit within the biophysical limits of the planet, we need to escape from the growth-mania and build up a society able to live better with less (Latouche). For this reason, the transition towards degrowth would require a change of the sustainability paradigm. As highlighted in Kemp and van Lente, a dual challenge would be needed. If on the one side the socio-economic system should evolve so as to reduce the overall impacts on the environment, on the other side the criteria used to judge the appropriateness of socio-economic decisions should change according to the new paradigm of degrowth. Within this context, policy and decision-makers are particularly important (Nill and Kemp).

The main pillars of degrowth economy and its main critiques to the present socio-economic system will be illustrated. Degrowth principles and initiatives will be enunciated in order to uncover research gaps and delineate future projects, developments and policy implications.

The concept of degrowth, originating at the intersection of different disciplines and theoretical backgrounds, is based on the idea that the paradigm of an infinite economic growth has had negative impacts on society and the environment. The paradigm of degrowth involves four interrelated research areas: Sustainability, Wellbeing, Ethics, and Democracy.

II Sustainability

Since the end of the Second World War, the environmental effects of economy have received increasing attention from economists. The worldwide deterioration of the environmental quality and the ecological requirement of economic systems have become an important issue both in the political debate and in academic literature.

In the early 1960s publications such as *Limits to Growth* (Meadows *et al.*), *Silent Spring* (Carson) or *The Entropy Law of Economic Process* (Georgescu-Roegen) focused on the negative environmental impacts of economic activities and highlighted the impossibility of an infinite economic growth in a limited and finite environment.

At the same time, however, the rapid technological development and improvements in efficiency and productivity fostered optimistic perceptions of the possibility to overcome the environmental constraints through development and innovation. The work on dematerialisation and depollution by Malenbaum and later Panayotou, generally known as the Environmental Kuznets Curve, linked economic growth to the declining rate of the material used and pollution. In 1987 the World Commission on Environment and Development defined the concept of sustainable development as a form of "development that meet the needs of the present without compromising the ability of future generations to meet their own needs" (43). Based on the idea that economic growth and technological progress are the best way to improve environmental quality and reduce poverty, sustainable development policies promoted economic growth and consumption increase both in developed and developing areas.

In recent years, different studies have investigated the depollution hypothesis and the environmental Kuznets curve in a large set of countries and with regards to different pollutants, materials and economic sectors. The main evidence shows that, if technological progress is able to reduce the quantity of resources used and pollution generated per unit of goods and services produced, the reduction in the price related to the efficiency improvements generates an overall increase of the aggregate demand, with consequent negative impacts on natural resources. This effect, generally known as Jevons Paradox, highlights that economic growth and consumption increase are not able to reduce the impacts on the environment.

As a consequence, recent developments of technologically oriented approaches to sustainability have been devoted to investigating the effectiveness of different strategies generally summarised under the name of "sustainable consumption and production". The "eco-efficient production", the "green-supply chains" and the "labelled products" are examples of activities aimed at reducing the waste and energy used, in order to produce greener products and to incentivise consumers to buy responsibly (Geels *et al.*).

All these approaches rely on the idea that the production and consumption activities taking place in a market-oriented environment are largely efficient and that marginal public interventions are needed to fix the market failures associated with pollution and resource depletion. The EU Sustainable Consumption and Production Action Plan, the OECD Green Growth document and the emission trading scheme are examples of policy interventions aimed at stimulating investments in eco-innovation through regulation, R&D subsidies, environmental taxes and cap-and-trade schemes.

Following the large economic and policy implications related to the sustainable production and consumption approach, various issues have been recently raised. The main concerns are related to the fact that market instruments, being based on profit and short-term perspective, are unable to fully address the scale and urgency of the environmental problems. According to this view, the degrowth theory advocates that a green transformation of production and consumption will not be enough to address the overexploitation of the natural environment. For this reason, a scale reduction of the present economic system would be needed to fit the biophysical limits of the planet.

III Wellbeing

Mainstream economic theory usually approximates wellbeing to the GDP level. Based on the utilitarian approach that considers people's utility dependent on consumption, the standard economic hypothesis reduces the definition of wellbeing to income and consumption. According to this approach, economic growth is considered as the best way to increase the wellbeing of people and societies. In reality, wellbeing is difficult to define mainly because it involves subjective perceptions of values and is subject to changes over space and time based on cultural variables, expectations and human adaptation (Andreoni and Galmarini, "How to Increase Wellbeing").

Over the last decades, numerous studies have been devoted to identifying the main variables that characterise wellbeing and also to investigating the negative or non-increasing relationships between income and wellbeing (Easterlin). Evidence provided by different studies highlights that wellbeing is a multidimensional concept generated by a set of different and interrelated variables. Table II.2.1 shows some of the most important theories and hypothesis related to the definition and quantification of wellbeing.

Based on this evidence, the theory of degrowth proposes to redefine the concept of wellbeing according to a set of multidimensional and subjective values.

Table II.2.1 Wellbeing theories and hypothesis

Easterlin Paradox	Once a certain absolute level of income is reached, gains in wellbeing are obtained by having higher income relative to other people, or by having benefits generated by social relationships or environment (Easterlin)
Critical level of GDP	Above a certain level of GDP, income inequalities, environmental degradation and reductions in leisure time tend to increase and the overall level of wellbeing tend to decrease (Clark *et al.*)
Relative Income Hypothesis	The level of wellbeing is influenced by comparisons with other individuals of society (Corazzini *et al.*)
Personality	Only around 10% of individual wellbeing is attributable to income. The remaining 90% is determined by personality and activities that people choose to engage in (Headey *et al.*)
Set point theory	Since people tend to adapt their aspirations to their changing circumstances, an increasing level of income does not generate an increasing level of wellbeing in the long term (Frey and Stutzer)
Expectation and consumerist values	It is generally recognised that expectation and consumerist values are an important causal factor of unhappiness and mental illness such as depression, anxiety and narcissism (Aydin)

Individual preferences, satisfaction of basic human needs, environmental variables, social relationships and expectations should be considered as important elements of wellbeing. Based on that, since human satisfaction is not strictly related to increasing levels of consumption, the need for an infinite economic growth is no longer justified (Andreoni and Galmarini). According to this approach, subjective elements and multidimensional perspectives should be taken into account in the decision-making process both from an individual and a societal perspective. The recognition of wellbeing as a complex element of life satisfaction also contributes to promoting processes of participative democracy and collaborative decision-making approaches such as for example *Agenda 21*, launched in 1992 after the UN conference on Environment and Development, and the recent Sustainable Development Goals.

IV Equity

According to mainstream economic theories, economic growth is considered as the best way to reduce poverty and inequality through income generation and redistribution. This approach is based on the so-called Kuznets Theory, named after the Nobel Prize winner Simon Kuznets, who has identified an inverted U-shape relationship between development and inequality. The basic idea is that during the first stage of development the level of inequality is really low. As far as economies develop, the unequal distribution of income, mainly generated by the concentration of the most productive economic activities in the hands of small groups of people, leads to an increasing level of inequality. In developed societies, the level of equality tends to increase mainly because in periods of economic growth the spill-over effects across economic activities tend to generate redistribution of income across all the levels of society (Desborder and Verardi).

Recent studies, however, have suggested that economic growth cannot be necessarily considered as synonymous with inequality reduction. According to data provided by Brandolini, in spite of the sustained economic growth over the last 50 years, the level of inequality has increased both in the UK and US. Taxation, redistribution policies, welfare state and support, together with mutual help and voluntary activities, are considered by different authors as elements that can reduce inequality in periods of economic growth or crisis (Putnam). In addition, *Capital in the Twenty-First Century* by Thomas Piketty highlights how the inequality reduction reported by the Kuznets Curve is not purely related to economic growth variables. Technological factors, sectoral spill-over effects, taxation, power distribution and state intervention are also important elements that should be used to explain the relationship between economic trends and inequality levels.

V Democracy

One of the main issues associated with socio-environmental problems is the debate about decision making. As highlighted by authors such as for example Funtowicz and Ravetz, when uncertainty and complexity are high, decision-making processes cannot be delegated to a small number of people. On the contrary, deliberation and planning should result from a participative process able to consider the plurality of values and perceptions of the different stakeholders. The impact generated both on present and future generations should also be considered.

The lack of transparency and the feeling of imposed decisions that characterise large areas of the present socio-economic system are some of the main elements generating social discontent and protests. Examples are provided by the recent mobilisation movements associated with the global financial crisis. *Movimento 5 Stelle* in Italy, *Indignados* and *¡Democracia Real YA!* in Spain, *Occupy Wall Street* in the US are examples of social instabilities and protests asking for decision-making processes based on transparency and participation.

In addition to that, the catalysing effect of information and communication technologies has largely contributed to the creation of non-border virtual communities based on the sharing of information, data and public debate. The large use of social media as a space for distribution of knowledge and debate has spread dialectic processes across countries and communities. Participative democracy and stakeholders' involvement seem to be even more fundamental to reducing conflicts and instabilities as well as to promoting sustainable long-term development.

The four pillars reported above highlight the main critical elements that, according to the theory of degrowth, contribute to generating instabilities in the present socio-economic system. In order to overcome these problems, alternative socio-economic structures are proposed by degrowth. Some of the main degrowth ideas will be illustrated together with examples of practical degrowth initiatives.

VI Degrowth principles and initiatives

During the last decades and in particular after the recent economic crisis a large number of degrowth practices have taken place both in developed and developing countries. The largest parts of them are voluntary bottom-up initiatives that have been put in place by individuals or small groups of people. This section presents an overview of the main degrowth principles, namely: (1) Sobriety and voluntary simplicity; (2) Conviviality and reciprocity; (3) Subsidiary and sustainable production principles; (4) Participation and transparency; (5) Redefinition of wellbeing and accountability system.

1   Sobriety and voluntary simplicity. Defined as a voluntary reduction of production and consumption activities, this principle is aimed at reducing the environmental impact and emphasising quality of life rather than quantity of consumption. It includes a set of strategies to reduce consumption and make production more ethical and sustainable. Examples are provided by product-life extension, self-production and exchange, reuse and recycling, local cooperative of production and consumption and leisure based on social and environmental activities (Latouche).

2   Conviviality and reciprocity. This principle includes activities devoted to society without monetary compensation and based on mutual support, voluntary activity and community exchange. Conviviality and reciprocity help decrease consumption while increasing social relationships, networking and social capital. Examples are provided by Time Banking, HOURS Currencies, Cohousing and non-paid jobs (Andreoni and Galmarini, "How to Increase Wellbeing").

3   Subsidiary and sustainable production. When possible, consumption needs to be satisfied by local production. This approach will reduce the global environmental impact by reducing the emissions generated by transport and will contribute to support local economy and sustainability. In addition, since consumers will have direct relationships with producers, it will be easier to track ethical and sustainable behaviours. Local currencies, cooperatives of production and consumption, urban gardening and the Transition Town Movement exemplify this principle (Garcia-Olivares and Solé).

4   Participation and transparency. Extended participation in decision making, involvement of local communities, consideration of short- and long-term impacts, attention for present and future generations are some of the principles that degrowth economy considers fundamental to implementing a

real, sustainable and participative democracy. When high uncertainty, complexity and plurality of values are involved, decision-making processes need to consider different parameters and perspectives together with the possible impacts generated both in the short and long term. *Agenda 21*, public debates and referendums are some of the practices used to achieve participative and transparent decision-making processes (Funtowicz and Ravetz).

5   Redefinition of wellbeing and accountability system. The reduction of the emphasis that our society has laid on materialistic values is a fundamental step to decrease the level of consumption and lower expectations of economic growth. Reducing the emphasis on GDP by introducing a set of indicators able to account for different environmental, social and wellbeing variables is a fundamental step towards changing perception about what really matters. The Material Flow Accounting, the Ecological Footprint, the different GDP adjustments and the multidimensional description of wellbeing are examples of this principle (Andreoni and Galmarini, "How to Increase Wellbeing").

The five categories and the examples reported above do not intend to be an exhaustive summary of the main bottom-up degrowth initiatives. Many others are implemented and different degrowth debates are taking place to explore the feasibility of a degrowth society. In addition to that, a large number of international conferences, meetings and publications have also been devoted to investigating the concept of degrowth from an academic perspective.[1] The concept of degrowth, the identification of different implementation mechanisms and the analysis of the possible impacts have been largely debated during the last decade. However, many research areas still need to be investigated.

VII Research gaps, future developments and policy implications

The degrowth principles and initiatives reported in the previous section represent a set of bottom-up voluntary practices generally implemented on a small and local scale. For this reason, limited effects have been generated on the global economic system. However, the implementation of degrowth practices on a larger scale could have massive impacts both on developed and developing countries. A clear understanding of the possible consequences on economy and society is then needed to investigate the feasibility of degrowth. In particular, further research should be devoted, but not limited, to investigating:

1   The socio-economic impacts generated by reduction of import-export activities v/local production increase. Within this context, the impacts on the development paths of developing countries, particularly if export-oriented, could be considered together with the analysis of the possible consequence on the global supply chain.

2   The impacts on existing business activities and the opportunities for ethical and sustainable businesses.

3   The cascading effects produced by GDP reduction on economy and society. Possible areas of research could be related to income distribution and taxation.
4   The impacts that a decreasing level of GDP could have on investment on research development, innovation and technology.
5   The impact that a reduction of market-based activities and an increase of voluntary work could generate on tax and the welfare state.
6   The impacts on wellbeing generated by reduction of consumption and an increasing level of voluntary activity, mutual support and co-operation.
7   Co-operation and mutual support between individuals as a way to increase social capital and trust among citizens.
8   Ways of implementing a system of real and participative democracy.
9   Ways of making these changes voluntary and acceptable to the majority of the population.
10 The impacts that the implementation of degrowth principles could have on environmental elements such as emissions, water or land availability.

The feasibility of degrowth should also be considered in a policy context. Up to now, very few political initiatives have been specifically aimed at promoting degrowth as a new economic paradigm. However, some existing policies and policy proposals already include elements that are relevant for the context of degrowth.

Work sharing and working time reduction

During the last decades, socio-political debates and research initiatives have been devoted to investigating the benefit of working-time reduction. In European countries, such as Belgium and the Netherlands, working-time policies based on the "life course approach" have proven to be effective towards increasing productivity and life-working balance. The basic idea is to have flexible working arrangements based on the change of working time preferences across the different stages of life. The Life-course Savings Scheme is an example of an existing policy aimed at providing options for part-time working, career breaks, and childcare options.

Minimum (and maximum) wages and job guarantees

In different countries, minimum wage policies have been introduced to reduce income inequalities, improve people's economic security and reduce public spending for assistance programmes. Subsided jobs are also offered to promote social inclusion, skill development, social stability and wellbeing. Examples are provided by the "welfare-to-work" scheme in the UK, or by the Jefes de Hogar Programme in Argentina. Degrowth analyses also suggest that the introduction of a maximum wage could be useful to further increase social equality and redistribution.

Support for sustainable production activities

Different kinds of public financial support, such as subsidies, tax exception, grants and provision of public buildings and spaces already exist for the non-profit as well as the co-operative economic sector. The extension of these benefits to activities and groups oriented to promote mutual support, conviviality and reciprocity could contribute to further increase the non-market activities encouraged by degrowth.

Reduce the focus on consumption

Restrictive criteria for advertising in public spaces such as the ban on commercial street advertising in Grenoble (France) or the Clean City Law in São Paulo (Brazil) can contribute to reduce the level of consumption and consequent impacts on resource depletion and pollution.

Establish an accountability system based on a plurality of variables

The focus on GDP as the main indicator of the wellness of societies can be changed by adopting an accountability system able to include a set of different socio-environmental variables. The Sustainable Development Indicators proposed by Eurostat are an example of how alternative descriptions of reality can be provided by adopting a multidimensional approach.

The few examples above show how different kind of policies can be designed to promote a transition toward degrowth. However, a large political debate across developed and developing countries would be needed to discuss the feasibility of degrowth and possible implementation policies.

VIII Final remarks

In this chapter, an overview of the main elements characterising the concept of degrowth is presented together with some of the main research gaps existing in academic literature. The theoretical background to degrowth comprises four main pillars: Sustainability; Equity; Wellbeing; and Democracy. In spite of the large number of academic contributions on these four research areas, the degrowth approach identifies socio-environmental unsustainability as a whole. Policies oriented towards economic growth and the consequent maximisation of consumption in a market-oriented system are identified by degrowth as the main causes for environmental impacts, social inequality and conflicts. A radical change in the system of values along with a redefinition of needs and values are fundamental to the promotion of a sustainable-oriented society.

In addition to the theoretical studies of degrowth, a large set of bottom-up initiatives have also been put in place, particularly after the global financial crisis. Time banking, local currencies, cohousing, reciprocity jobs and transition town movements are just some examples. In spite of the increasing involvement of local communities and the large debate taking place both in civil society and academia, large research gaps still exist with regards to the identification of the main impacts that a degrowth-oriented society could generate when applied on a larger scale. As argued in the previous paragraph, increasing efforts should be devoted to investigating the possible impacts that a degrowth economy could generate both at local and global level, particularly considering that the present socio-economic structure is characterised by globalisation and inter-sectoral connections. The feasibility of degrowth must be urgently studied in order to make degrowth a feasible alternative for a sustainable transition.

Note

1 See the special issues on Degrowth published by many prominent journals or the Degrowth conference in Paris (2008), Barcelona (2010), Venice (2012), Montreal (2012), Leipzig (2014) and Budapest (2016).

Bibliography

Andreoni, Valeria. "Can Economic Growth be Sustainable? The Case of EU27". *Journal of Global Policy and Governance*, vol. 1, no. 2, 2013, pp. 185–195.

Andreoni, Valeria, and Stefano Galmarini. *Mapping the Distribution of Wellbeing in Europe beyond National Borders.* European Commission – JRC Scientific and Policy Report, 2015.

Andreoni, Valeria, and Stefano Galmarini. "How to Increase Wellbeing in a Context of Degrowth". *Future*, vol. 55, no. 1, 2014, pp. 78–89.

Andreoni, Valeria, and Stefano Galmarini. "On the Increase of Social Capital in a Degrowth Economy". *Procedia – Social and Behavioural Science*, vol. 72, no. 1, 2012, pp. 64–72.

Aydin, Necati. "Subjective Wellbeing and Sustainable Consumption". *The International Journal of Environmental, Cultural, Economic, and Social Sustainability*, vol. 6, no. 1, 2010, pp. 133–148.

Brandolini, Andrea. *A Bird's Eye View of Long-run Changes in Income Inequality.* Bank of Italy Research Department, Rome, 2002.

Carson, Rachel. *Silent Spring.* Houghton Miffin, 1962.

Clark, Andrew, Paul Frijters, and Michael A. Shields. "Relative Income, Happiness and Utility: An Explanation for the Easterlin Paradox and other Puzzles". *Journal of Economic Literature*, vol. 46, no. 1, 2008, pp. 95–144.

Corazzini, Luca, Lucio Esposito, and Francesca Majorano. "Reign in Hell or Serve in Heaven? A Cross-country Journey into the Relative vs Absolute Perceptions of Wellbeing". *Journal of Economic Behavior and Organization*, vol. 81, no. 3, 2012, pp. 715–730.

Desborder, Rodolphe, and Vincenzo Verardi. "Refitting the Kuznets Curve". *Economic Letters*, vol. 112, no. 2, 2012, pp. 258–261.

Easterlin, Richard A. "Does Economic Growth Improve the Human Lot? Some Empirical Evidence". *Nations and Households in Economic Growth. Essays in Honor of Moses Abramovitz*, edited by Paul A. David and Melvin W. Reder, Academic Press, 1974.

European Commission. *Sustainable Production and Consumption Plan*, 2008, http://ec.europa.eu/environment/eussd/escp_en.htm.

Frey, Bruno, and Alois Stutzer. "The Economics of Happiness". *World Economics*, vol. 3, no. 1, 2002, pp. 25–41.

Funtowicz, Silvio, and Jerry Ravetz. "Science for the Post-normal Age". *Future*, vol. 25, no. 7, 1993, pp. 739–755.

Garcia-Olivares, Antonio, and Jordi Solé. "End of Growth and the Structural Instability of Capitalism – From Capitalism to a Symbiotic Economy". *Future*, vol. 68, 2015, pp. 31–43.

Geels, Frank W., Andy McKeekin, Josephine Mylan, and Dale Southerton. "A Critical Appraisal of Sustainable Consumption and Production Research: The Reformist, Revolutionary and Reconfiguration Positions". *Global Environmental Change*, vol. 34, no. 1, 2015, pp. 1–12.

Georgescu-Roegen, Nicholas. *The Entropy Law and the Economic Process.* Harvard University Press, 1971.

Headey, Bruce, Ruud Muffels, and Gert G. Wagner. "Long-running German Panel Survey Shows that Personal and Economic Choices, Not Just Genes, Matter for Happiness". *Proceedings of the Academy of Sciences*, vol. 107, no. 42, 2010, pp. 17922–17926.

Kemp, Rene, and Harro van Lente. "The Dual Challenge of Sustainability Transitions". *Environmental Innovation and Societal Transitions*, vol. 1, no. 1, 2011, pp. 121–124.

Latouche, Serge. "Degrowth". *Journal of Cleaner Production*, vol. 18, no. 6, 2010, pp. 519–522.

Malenbaum Wilfred. *World Demand for Raw Materials in 1985 and 2000.* MacGraw-Hill, 1978.

Meadows, Donatella, Dennis Meadows, Jørgen Randers, and William W. Behrens III. *The Limits to Growth.* Universe Books, 1972.

Nill, Jan, and René Kemp. "Evolutionary Approaches for Sustainable Innovation Policies: from Niche to Paradigm?" *Research Policy*, vol. 38, no. 4, 2009, pp. 668–680.

OECD. *Toward a Green Growth*, 2011, www.oecd.org/env/towards-green-growth-9789264111318-en.htm.

Panayotou, Theodore. *Empirical Tests and Policy Analysis of Environmental Degradation at Different Stages of Economic Development.* Working Paper WP238, Technology and Employment Programme, International Labour Office, Geneva, 1993.

Piketty, Thomas. *Capital in the Twenty-First Century.* Harvard University Press, 2013.

Putnam, Robert D. "Bowling Alone. America's Declining Social Capital". *Journal of Democracy*, vol. 6, no. 1, 1995, pp. 65–78.

Schneider, François, Giorgos Kallis, and Joan Martinez-Alier. "Crisis or Opportunity? Economic Degrowth for Social Equity and Ecological Sustainability". *Journal of Cleaner Production*, vol. 18, no. 6, 2010, pp. 511–518.

World Commission on Environment and Development. *Our Common Future.* Oxford University Press, 1987.

II.3 Institutionalist climate governance for pleasant cities and the good life

Gjalt Huppes and Ruben Huele

I Effective climate policy in cities: Leviathan or decentral creativity?

I.1 Decentralised power vs deconcentrated tasks of cities

Cities must play a key role in effective climate policy, a role that can be specified in two very different ways. They can have decentralised autonomy regarding many subjects, including climate emissions (decentralisation), or they have to implement centrally defined tasks, including centrally decided climate policies (deconcentration). Decentralisation relates to local autonomy while deconcentration refers to regional branches of central government. Both elements are present in city government, often in mixed forms, with some autonomy in the implementation of centrally defined tasks. Climate policy under the Paris Agreement (2015) sets a goal of a maximum of 2°C global temperature rise, or preferably only 1.5°C. This will require emission reductions of well over 90 per cent by 2050, especially in developed countries. Such reductions will have a massive impact on virtually all activities in society, including those at city level. If climate policy forces cities to implement centrally decided policies, they will move towards a dependent, deconcentrated role, focusing on technical issues. Current climate policy tends towards such central planning: central emission reduction targets are translated into sectoral tasks and subsequently into more technology-specific implementation. An example is the EU plans for refurbishing the housing stock to achieve low emission, as laid down in the Energy Efficiency Directive, with around 3 per cent of the stock to be refurbished each year (European Union).[1] National governments are to enact this task in national emission reduction plans, and then translate these into tasks for more local governments and organisations. The logic of effective administration also requires performance reporting, as a key ingredient of control. At the local level, choices are predefined and limited to details. In this situation, deconcentration of national tasks becomes dominant over decentralisation, and city autonomy, as a key factor for a good life in pleasant cities, is thereby substantially reduced, for apparently sound global climate policy reasons.

I.2 Proactive decentral action

There is an alternative to this type of planning for climate policy instrumentation. It does not set targets for the world, translated ultimately into specific tasks for cities, businesses and citizens. Instead, it creates incentives for emission reduction by redesigning central institutions according to policies to some extent already established in other environmental domains such as nature conservation and waste management and recycling (Mazzanti and Montini; Dong *et al.*). One main institutional element is the internalisation of external climate effects, the other is the correction of the currently highly deficient energy markets, especially regarding electric power. Cities then have an incentive to adjust their actions so as to take climate consequences into account, for one thing because all other decision-makers function under the same dynamic incentive. For example, high energy prices and reduced car ownership will entail spatial plans to abandon shopping malls in the countryside and create better options for sound local reasons, not requiring central policy. Whereas passive adjustment to new climate policies is one option, as in any other domain, proactive creativity is a better option, for cities and for the citizens and the businesses involved.

I.3 Core question and chapter outline

The central question in this chapter is how climate policy instrumentation can be redirected, in order to create decentralised autonomy for cities, while moving towards a near-zero emission society. Two schools of thought on governance are described, each with climate policy instrumentation strategies linked to them, and the consequences for cities. Before going into the climate policy details, Section II first discusses cities in the context of their wider government structure, involving other layers of government, other public organisations, and also private organisations, including NGOs and businesses, from local to global. Next, Section III describes the currently dominant approach in climate policy instrumentation, with its pros and cons. This system sets emission reduction targets for each year and specifies how these targets can be reached in a most effective and efficient way. This logic permeates almost all current political theory, aimed at effectiveness and optimisation. Optimisation goes beyond the purely economic, integrating moral and ethical issues in a welfare-theoretical framework. The *Planning and Control School* is described in terms of its general and historical background as well as of the ensuing climate policy instrumentation. Similarly, the *Institutionalist School* is illustrated with regard to its more general and historical background as well as to a strategic logic for climate policy instrumentation of a very different kind, as described in Section IV. It relates to the same institutional development which also formed the basis of the Industrial Revolution, with some fundamental adaptations now required, due to the latter's success. Shifting climate policy to the institutionalist approach means shifting responsibility for emission reduction to a decentralised level, involving cities. Being autonomously responsible requires a different attitude to local policy and to decentralised actions in general, the necessary reorientation being described in Section V. Passive adjustment will hardly lead to resilient and pleasant cities. Local autonomy requires local initiative with local strategies, and some indications of the sort of actions that may develop, not as prescriptions, are presented. Next, Section VI sketches how in institutionalist governance the top-down approach to cities may be replaced by bottom-up actions by cities. How could cities relate to the wider world, how can they connect with other cities and other public bodies, and with NGOs and businesses? Ultimately, cities could become active partners in global climate policy negotiations (Amen *et al.*). But should they? Finally, in Section VII the planning and control and institutionalist approaches are reviewed from broader perspectives in society, including central-isation, democracy, nationalism, populism, and legitimacy.

II Cities as part of public governance in climate policy

II.1 What are cities?

Cities are here taken to mean the local to regional government level. Cities may be seen more generally as socio-economic and cultural entities, even if they do not have one city government. London is an example: it did not have a city gov-ernment or a Mayor until a few decades ago. Paris is a different example, the Paris city government being restricted to Paris within the ring road (Boulevard Périphérique), with all suburbs functioning as independent cities with their own local governments. The city of Paris covers only 20 per cent of the urban Paris population. Expansion of urban areas in the socio-economic sense mostly does not coincide with the area expansion of city government. In the US, sub-urbanities can be split off into independent cities. In a socio-economic sense, globalisation has created global interrelations, even leading to ideas about global cities (Curtis). The broader term "urban" may cover such larger socio-economic entities. A more physical definition is linked to the urban metabolism of energy and materials, also defined independently of city boundaries (Weisz and Stein-berger). However, as public policy is our focus, the government level of cities as local governments is the reference, even if some further decentralisation is present. Metropolitan governments, if present, will have a city role as well. The city level is part of broader urban relation, from local cities, to metropolitan cities of cities, to international cities (McCarney and Stren). National govern-ments may effectively have decentralised their climate policy instrumentation to the level of provinces and states, as is largely the case in the US. City govern-ments may differ considerably between countries as regards organisation and tasks. In the Netherlands, mayors are not elected but appointed by central gov-ernment. Even though they have limited direct tasks, only ensuring public order, they play an important role by chairing local government, including safeguard-ing against misconduct of local government and its officials. Within this diver-sity of city arrangements there is some unity. Typically, the cities considered here, including metropolitan governments and decentralised city sub-areas, are in charge of detailed spatial planning, implementation of infrastructure, public transport, public housing, economic development, and much more, but are less directly linked to climate performance. Views on pleasant cities refer to local units with a substantial government structure, important tasks, and links to several layers of a more centralised government.

II.2 Emission reduction task of cities

Climate policy under the Paris agreement requires steep emission reductions, starting soon. The emission budget for keeping global warming to below 2°C was around 800 Gt in 2015 (Rogelj *et al*. 251). With current yearly emissions at around 40 Gt, this leaves 20 years of non-rising emissions, and then a jump towards zero emissions. Such a jump is impossible. With emission reductions spread out evenly, and starting now, near zero emissions can be realised by 2055, roughly corresponding to 90 per cent reduction by 2050. This reduction task is set in the Paris agreement. It is the same for any climate governance option. Emissions are to fall to near zero in most energy applications, with few exceptions. Cities will therefore also have to be nearly emission-free. This is the task ahead.

Moving to a low-carbon economy is not just a matter of developing, adding, and implementing low-emission energy technologies. This is an additivity view according to which adding renewables like wind and solar could solve all problems. Though such urban technology developments are essential, cities hardly play a role in their development and a limited role in their direct implementation. But merely adding renewables will not work. If cars are to drive on electricity, they need charging stations. If road traffic is to be reduced, new spatial planning and infrastructure must be implemented. Low-emission public transport, walking and cycling require investments in infrastructure, and rules regarding the use of public space. A main route toward refurbishing high-emission buildings appears to be planning by cities. The intermittent nature of solar and wind creates fluctuations in electricity supply not linked to the intermittency of demand, with local storage options being one part of the solution. If such adjoining local developments do not take place, renewables introduction may grind to a halt for technical and economic reasons as well as due to opposition by those on which the renewable systems are imposed. Hence, the autonomous city level is unavoidable in moving towards a low-emission society.

II.3 Cities as part of current climate policy architecture

Currently developing climate policy defines how technologies are to perform and how behaviour is to be adjusted. Several policy instruments are in place and further instruments must emerge to meet the challenge. There is substantial literature on how to better integrate climate policy at the different levels of government involved, ranging from cities, provinces, and states to the national and supranational levels. This integration is seen by the OECD as a technical issue (Corfee-Morlot *et al*.) and described by them in much detail (OECD/IEA/NEA/

ITF 185ff.). This climate policy role requires city government to be substantially strengthened (McCarney 85), including more active involvement of its citizens. City performance must be measured on the basis of international standards being developed, the first being ISO37120, which details 100 indicators for measuring city climate performance (Hurth and McCarney). Focusing on effectiveness in such detailed terms is a logical step, especially as climate policy is not yet delivering the extreme reductions required in the Paris Agreement (2015).

II.4 Governance choice based on effectiveness and efficiency alone?

There is a dominant view in policy analysis that effectiveness and efficiency (as cost-effectiveness) are core goals in designing and redesigning policy instrumentation (Sunstein). Better regulation also means more integrated regulation, making instrumentation *simpler*, to borrow a term from Sunstein's book title. The goal in the climate domain would be less overlap, fewer instruments, covering all relevant emissions, with clear tasks for all government levels involved. This implies sparseness in terms of instrument numbers; non-overlap of instruments to avoid complexity; completeness, as covering all aspects relevant to emission reduction; and equal distribution of burdens of emission reduction, for reasons of efficiency. These technocratic goals are laudable goals, but they are not sufficient. They presuppose that a major architecture of instrumentation is already in place, as a result of governance choice. Moreover, the role of cities is not just a technical issue of better policy alignment and better implementation. Basic governance choice first determines the role of cities. This role may concern the implementation of policy elements resulting from higher-level policy decisions, ultimately based on internationally set targets for emission reduction. In this form of governance, cities are the recipients of policy, implementing what is needed to achieve the climate policy target, in ever better and more relevant detail. There is a different option for national and international climate policy instrumentation, however, according to which cities, far from being merely recipients, are generically incentivised and autonomously enabled. Prime central instruments then do not relate to planning, but to incentivising as well as creating options, such as a generically applied high and rising carbon tax; a well-developed real-time electricity market with equal prices for all primary and secondary producers and users; public supply of naturally monopolistic infrastructure; and support for longer-term high-risk technology development. Such an institutional policy allows cities to develop and use their creativity for a common purpose: climate policy. While developing cultural independence and vitality, cities also improve their wellbeing.

II.5 Historical developments in governance

These two governance views have a long history, starting in the city states of ancient Greece. The dividing lines between Plato's and Aristotle's views include a focus on central power for the common good versus procedural controls on central power, taking decentralised views and interests into account. Plato accepted the poison cup for Socrates because he accepted absolute power (Popper, vol. 1), while Aristotle fled when a similar prospect arose for him in 322 BC. Political theory in the West has been polarised between centralised effectiveness and institutionalised restrictions on central power, as exemplified by the Magna Carta (1215), the Bill of Rights in England (1689), the Bill of Rights in the USA (1789), and depicted in Hobbes' *Leviathan* (1651). The American and French revolutions have brought systems of democracy with safeguards on abuse of power but not limiting the domain of power. Absolute democracy, especially forms of direct democracy, tends toward absolutism. "We the people" can be specified in different ways, but always stating absolutely what is right (Morris). If cities are to survive independently, this tendency must be reversed. At the same time, the problems of industrialised and globalised society, like climate change, need to be resolved somehow, at the institutional level. Cities should not submit to a centralised dirigiste power ultimately vested in global governance, setting legally binding emission caps for countries as nation states, and thereby dictating the actions of cities, businesses and populations.

II.6 Co-ordination by hierarchy or by incentivised autonomy and networks

Co-ordination sounds like a neutral term, hiding the fact that power is always involved, in the sense that others do what someone wants. Co-ordination in society is always based on power, directly by controlling others or indirectly through rules changing behaviour of others in subtler ways. In this sense, power in climate policy is the power to achieve climate results, focusing on command and control or on adapted institutions, and also involving co-operative relations, including economic relations. For the social system as a whole, it is not just governments at different levels but also businesses that play their part, not freely but within the publicly developed and maintained market structure, and partly developing spontaneously as well (Boulding; Thompson *et al.*). In the hierarchical approach, what cities should do is decided centrally, ideally leading to an optimised overall structure of policy instrumentation and implementation. Such central co-ordination is one option, viewing the government of countries like the management of large companies, which also have decentralised departments and local branches. Businesses also have to incentivise at least some decentralised creativity, based on some level of decentralised autonomy. The central management, however, will always control what is happening at the decentral level, with specific carrots and sticks, and within the wider order of the market structure. The limitations of co-ordination by hierarchy relate to the limits of information and information processing, especially in a dynamic context. The equivalent autonomy model of business would be one of locally independent companies. But again, they are autonomous only within the given market structure. They have to build up relations with clients and competitors, with regulatory bodies and with their investors, and with research and development organisations and others for their dynamics. They are autonomous within institutional constraints.

They have to develop complex networks, in which hierarchical power over others is not fully absent but limited.

Similarly, autonomous cities are never fully autonomous and to the extent that they are autonomous, they can hardly act alone. Novel technologies arrive through contacts with potential providers; ideas on infrastructure develop in discussions with others, including other cities; what is acceptable for citizens depends on views developing through wider discussions in civil society, in an economic context that is more centrally structured, as market institutions are mostly not local. In this sense, autonomous cities have a much broader task than those directed hierarchically, involving dynamics of complex networks. Maintaining effective co-ordination requires some hierarchy within cities, with internal and external tasks of city departments. Even autonomous cities can kill local initiatives by following rigid local planning. In terms of local climate policy instrumentation, there is again a choice between planning and control and "softer" institutional arrangements. In institutionalist governance, both central and local policy leave the actual decisions to decentral parties, not as inactive laissez faire but by creating conditions and processes under which climate-relevant developments can, and are likely to, emerge.

III Planning and control climate governance and cities

III.1 Planning and control governance

Planning and Control is the most common political science model of public government. It also has a long philosophical tradition, going from Plato to Bacon to Hobbes, and in modern times Rousseau. Before democracy, sovereignty resided with a central body or in a person, the monarch, increasingly with absolute power. After Rousseau and the American and French Revolutions, the monarch has been replaced by *the people*. This justifies centralised decision-making with power to implement whatever is decided democratically. Plans for specific issues are developed, decided upon democratically, implemented, and checked for sufficient progress of implementation, and effects resulting. The same model is used in larger companies, which have a central decision-making unit responsible towards the shareholders, with a function like *the people*, with the power to overrule all lower-level departments. Autonomy is highly valued, as long as it delivers what is supposed to be delivered, in the judgment of the central authority. Similarly, the place of cities in public governance is to contribute to what the nation state wants to realise, in this case deep emission reductions. In terms of Hobbes' famous Leviathan, cities are body parts, guided by the head.

III.2 Planning and control instrumentation for effective climate policy

The Planning and Control strategy offers the simplest links to improve on what is there already, setting quantified targets and using performance measurement.

In a governance sense, it relates to an incrementalist approach, focusing on political and administrative feasibility, and on avoiding big mistakes by means of stepwise policy development (Dahl and Lindblom; Lindblom). The European Commission has its Strategic Planning and Programming Cycle, akin to the US PPBS (Planning Programming Budgeting System) starting in the 1960s, with yearly checks on achieving objectives. Longer-term issues also tend to be stated in terms of medium-term targets, with instruments adapted in the process. Explicit higher-order strategic planning of long-term instrumentation is absent, as this does not fit within the quantified effectiveness and efficiency framework. In the energy and climate policy domain, the time horizon issue is described as a core element (Grubb *et al.*). The short run is dominated by what can be done practically, now. The medium term is based on optimisation, taking care that all objectives are included in a balanced way. The optimisation process starts now and covers the predictable future, somewhat over a decade. The long term, however, requires deep transformations, which are not set in motion by adjusting and optimising. Deep transformations require sweeping new policies, actively transforming society towards near-zero emissions. Grubb, Hourcade and Neuhoff acknowledge the long-term problem but do not look into the new policy instrumentation required. Their logic of action is to do the right things, making climate policy part of wider nation-wide welfare optimisation, taking dynamics into account independently.

III.3 Planning for wider welfare optimisation

The focus on optimisation leads to climate policy being placed in a wider context, linked to wider policy domains like economic growth, energy supply, housing, and income distribution, and other environmental issues. Optimisation cannot just be a climate affair. Applied social welfare functions have been at the basis of this approach, which developed theoretically and empirically after the Second World War. A few big names are Samuelson, Arrow, Sen and Stiglitz, Nobel Prize winners, with Arrow, Sen, and Stiglitz active in the climate policy domain, using a top-down optimisation view. The core reasoning, generalising Jeremy Bentham's principle of the greatest happiness for the greatest numbers, is: *what is best collectively must be realised.* The derived actions for cities are relatively policy-free technical adjustments aimed at implementing their tasks. The social welfare function started mainly in economic terms, restricting its domain to the analysis of economic activities such as the public supply of roads, airports, and education. Cost-benefit analysis originated as a most practical tool, also telling cities what to do and what not to do. After around 1970, the welfare approach was broadened to cover ethical and normative issues as well. The role of cities in this integrated approach was then reduced further. Social welfare becomes more difficult to define operationally at city level; actions come from "above", and tend toward fragmented local adjustments.

This integrated governance approach now also dominates the climate policy domain. Eighteen post-carbon strategies have now been surveyed (Wiseman *et al.*).

Most of them use an incremental and optimising approach, all in a Planning and Control mode of governance, even if they have transformational change as their ultimate goal. In a most practical sense, this governance approach may, for example, guide the mix of renewables to be introduced, not just for climate and cost reasons but also to cover energy, water use, emissions, land use, employment and income distribution (Shmelev and van den Bergh). What cities must do follows from this analysis, mediated through national and regional plans. Quantification of climate damage per unit of emission is required to place climate policy in this overall welfare perspective. This quantification has proven elusive, owing to time delays in effects and low-chance catastrophic effects (Weitzman). Though historically linked to somewhat restricted economic reasoning, the definitions of collective welfare have broadened since the 1970s, as in the UN Human Development Indicator, first taking into account life expectancy and schooling. More recently (Jones and Klenow) included not only consumption level but also leisure, mortality, and inequality in an adjusted quantified GDP measure. Global public goods like climate stability are part of the equations, with climate policy to be mainstreamed in all policy domains. The report for the French government (Stiglitz *et al.*) comes close to being an action-oriented handbook, covering integrated climate policy:

> these dimensions [of wellbeing] should be considered simultaneously:
>
> i   material living standards (income, consumption, and wealth);
> ii   health;
> iii education;
> iv   personal activities including work;
> v   political voice and governance;
> vi   social connections and relationships;
> vii environment (present and future conditions);
> viii insecurity, of an economic as well as a physical nature.

$$(7)$$

Any option increasing overall welfare in society is useful, priority being given to the most useful ones. But what can cities do with this mostly abstract and macro-level analysis, which fails to show the causal relations between these overlapping elements? And what can businesses and consumers do with this analysis, functioning at city level?

III.4 City role under planning and control climate policy

For decentral decision-makers, public and private, there is more to life than greenhouse gas emissions, invisible to the eye and with negligible individual contributions to the global problem of climate change. Forcing upon society the technologies required for emission reductions is an option which entails setting goals and targets in ever more binding fashion, translating them to lower administrative levels, and ultimately to the local level of cities, where many behavioural choices reside: how to shop, what to eat, how to heat, and where to go on holidays. Billions of individuals make decisions, as private consumers, in businesses and governmental organisations, and as citizens. Controlling these aspects in order to achieve deep emission reductions is a challenging task. It requires action for change in virtually all domains of private life and in all economic sectors, against a background of other dynamics, with conflicting goals. Leviathan would be among us. But such strong actions, ultimately at the urban level of metropolises and cities, seem hardly possible effectively. This direct version of Leviathan (Hobbes, Chapters 17–18) has lost out in the highly complex industrialised and globalised world we inhabit. It still is, however, the main line for discussing the planning and control mode of governance, legitimised by reasoning from the people's sovereignty, of which many versions exist (Morris). Optimising the emission-reducing actions centrally, from above, is arduous as well as detrimental to creative climate action. Its technocratic nature is ultimately detrimental to democracy and legitimacy, and to the good life, which is mostly embedded in local life in cities. It may also be ineffective in the long term, moving to an unacceptable tangle of behavioural and technological details in policy.

Seemingly decentralised autonomy may also become part of centralised policy, shifting to deconcentration. The Covenant of Mayors for Climate and Energy in the EU (www.covenantofmayors.eu) is an example of a horizontal organisation channelling information exchange between cities. Such an organisation may easily be transformed into a vertical one, in this case with the power from the European Commission reaching to street level. In the EU Strategy on Heating and Cooling, for example, the Commission sees the Covenant of Mayors as a tool to support local authorities.[2] This tool involves conditional funding instruments for specific technologies, partly restricted to signatories to the Covenant.[3] The vertical carrots and sticks are clearly visible. This may be most useful from a partial effectiveness and efficiency point of view, making cities do what is deemed necessary and best. But the Covenant is then not a horizontal structure any more. It becomes part of the vertical Command and Control strategy of the EU, the EU in turn being forced by internationally agreed reduction targets it helped shape. Also, it assumes that mayors represent the central democratic power in cities. This may not be the case, as in the Netherlands for example, where mayors are appointed by central government. The EU Covenant of Mayors interferes with such national constitutional arrangements.

It might be better for longer-term climate policy instrumentation to deviate from the planning path, assigning substantial autonomy to others than the central authority of the nation state. There may not be a single direction for such non-hierarchical autonomy, which will involve much more than the autonomy of cities. But it will certainly require real autonomy of cities. Institutional arrangements must both safeguard the goals of climate stability and the autonomy of other actors, here focusing on cities. This challenge can be defined in terms of central institutions and only public supply of natural monopolies, as in large-scale infrastructure. They open up options for creative local action, public and private.

IV Institutionalist climate governance and cities

IV.1 *Institutionalism for decentral innovation*

The institutionalist strategy is linked to the more empirically oriented institutionalist school, studying long-term societal development (Acemoglu *et al.*; Mokyr; North). Critical approaches focus on institutional economics (Coase), institutional processes in the social sciences (Immergut and Anderson; Munck af Rosenschöld *et al.*), political economy (Chang), and the use of soft instruments (Carrigan and Coglianese) in political science. Ha-Joon Chang argued against the simplistic non-institutional neo-liberal view of "the" market, which erroneously abstracts from the institutions capable of creating and regulating markets in very different ways (Chang). There is a direct link to creative bottom-up processes, which are applicable mainly at a more local level (Ostrom). Current economic institutions lack internalisation of external effects on climate, creating an uphill struggle for decentral climate action. If the climate fervour passes, behaviour may well fall back on what is dictated by poorly developed markets, regarding both cities and private actors to be served by cities. Internalisation of external climate effects, and further institutional adaptations, create a more level playing field, with cost reduction then also favouring emission reduction.

IV.2 *Institutionalism and monopoly*

Institutions constitute the basic, relatively stable, fabric of society, including constitutional government arrangements, the legislative system, the judicial system, the educational system, public-private delineations, and more or less fixed normative and ethical principles. Many of these are vested in specific organisations like public bodies, research organisations and universities, as well as private organisations. Here, institutions are taken as the more abstract rule systems. One core element of Institutionalism is that no organisation in society should have absolute authority. The source of sovereignty is not acquired Power, God, or the People; such framing of options is wrong, leading to absolutism. Authority must not be monopolised, including decentralised monopoly, as may develop in cities. An analogy with the patent system may clarify the issue of autonomy without absolute power. Patents used to be the prerogatives of royal sovereignty, given to persons or organisations, as derived monopolies. The patent system on inventions is an example of a more specific legal rule system, with a rapid domain expansion in recent decades. The patent system belongs to governance; its development is part of strategic governance development. The argument for a publicly created technology monopoly, albeit a temporary one, is utilitarian: it helps develop technologies, improving welfare in society dynamically by financially supporting innovation. Forcing climate policy measures on cities similarly reflects the central monopoly, using its power at the city level for utilitarian reasons.

Critics of the patent monopoly might use the same welfare reasoning as used in supporting the patent system, but in their case to abandon patents, based on more empirical analysis. Like central technology enforcement for climate reasons, the patent monopoly may well run counter to innovative dynamics, like those required to reduce climate emissions. Three lines of empirical research support this position. Historically, the patent system did not contribute to innovation in the last century and a half, as innovation was faster where patents did not apply (Moser). Patents reduce the publicly available pool of knowledge, which may then be the overarching effect (Stiglitz) based on modelling with reasonable empirical assumptions. Many cases of detrimental effects of patent monopolies have been surveyed (Boldrin and Levine), including James Watt holding up the development of the steam engine for over a decade. The planning type of climate policy instrumentation may place the power of implementation at decentral level, deconcentrated, effectively creating a monopoly at local level for the technologies prescribed and reducing dynamics. Command and control instruments for environmental policy tend to be slower and costlier than those resulting from more generic financial incentives (Romstad).

The institutionalist type of instrumentation refrains from specifying technological and behavioural requirements, including decentral ones. Stable long-term change comes from creating the right institutions, as economic history has extensively shown (Acemoglu *et al.*; Mokyr; North, "Economic Performance"; North, *Institutions, Institutional Change*), avoiding local monopolies. The generic non-planning structure allows for bottom-up developments both in the private and public domains, the power of which has been shown by Ostrom, with many of her examples focusing on the local domain. At the decentralised level, technologies are developed privately but also publicly, as by cities, if there is good reason to do so. The central issue is that neither party has an effective monopoly. A very substantial subsidy for one party, as often seen in public-private partnerships, also in effect creates a monopoly. Where monopolies are unavoidable, as with natural monopolies in infrastructure, they must be public and comply with suitable rules in order to avoid misuse of monopolistic power.

Regarding climate changing emissions, decentral public and private parties can be followers or they may be pro-active leaders with longer-term strategies. They act from their more individual perspectives, integrating cost considerations, climate considerations, justice considerations and individual knowledge at that level, to their discretion. How should cities act, what should they do, and what not, when acquiring decentralised autonomy and hence responsibility?

V Taking decentralised responsibility

V.1 From hierarchy to autonomy and networks

The development of instrumentation is a strategic governance issue, certainly with regard to deep emission reduction. The two styles of climate policy governance lead to two very different sets of instruments, with very different styles of climate governance at the local urban level. The urban level of cities and the reduction of actual emissions are paramount in both strategies. Moving from central planning and control to decentralised autonomy is, however, not just an issue of creating some new institutionalist central instruments. Without direct executive power, emission reductions are to be brought about not by plan but by incentivising and enabling. There is a clear public task in designing spatial plans fit for low emissions activities. There is also a clear task regarding publicly owned monopolistic infrastructure, like grids for charging electric cars, as well as a very substantial task in sifting through existing regulations to be abolished such as those meant to improve energy efficiency and reduce emissions, which however tend to obstruct deeper dynamics. Forcing refurbishment to obtain high insulation by means of standards, procedures, and subsidies may well reduce emissions but also makes the shift to electric heat pumps with heat storage unattractive, creating a lock-in. The long term is then jeopardised to achieve shorter term improvements. Such regulations, now being forced on cities, will mainly disappear in Institutionalism, not to be replaced by well-intentioned similar city-level regulations. With all pricing adapted to substantial emission taxes, all parties will become more interested in novel technologies, a process that is to be supported by a networking role of cities.

The stimulating networking role of cities is a type of policy to be developed, closely linked to innovation. Public space is merely a starting point. Participation must be attractive for discussion to become self-sustained. Just linking internal and external persons and organisations is not enough. The number of possible linking relations with other cities and with public and private organisations is very large. The task of co-ordinating the many millions of actors involved seems quite unsurmountable. Networks in a city must have a focus, connecting only the most relevant participants. With natural gas heating becoming more expensive owing to the central emissions tax, other heating systems must emerge, like electric heat pumps with heat and cold storage. What is attractive, when and where? In the new context of relatively low-priced electricity compared to natural gas, such questions require using the network in order to stimulate serious discussion at city level. How to accommodate electric cars with self-driving capability? This does not only involve providing charging stations in the new variably priced electricity market. An attractive new city infrastructure must include walkability and cyclability where locally relevant, and new and different types of parking facilities. Taking responsibility for the design processes involved, based on autonomy, makes for an interesting and good life, as well as being commercially attractive for businesses.

Such decentralised debate shares some characteristics with certain study groups in the field of agriculture in the Netherlands and UK, where businesses co-operate for innovation, presenting ideas and options to each other, including pros and cons, followed by private commercialisation. There is renewed interest in this subject, as innovation for transformative change involving disparate partners requires co-operation before competition, coopetition being the new subject name (Barney *et al.*; Ritala *et al.*). Social proximity is one key feature in building up trust (Jakobsen and Steinmo), in the domain of city support. In the more top-down policy approach, the link to the local level is experienced as difficult to establish, see for example the survey of literature regarding the implementation of solar cell technologies (Margolis and Zuboy). The barriers to implementation include lack of information dissemination; difficulty of overcoming established energy systems; failure to account for all costs and benefits of choices; and lack of community participation. These barriers have a substantial local reference, with local differences, and can hardly be resolved top down. Of course, not all issues can be resolved at city level; they require participation in higher-level networks, such as those for establishing effective codes, standards, and interconnections, and linking to larger firms. The study group model, involving citizens, businesses, schools, and research bodies, seems an adequate option for organising autonomous responsibility. It is substantially culture-dependent however, requiring mutual trust. Their development does not come automatically.

V.2 Why take active responsibility?

Why should cities become proactive in emission reduction? There are various answers, like prime mover advantages for the city and its citizens, reducing costs in the newly created price environment of the emissions tax, and using the advantages of the open electricity market. Helping to actively create the future is also inherently attractive in the psychological sense of being master of one's own destiny. In negative terms, without the open guidance of civil society discussions, the context for public and private decisions is not clear. Adjoining public policies will then not develop adequately, with private actions delayed by uncertainty.

In preparing for the future, one wants to be on the right side of history. Some examples are clear. Building a coal harbour currently seems less than wise. Supporting deep sea oil drilling and out-of-town shopping malls probably means investing in a lost cause. But how to deal with natural gas at city level? Should new city expansions not have natural gas, nor butane and propane? How to disconnect existing natural gas connections? What may cities decide?

A case in Amsterdam may illustrate some governance issues involved. In a newly developed area, Zeeburg, no natural gas grid is supplied. The city waste incineration plant AEB[4] and the Swedish energy supplier Vattenfall are building a subsidised district heating grid, partly based on natural gas. The local authorities made linking up to this district heating system obligatory at Zeeburg. This is the old system of top-down policy. A group of 43 people building their own apartments and some commercial premises, called Nautilus,[5] decided to go for a heat pump with an underground seasonal heat and cold storage system. They went to court to fight the city's monopoly, and won. Things could have been different in the institutionalist governance version, without public and private monopolists in the market. The capable persons and organisations involved would have looked into combinatory options, with district heating offering some heat at low-demand periods for storage and the heat pump system supporting district heating at times of high demand in winter. Of course, this requires new designs, new contracts, new technologies, and some additional infrastructure. But it could have paid off for all parties, already resulting in reduced emissions. It would also have been a learning experience open to improvements.

V.3 Climate effectiveness of cities

The question remains if under institutionalist central government, city policies and urban developments will deliver what is needed in terms of deep emission reduction. As the Amsterdam example indicates, the institutionalist approach may be faster and more effective than planning and control. Whether emission reduction is fast enough for the 2- or 1.5-°C goal is yet unclear. Of course, higher-level institutional development, especially the level of the emissions tax, will play a role. But reducing opposition by means of an open voluntary process seems a good approach to faster reduction. Cities can bring together private parties with an economic interest and with a drive, on topics that fall under the direct responsibility of the city, like spatial planning and public infrastructure.

Is it relevant to ask whether cities reduce emissions sufficiently, and can there be an answer? In Institutionalism, there are no emission targets for cities, nor for countries. Also, technical targets, like the percentage of dwellings refurbished, do not exist. The rules of the social order have been changed to induce emissions reduction generically. In institutionalist governance the rules are changed, leading to overall reduction, not at completely predictable speeds and locations, but overall not necessarily slower than in Planning and Control, and probably faster in the long term. The main role of cities is in accommodating the energy transformation in the energy use chains. The fossil emissions currently left will be concentrated at the most efficient production locations, such as those close to deep water for cooling. There, emissions will then remain higher than in inland cities. Are coastal cities then not on track? That is a wrong question. Also, inland cities stating they are on track give irrelevant information. The open nature of city actions under general incentives and options creation may well help solve deep emission reduction transformation more effectively than Planning and Control, but not in a localised quantifiable way.

VI Autonomous cities in a wider world

Cities are not independent city states engaged in international relations, like the Greek city states once were. Cities are part of more central administrative entities, including nation states, with regional and provincial intermediate layers. Nations in turn engage in international relations, substantially binding as in the EU or more open as in global international agreements. Can autonomous cities follow suit in international climate policy, co-ordinating their actions? This seems well beyond the capacities of city governments. Connections are possible with only a few others, because of the costs of interacting with many partners from different cultures, climates, and regulatory structures. Learning from others would be the main goal, as more binding co-operation like those in international treaties is not in the domain of city governments. There is no hard logic for setting up structures for international co-operation by cities aiming to achieve a low emission society. Already, there exist many more general frameworks, like conferences, workshops, visits by relevant persons, and contacts between companies. What else could be useful?

VI.1 Horizontal connections and competition

Horizontalism and competition are intertwined: exchange of information strengthens all those participating in development and debate, but there is also a point where competition starts. Autonomous cities can have an initiating role in such networked co-operation arrangements between cities, as public coopetition. In the competition stage, being a prime mover is essential. Ultimately, prime mover motives must be strongly supported to achieve success. Remaining outside of information exchange precludes any chance of wider success. The main city task is however in enticing private developments in the city. Hosting meetings of scientists, politicians, business organisations, NGOs (non-governmental organisations) and cultural organisations, and boosting high-level tourism with support and interesting cultural events are all part of the prime mover game. Focusing on strengths is a key part of the city strategy. Not all regions can be global players in IT, or life sciences, or renewable energy systems. One general requirement is to be at the frontier: be broad, diverse, and attractive, more so than others. This opens up options and benefits. But what should specifically be done in relation to climate? Initiating and supporting coopetition alliances in key low-emission domains is one prime task (Jakobsen and Steinmo). Especially if city actions are a necessary part of innovative developments, a city role in horizontal coopetition alliances is to reduce uncertainty for private participants in the innovation processes. Such alliances may have a core in a city but must also expand beyond the city level.

VI.2 Institutionalist international governance

The World Trade Organization (WTO) is based on a strategy to build generic institutions, enabling international trade developments but not guiding them in content. By contrast, the UNFCCC climate agreements are based on a planning strategy leading to permits, standards, and subsidies in national climate policy. The generic WTO-type strategy might be developed for climate policy as well, replacing the top-down approach. An international climate agreement should first seek the same set-up and level of the national emission tax, dealing with countries without a similar climate policy, and at more regional levels linking high-tension electricity grids, organisationally and with super-grid interconnections. The decentral role of cities and private organisations then can be very different, shifting emphasis from implementation to taking initiatives (Figure II.3.1).

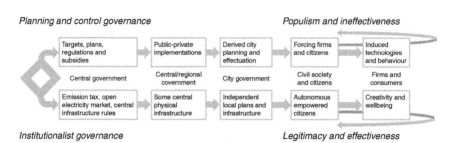

Figure II.3.1 Different roles of decentral actors under planning and control and institutionalism.

VII Decentralisation, democracy and legitimacy, and effectiveness of climate policy

Current climate policy instrumentation is at the crossroads. The current strategic entry to climate policy involves top-down processes through Planning and Control of emission reduction, based on centralised democratic sovereignty. Detailed regulations and targets make local initiatives more complex in terms of technical restrictions and information requirements, obliterating local democracy. Forcing change in the Planning and Control way has not yet led to a reduction of global CO_2 emissions, with atmospheric concentrations rising at the highest speed ever. Reconsidering this organisation of climate policy seems wise from this global point of view, with Institutionalism as the main alternative. Institutionalism also covers effectiveness and efficiency in climate policy, usually linked to Planning and Control, but without real justification (Grubb *et al.*): optimisation works in the medium term at best, whereas deep reductions require deep transformation, not guided by quantified optimisation. By creating dynamics, Institutionalism may inherently be more effective and efficient in the long term, as historical analysis of the role of institutions in economic growth has shown. The case for cities being empowered in a decentralised fashion, as promoted by Institutionalism, can be argued from effectiveness and efficiency considerations alone already. The role of cities would then become very different, moving from guided climate policy implementation to decentralised autonomy with responsibility. Autonomy, responsibility, and wellbeing are closely connected, even in the good life described by Aristotle, in turn closely connected to political autonomy: "[man] has the perception of good and bad, and right and wrong, and the other moral qualities, and it is partnership with these things that makes a household and a city-state" (*Politics*, Book 1, section 11). Autonomy pertains to people within the city state as well, against Plato.

Current psychological considerations, in a similar vein, relate autonomy to intrinsic motivation, self-regulation, and wellbeing (Ryan and Deci). The thwarting of autonomy and competence diminishes motivation and wellbeing. Discontent was not yet politically dominant in 2000, but now seems on the rise.

Empirical analysis of educational success also shows that students improve less when subjected to direct control than when given active autonomy (Fei-Yin Ng *et al.*). Discontent and passiveness among substantial parts of the population are a reality, closely linked to a lack of legitimacy of governments and rulers, and endangering governmental institutions. Currently proposed solutions, involving more voting as in referenda, do not ensure autonomic competence, certainly not if voting with millions of others; "we the people" is a dangerous concept (Morris). The importance of democracy as majority rule may reside in a very different domain. Avoiding usurpation of government power by institutional constraints and regular change of rulers is a precondition for the lasting autonomy of cities and the good life. It is also a condition for Institutionalism, but not a sufficient one.

Climate policy instrumentation, happiness and good life are interrelated. Institutionalist instrumentation is related to decentralised autonomy within generic conditions of incentives and options creation. Decentralised autonomy is related to personal motivation and development, to competence, and to taking one's life in one's own hands; happiness and wellbeing may then follow. Happiness is also a matter of external factors, of luck, Aristotle stresses, which is difficult to organise directly. Effective climate policy will certainly reduce the risk of climate catastrophes, avoiding bad luck. Autonomy in politically autonomous cities is part of Institutionalism for effective climate policy, reducing the risk of other disasters as well, those of a political origin, with social and economic disasters following. The Planning and Control approach may therefore be a dangerous one, with populism lurking. More direct voting for ill-considered democracy reasons may deal a blow to any effective climate policy, and to other policies as well.

Local legitimacy creates a link back up to the legitimacy of central government, to support national climate policy. Trust, support, and legitimacy are here taken broadly and extend to informal processes beyond legal-administrative ones, including concepts like cultural legitimacy (Kailitz 43–44). These processes are substantially rooted in the public space at the city level, in autonomous cities. Legitimacy in this general sense, used also in public parlance, is related to views such as those by Greene. With civil society involved in the preparation of policy instrumentation and implementation, trust can be created in layered processes (Luhmann), starting at the city level. Legitimacy, effective climate policy and wellbeing in cities are directly connected to institutionalist climate policy instrumentation.

Notes

1 See the European Commission, https://ec.europa.eu/energy/en/topics/energy-efficiency/energy-efficiency-directive.
2 See the Covenant of Mayors for Climate and Energy, www.covenantofmayors.eu/news_en.html?id_news=713.
3 See the Covenant of Mayors for Climate and Energy, www.covenantofmayors.eu/Funding-Instruments,87.html.

4 See AEB Amsterdam, www.aebamsterdam.com/.
5 See Woonwerkgebouw Nautilus, http://nautilus-amsterdam.nl/.

Bibliography

Acemoglu, Daron, Siomon Johnson, and James A. Robinson. "Institutions as a Funda-
   mental Cause of Long-Run Growth". *Handbook of Economic Growth*, vol. 1a, edited
   by Philippe Aghion and Steven M. Durlauf, Elsevier, 2005, pp. 386–472.
Amen, Mark, Noah J. Toly, Patricia L. McCarney, and Klaus Segbers. *Cities and Global
   Governance: New Sites for International Relations*, Ashgate, 2013.
Aristotle. *Politics*. Translated by Harris Rackham, Harvard University Press, 1934.
Barney, Jay B., Giovanni Battista Dagnino, Valentina Della Corte, and Erik W. K. Tsang.
   "Special Issue on Coopetition and Innovation in Transforming Economies". *Manage-
   ment and Organization Review*, vol. 12, no. 2, 2016, pp. 417–420, doi:10.1017/
   mor.2016.15.
Bentham, Jeremy. *A Fragment on Government: Being an Examination of What Is
   Delivered, on the Subject of Government in General.* Payne, Elmsly and Brooke, 1776.
Boldrin, Michele, and David K. Levine. *Against Intellectual Monopoly*. Cambridge Uni-
   versity Press, 2008.
Boulding, Kenneth E. *Three Faces of Power*. Sage, 1990.
Carrigan, Christopher, and Cary Coglianese. "The Politics of Regulation: From New
   Institutionalism to New Governance". *Annual Review of Political Science*, vol. 14,
   no. 1, 2011, pp. 107–129, www.annualreviews.org/doi/abs/10.1146/annurev.
   polisci.032408.171344. Accessed 13 March 2017.
Chang, Ha-Joon. "Breaking the Mould: An Institutionalist Political Economy Alternative
   to the Neo-Liberal Theory of the Market and the State". *Cambridge Journal of Economics*, vol. 26, no. 5, 2002, pp. 539–559.
Coase, Ronald. "The Task of the Society". Opening Address to the Annual Conference of
   the International Society for New Institutional Economics, 17 September 1999. *News-
   letter of the International Society for New Institutional Economics*, vol. 2, no. 2, 1999,
   pp. 1–6.
Corfee-Morlot, Jan, Michael G. Donovan, Ian Cochran, Alexis Robert, and Pierre J. Teas-
   dale. "Cities, Climate Change and Multilevel Governance". OECD Environment
   Working Paper no. 14, 2 December 2009. OECD Publishing, 2014, doi:http://dx.doi.
   org/10.1787/220062444715.
Curtis, Simon. "Global Cities and the Transformation of the International System".
   Review of International Studies, vol. 37, no. 4, 2011, pp. 1923–1947.
Dahl, Robert Alan, and Charles Edward Lindblom. *Politics, Economics and Welfare:
   Planning and Politico-Economic Systems, Resolved into Basic Processes.* Harper &
   Brothers, 1953.
Dong, Liang, Yutao Wang, Antonio Scipioni, Hung-Suck Park, and Jingzheng Ren.
   "Recent Progress on Innovative Urban Infrastructures System towards Sustainable
   Resource Management". *Resources, Conservation and Recycling*, forthcoming, http://
   dx.doi.org/10.1016/j.resconrec.2017.02.020.
European Union. "Energy Efficiency Directive 2012/27/Eu of the European Parliament
   and of the Council". Official Journal of the European Union, 2012.
Fei-Yin Ng, Florrie, Gwen A. Kenney-Benson, and Eva M. Pomerantz. "Children's Achieve-
   ment Moderates the Effects of Mothers' Use of Control and Autonomy Support". *Child
   Development*, vol. 75, no. 3, 2004, pp. 764–780, doi:10.1111/j.1467-8624.2004.00705.x.

Greene, Amanda R. "Consent and Political Legitimacy". *Oxford Studies in Political Philosophy*, edited by David Sobel, Peter Vallentyne, and Steven Wall, 2 vols., Oxford University Press, 2015, pp. 71–97.

Grubb, Michael, Jean-Charles Hourcade, and Karsten Neuhoff. *Planetary Economics: Energy, Climate Change and the Three Domains of Sustainable Development.* Taylor and Francis, 2014.

Hobbes, Thomas. *Leviathan.* 1651. Edited by C. B. Macpherson, Penguin Books, 1968.

Hurth, Victoria, and Patricia McCarney. "International Standards for Climate-Friendly Cities". *Nature Climate Change*, vol. 5, no. 12, 2015, pp. 1025–1026.

Immergut, Ellen M., and Karen M. Anderson. "Historical Institutionalism and West European Politics". *West European Politics*, vol. 31, nos. 1–2, 2008, pp. 345–369.

Jakobsen, Siri, and Marianne Steinmo. "The Role of Proximity Dimensions in the Development of Innovations in Coopetition: A Longitudinal Case Study". *International Journal of Technology Management*, vol. 71, no. 1–2, 2016, pp. 100–122, doi:10.1504/ijtm.2016.077976.

Jones, Charles I., and Peter J. Klenow. "Beyond GDP? Welfare across Countries and Time". *American Economic Review*, vol. 106, no. 9, 2016, pp. 2426–2457.

Kailitz, Steffen. "Classifying Political Regimes Revisited: Legitimation and Durability". *Democratization*, vol. 20, no. 1, 2013, pp. 39–60.

Lindblom, Charles E. "The Science of 'Muddling Through'". *Public Administration Review*, vol. 19, no. 2, 1959, pp. 79–88, doi:10.2307/973677.

Luhmann, Niklas. 1968. *Trust and Power.* John Wiley and Sons, 1979.

Margolis, Robert, and Jarett Zuboy. "Nontechnical Barriers to Solar Energy Use: Review of Recent Literature". National Renewable Energy Laboratory, 2006.

Mazzanti, Massimiliano, and Anna Montini. "Waste Management Beyond the Italian North-South Divide: Spatial Analyses of Geographical, Economic and Institutional Dimensions". *Handbook on Waste Management*, edited by Thomas C. Kinnaman and Kenji Takeuchi, Edward Elgar, 2014, pp. 256–285.

McCarney, Patricia. "Cities and Governance: Coming to Terms with Climate Challenges". *Climate Change Governance*, edited by Jörg Knieling and Walter Leal Filho, Springer, 2013, pp. 85–103.

McCarney, Patricia, and Richard E. Stren. "Metropolitan Governance: Governing in a City of Cities". *State of the World's Cities Report*, edited by United Nations-Habitat, 2008, pp. 226–237.

Mokyr, Joel. *The Gifts of Athena: Historical Origins of the Knowledge Economy.* Princeton University Press, 2004.

Morris, Christopher W. "The Very Idea of Popular Sovereignty: 'We the People' Reconsidered". *Social Philosophy and Policy*, vol. 17, no. 1, 2000, pp. 1–26.

Moser, Petra. "Patents and Innovation: Evidence from Economic History". *The Journal of Economic Perspectives*, vol. 27, no. 1, 2013, pp. 23–44.

Munck af Rosenschöld, Johan, Jaap G. Rozema, and Laura Alex Frye-Levine. "Institutional Inertia and Climate Change: A Review of the New Institutionalist Literature". *Wiley Interdisciplinary Reviews: Climate Change*, vol. 5, no. 5, 2014, pp. 639–648, doi:10.1002/wcc.292.

North, Douglass C. "Economic Performance through Time". *The American Economic Review*, vol. 84, no. 3, 1994, pp. 359–368, doi:10.2307/2118057.

North, Douglass C. *Institutions, Institutional Change and Economic Performance.* Cambridge University Press, 1990.

OECD. *Aligning Policies for a Low-Carbon Economy.* OECD Publishing, 2015, doi:10.1787/9789264233294-en.

Ostrom, Elinor. "Beyond Markets and States: Polycentric Governance of Complex Economic Systems". *American Economic Review*, vol. 100, no. 3, 2010, pp. 641–672, doi:10.1257/aer.100.3.641.

Popper, Karl. *The Open Society and Its Enemies.* 1945. Routledge, 2012.

Ritala, Paavo, Sascha Kraus, and Ricardo B. Bouncken. "Introduction to Coopetition and Innovation: Contemporary Topics and Future Research Opportunities". *International Journal of Technology Management*, vol. 71, no. 1–2, 2016, pp. 1–9.

Rogelj, Joeri, Michel Schaeffer, Pierre Friedlingstein, Nathan P. Gillett, Detlef P. van Vuuren, Keywan Riahi, Myles Allen, and Reto Knutti. "Differences between Carbon Budget Estimates Unravelled". *Nature Climate Change*, vol. 6, no. 3, 2016, pp. 245–252, doi:10.1038/nclimate2868.

Romstad, Eirik. "Theoretical Considerations Regarding the Effectiveness of Policy Instruments". *The Market and the Environment: The Effectiveness of Market-Based Policy Instruments for Environmental Reform*, edited by Thomas Sterner, Edward Elgar, 1999, pp. 50–65.

Ryan, Richard M., and Edward L. Deci. "Self-Determination Theory and the Facilitation of Intrinsic Motivation, Social Development, and Well-Being". *American Psychologist*, vol. 55, no. 1, 2000, pp. 68–78.

Shmelev, Stanislav E., and Jeroen C. J. M. van den Bergh. "Optimal Diversity of Renewable Energy Alternatives under Multiple Criteria: An Application to the UK". *Renewable and Sustainable Energy Reviews*, vol. 60, 2016, pp. 679–691.

Stiglitz, Joseph E. "Intellectual Property Rights, the Pool of Knowledge, and Innovation". NBER Working Paper 20014, March 2014. National Bureau of Economic Research, doi:10.3386/w20014.

Stiglitz, Joseph E., Armartya Sen, and Jean-Paul Fitoussi. *Report by the Commission on the Measurement of Economic Performance and Social Progress*. OFCE – Centre de recherche en Économie de Sciences Po, 2009, www.ofce.sciences-po.fr/pdf/dtravail/WP2009-33.pdf. Accessed 8 March 2017.

Sunstein, Cass R. *Simpler: The Future of Government*. Simon and Schuster, 2014.

Thompson, Grahame, Jennifer Frances, Rosalind Levacic, and Jeremy Mitchell. *Markets, Hierarchies and Networks: The Coordination of Social Life*. Sage, 1991.

Weisz, Helga, and Julia K. Steinberger. "Reducing Energy and Material Flows in Cities". *Current Opinion in Environmental Sustainability*, vol. 2, no. 3, 2010, pp. 185–192.

Weitzman, Martin L. "Fat-Tailed Uncertainty in the Economics of Catastrophic Climate Change". *Review of Environmental Economics and Policy*, vol. 5, no. 2, 2011, pp. 275–292, doi:10.1093/reep/rer006.

Wiseman, John, Taegen Edwards, and Kate Luckins. "Post Carbon Pathways: A Meta-Analysis of 18 Large-Scale Post Carbon Economy Transition Strategies". *Environmental Innovation and Societal Transitions*, vol. 8, 2013, pp. 76–93, doi:10.1016/j.eist.2013.04.001.

II.4 Assessing public awareness about biodiversity in Europe

Anna Kalinowska

Biological diversity is broadly understood as the variety of life on Earth. This includes diversity within species, between species and of ecosystems. Biodiversity underpins human wellbeing. It provides the food we eat and materials for the homes in which we live, and supports jobs, economic security and development. However, many of the important roles of biodiversity often go largely unrecognized and are not widely understood.

Braulio Ferreira de Souza Dias, *Convention on Biological Diversity* (29–33)

In myriad ways humanity is linked to the millions of other species on this planet. What concerns them equally concerns us. The more we ignore our common health and welfare, the great are the many threats to our own species. The better we understand and the more we rationally manage our relationship to the rest of life, the greater the guarantee of our own safety and quality of life.

Edward O. Wilson, "Foreword" (VIII–IX)

I Introduction

The dramatic increase in human population beginning in the late seventeenth century and further accelerated by the Industrial Revolution set in train processes leading to changes in the environment that human beings had never encountered or experienced before. The natural process by which species head for extinction accelerated dramatically and indeed the danger of whole ecosystems being lost became a reality. It was not until the mid-twentieth century that the need for international action came to be signalled by cumulative documentation and analysis regarding consequences, not only for nature, but for ourselves. This change in fact went hand in hand with new insights from ecological science. The concept of sustainable use was both associated with, and generated by, an emerging concept of biological diversity, which had both biological and political significance. The approach stressed – and still stresses – the importance of the entire range of forms of life on Earth, seen as the heritage of all humankind, but also as a *sine qua non* condition underpinning our species' success. For that reason, people ought to be taking every care to ensure the preservation of that natural heritage. Side by side, biodiversity and sustainable development were the key philosophies, as Rio de Janeiro 1992 played host to the Earth Summit or UN Conference on the Environment and Development. It was in the same spirit of sustainability that the Summit opened for signature the United Nation Convention on Biodiversity. Each of the Earth Summit 1992 documents also made it clear how necessary it was to raise the public's level of awareness about how to achieve sustainable development and how to halt biodiversity loss.

Biodiversity matters for all spheres of human life and activity and yet biodiversity loss has increased to an unprecedented level, both in Europe and worldwide. Scientists agree that the current global rate of species extinctions is on average somewhere between 100 and 1,000 times greater than the natural background extinction rate in the pre-human era (Pimm *et al.* 18; Reichholf 88). The mismanagement and destruction of ecosystems and over-exploitation of species is ongoing all over the world. It lowers the quality of the planet's global and local resources and destabilises the physical environment. In addition, the exploitation of biodiversity is affecting ecosystem services, i.e. benefits that people can obtain from ecosystems such as those which are essential for life (e.g. food, clean air and water) or those which improve our quality of life (e.g. recreation and beautiful landscapes).

The alarming biodiversity loss, among the challenges faced by contemporary civilisation, seems to be one of paramount importance, but what should be a cause for deep concern is very much underestimated in public perception. It is impossible to save biodiversity and enhance its benefits for people without public awareness and knowledge of biodiversity issues. Public opinion is essential to influencing politicians and decisions-makers, it provides a barometer for public support and interest and motivates individuals at all levels.

What is the level of public awareness about biodiversity in Europe? Do Europeans agree or disagree that the European Union should better inform citizens about the importance of biodiversity? This chapter presents the results of the European Environment Agency consecutive surveys on the attitudes of Europeans towards issues of biodiversity. It discusses the efforts undertaken to raise the level of public awareness on biodiversity through education and communication. It illustrates selected examples of good practices in the promotion of biodiversity and the best tools and approaches relevant to the Strategic Plan for Biodiversity 2011–2020 launched by the Convention on Biological Diversity in 2011.

II The state of biodiversity in Europe

As the signatories to the United Nation Convention on Biological Diversity (United Nations), all European countries individually and the European Union committed themselves to the protection of biological diversity. In order to discuss the status of biodiversity in the European Union and the perception of Europeans, a general picture of nature in Europe and its threats is needed.

The EU Member States stretch from the Arctic Circle in the North to the Mediterranean in the South and from the Atlantic coast in the West to the Pannonian Steppes in the East – an area characterised by a great diversity of landscapes and habitats and wealth of flora and fauna. European biodiversity includes 488 species of birds (International Union for Conservation of Nature 2010), 260 species of mammals (Temple and Terry), 1,151 species of reptiles, 85 species of amphibians, 546 species of freshwater fishes (Kottelat and Freyhof, 2007), 20,000–25,000 species of vascular plants (*Euro + Med PlantBase*) and well over 100,000 species of invertebrates (*Fauna Europaea*). Europe has arguably the most highly fragmented landscape of all continents, and only a tiny fraction of its land surface can be considered as wilderness. For centuries, most of Europe's land has been used by humans to produce food, timber and fuel and provide living space and currently more than 80 per cent of land in Western Europe is under some form of direct management by the European Environment Agency. Consequently, European species are to a large extent dependent upon semi-natural habitats created and maintained by human activity, particularly traditional, non-invasive forms of land management. These habitats are under pressure from agricultural intensification, urban sprawl, infrastructure development, land abandonment, acidification and eutrophication. Many species are directly affected by overexploitation, persecution and impact of alien species, as well as climate changes posing an increasingly serious threat in the future (van Swaay *et al.*). Europe is a huge, diverse region and the relative importance of different threats varies widely across its biogeographic regions and countries.

Although considerable efforts have been made to protect and conserve European habitats, decline and the associated loss of vital ecosystem services such as water purification, crop pollination and carbon sequestration continue to be a major concern in the region (van Swaay *et al.*). Currently almost 25 per cent of European animals face the risk of extinction. For example, nearly one-sixth (15 per cent) of Europe's mammalian species is threatened and a further 9 per cent are close to qualifying for the threatened status (Temple and Terry). By comparison, despite the lack of good trend data from many countries, the results show that about a third of European butterfly species suffered a decline in their populations over the last ten years and 9 per cent are threatened (van Swaay *et al.*). According to the International Union of Conservation of Nature – IUCN, some of 21 per cent of Europe's vascular plants species and half of the continent's vascular endemic plants are in danger of extinction (European Commission, *LIFE and Endangered Plants*).

Published by the European Environmental Agency in 2015, the State of Nature in the EU report results from the largest collaborative data-collection and assessment of nature ever undertaken across the EU Member States in the period 2007–2012 and provides the most comprehensive picture of European habitats and species. Looking at birds, the report concludes that 17 per cent of European bird species are threatened and 15 per cent are near threatened, declining or depleted. At present only 16 per cent of the assessed habitats is in a favourable conservation status. Most European ecosystems are now defined as degraded and the conservation status of more than 30 per cent is bad (European Environmental Agency, *State of Nature in the EU*). As Europeans currently consume more than twice what the EU's land and sea can deliver in terms of natural resources, 88 per cent of the fish stock is over-exploited or significantly depleted (European

Environmental Agency, *European Environment*). The condition of most eco-system services has showed either a degraded or mixed status across Europe (Harrison *et al.*). Although action to halt biodiversity loss requires money, the cost of inaction is expected to be even higher. Biodiversity loss is in fact very costly for society, particularly for sectors that depend heavily on ecosystem services. For example, within the EU as a whole, the estimated economic value of insects' pollination is €15 billion annually (European Commission, *The EU Biodiversity Strategy to 2020*).

There is enough evidence to certify that biodiversity loss is an enormous challenge for the European Union. The EU specific answer for such loss is the European network of nature protection areas (over 26,000 sites across Europe) known as the *Natura 2000* Network.

III Are Europeans familiar with the term "biodiversity" and the *Natura 2000* network?

It is clear that biodiversity conservation, especially on the sites of *Natura 2000*, cannot be achieved without the widespread engagement of society as a whole, and engagement cannot be achieved without public knowledge of biodiversity issues even on a basic level. With this in mind the European Commission requested the Flash Eurobarometer survey on *Attitudes of Europeans towards the Issue of Biodiversity*, in which EU citizens were asked questions about biodiversity and *Natura 2000*. Do people agree or disagree that the EU should better inform citizens about the importance of biodiversity?

The European Commission published the results of *Attitudes of Europeans towards the Issue of Biodiversity. Analytical Report* Wave 1 in 2007, Wave 2 in 2010 and Wave 3 in 2013. The 2013 report presents comparative data from the three waves. The survey was carried out by the TNS Political and Social network throughout the 27 Member States of the EU. Some 25,537 respondents (1,000 per country) from different social and demographic groups were interviewed via telephone in their mother tongue on behalf of the European Commission.

As the results obtained from individual states are presented in the report without references to the regions, one question arises: is the level of awareness and knowledge about biodiversity and *Natura 2000* in Central European countries significantly different from the other EU countries? If so, what are the reasons for such differences? Considering the specific asset of Central Europe, to what extent do the characteristics of the regions influence the level of public familiarity with the idea of biodiversity and perceived seriousness of biodiversity loss? These questions invite a comparison between public awareness of the *Natura 2000* Network in Central European countries and the average awareness across the EU countries. The comparison will be based on data concerning seven Central European countries and on the analysis of those data against the background of all other countries covered by the report.

The Flash Eurobarometer survey interviewed EU citizens in order to assess how familiar they are with the term "biodiversity", the term "*Natura 2000*" and the concept of biodiversity loss. Thirteen questions were aimed to gain an insight into the perception of biodiversity loss at domestic, European and global levels. The report also dealt with several other aspects of biodiversity conservation, examining Europeans views on why preserving biodiversity is important and what EU measures and personal measures can be taken to prevent the loss of biodiversity. Question 1 in the Flash Eurobarometer survey is: Have you ever heard of the term "biodiversity"? The comparison between the 2007 and 2010 results showed that in 12 out of the 27 EU Member States the proportion of respondents who had never heard about the term biodiversity decreased by at least 5 percentage points. In 2010 more than 38 per cent of Europeans had heard of biodiversity and knew what it meant. The results of the survey requested by the Directorate-General for Environment in 2013 show that familiarity with biodiversity increased again in the majority of Member States compared with the survey in 2010. Across the EU, 44 per cent of Europeans have heard about biodiversity and know what it means.

The situation is different in Central European Countries. The Czech Republic initially saw the largest increase in the fraction of respondents who knew the meaning of the term. For example, in 2007 just 6 per cent of the respondents said they knew what biodiversity meant. In 2010 this proportion increased to 21 per cent, but the results of the 2013 survey show a decrease in number to only 17 per cent. A similar trend of initial increase from 18 per cent in 2007 to 23 per cent in 2010 and again decrease to 10 per cent in 2013 was observed in Hungary. In Poland, the results of the survey indicate a trend of permanent decrease of familiarity with the term. In 2007 more than 31 per cent of interviewees said they knew the meaning of the term, but six years later, in 2013, only 19 per cent did. It is worth mentioning that in 2012 national investigations were carried out in Poland at the request of the Polish Ministry of Environment (Ministerstwo Środowiska, *Research on Environmental Awareness*). Those resulted in a more optimistic assessment. According to the Polish report, more than 38 per cent of the respondents maintained they know and understand the term biodiversity. In Slovakia and Slovenia there has been a constant increase: from 6 per cent in 2007 to 15 per cent in the former in 2013 and from 25 per cent to 35 per cent in the latter. Altogether, even if we include the results from national research in Poland, in the five Central European countries mentioned above the average fraction of people familiar with the term is far less than the EU average of 44 per cent. This is in stark contrast with older Central European EU members, Austria and Germany, where this fraction is the highest in Europe, namely 80 per cent in both countries. This situation is rooted in the historical background. In those years the countries in transition and involved with the accession process were building the environmental infrastructure and implementing EU environmental policies in order to narrow the gap between the old and the new EU members. The "brown issues" dominated the content of education and the interest of the media. Problems of nature conservation were replaced by problems of recycling and renewable energy (Kalinowska, "Biodiversity Loss and Public Opinion"). This had an impact on the sphere of public environmental consciousness, causing a phenomenon clearly described by Edward O. Wilson:

Most people understand very well the dire effects of toxic pollution on their health. They also know that the ozone hole in the upper atmosphere is not a good thing, and that global warming, destruction of forests, and depletion of fresh water reserves are serious global threats. What has been harder to grasp, not only by the general public but also by most scientists, is the profound influence biodiversity has on human well-being. The reason is the prevailing world view that health is largely an internal matter for our species, and, with the exception of domesticated species and pathogenic microorganisms, the rest of life is something else.... For many reasons, not least our own well-being, we need to take better care of the rest of life. Biodiversity ... will pay off in every sphere of human life, from medical to economic, from our collective security to our spiritual fulfilment.

(VII–VIII)

The situation is somewhat different when we compare public knowledge about biodiversity and awareness of *Natura 2000*, a Europe-wide network of protected areas designed to halt biodiversity loss in the EU. It is the centrepiece of the EU nature and biodiversity policy. The aim of the network, established under the 1992 Habitat Directive, is to assure the long-term survival of Europe's most valuable and threatened species and habitats. It includes Special Areas of Conservation (SAC), designated by Member States under the Habitats Directive, and Special Protection Areas (SPAs), which they designate under the 1979 Birds Directive. The establishment of this network of protected areas also fulfils the Community obligation under the UN Convention on Biological Diversity.

Judging from the long time that has passed since the establishment of the *Natura 2000* Network, the number of sites and its high rank among the EU policies, the idea should be widely known. The answer to the question in the EU Flash Eurobarometer survey: "Have you heard of *Natura 2000*?" should thus be positive.

Instead, despite the importance of *Natura 2000* for conservation policy, European awareness of the Network is very low. Approximately eight in ten respondents had never heard of it. Even if the number of people who completely ignore it has decreased (from 80 per cent of the EU population in 2007 to 73 per cent in 2013), the percentage of people who declared themselves to be well informed still does not exceed 11 per cent. The lowest awareness is observed mainly in the older EU countries such as the UK (1 per cent), Ireland (1 per cent) and Denmark (6 per cent). In comparison with other countries, the level of awareness in Central Europe is relatively high. In most Central European countries, the fraction of respondents who had heard about *Natura 2000* and know what it is matches or exceeds the EU average of 11 per cent (Poland 34 per cent, Slovenia 32 per cent, Austria 19 per cent, Hungary 15 per cent, and the Czech Republic 11 per cent). Moreover, a steady increase in the awareness has been reported in all Central European countries since 2007. The improvement is observed even in Slovakia and Germany – the only Central European countries where awareness of *Natura 2000* is a little lower than average for the EU. Also, the percentage of citizens who have heard about it in all Central European countries (except Germany) is much higher than the average in all the 27 EU countries. The relatively high level of awareness in Central Europe could be explained by considering the comparatively short time elapsed from the start of the process of accession to the EU. The *Natura 2000* Network is still a fresh and sometimes controversial idea in all the Central European states, which became EU Member States in 2004. It regularly attracts media attention and causes the involvement of the environmental NGOs owing to the selection process of candidate areas and related local conflicts. For example, in Poland a very important role in raising public awareness of *Natura 2000* is played by the governmental financial support for educational programmes devoted to its introduction in the local communities. Several educational campaigns, information leaflets and programmes aimed at answering the "frequently asked questions about *Natura 2000*" are financed by the National Fund for Environmental Protection and Water Management, which is the institution managing the funds from obligatory fees and fines for the use of the environment (Kalinowska, "Biodiversity Loss and Public Opinion").

IV What is the level of public awareness of biodiversity in Europe?

In response to the *EU Biodiversity Strategy to 2020* (European Commission), the results of the 2007, 2010 and 2013 Flash Eurobarometer surveys were verified through new research in 2015. In contrast to the previous Special Eurobarometer telephone surveys, this study was conducted face-to-face due to the complexity of the topic, and the results show that the 60 per cent of respondents have heard of the term biodiversity. However, only half know what it means. Two thirds (66 per cent) of EU citizens do not feel informed of the loss of biodiversity with 22 per cent not feeling informed at all.

Overall, at least eight of ten respondents believe that the decline and possible extinction of animals, plants and ecosystems are serious problems at the national, European and global levels (TNS). In 2015, 60 per cent of EU citizens strongly agreed that it is important to halt biodiversity loss because our wellbeing and quality of life are dependent upon nature and biodiversity, but the survey also revealed that most Europeans are unaware of what the EU is doing to save biodiversity. Roughly three quarters of Europeans have not heard of the *Natura 2000* Network, and approximately half of those who have heard of it do not know what it is. On the other hand, nearly two thirds of Europeans claim that they are taking personal steps to protect biodiversity and nature, and half of them would like to do even more (TNS).

Similarly, according to *Special Eurobarometer 421. The European Year for Development – Citizen's Views on Development, Cooperation and Aid*, published in 2015 by the European Commission, there is strong support for increasing development aid in almost all Member States; respondents are more likely to believe that development aid should be increased. Not only do most respondents agree that assisting developing countries should be one of the main priorities of the EU, but they also believe that doing so would benefit Europeans. Therefore, providing aid is in the EU's own interest. Health (39 per cent), peace and security (36 per cent) and education (34 per cent) are considered to be the most pressing challenges for developing countries. Interestingly, these are more likely to be viewed as more important than meeting basic human needs such as food security and agricultural production (25 per cent) or water provision and sanitation (30 per cent). In contrast, very few respondents consider environmental protection and biodiversity (both 7 per cent) to be the most pressing challenges. The results of this survey must be considered in light of the six main targets of the *EU Biodiversity Strategy to 2020*, the last but not least of which is the EU making a larger contribution towards averting global biodiversity loss. Only if the public understands what the loss of biodiversity means for human wellbeing in developing countries can Europe reach such an ambitious goal.

V Information and education for the protection of biodiversity

Socio-demographic analysis of the Eurobarometer surveys and several other sociological studies and reports on environmental awareness have revealed that education most significantly influences the level of knowledge about biodiversity and the *Natura 2000* initiative (Bortłomiuk and Burger; Ministerstwo Środowiska, *Research on Environmental Awareness*; Pietrzyk-Kaszyńska *et al.*; European Commission, *Attitudes of Europeans*; Kalinowska, "Biodiversity Loss and Public Opinion"). Additionally, in socio-demographic terms, better-educated respondents are more positive about development aid (European Commission, *Special Eurobarometer 421*). Biodiversity cannot be conserved without widespread societal engagement, so the active involvement of stakeholders, key policy sectors and civil society will be essential. A higher level of education and access to information about biodiversity are the best guarantors of such engagement.

Question 8 in the Flash Barometer survey is whether Europeans would agree or disagree that the EU should take various measures to combat biodiversity loss, and nearly three quarters of the respondents in all EU countries completely agreed that the EU should better inform its citizens about the importance of biodiversity. Moreover, none of the seven Central European countries were in the group of states whose citizens were the least likely to affirm a need for better information (European Commission, *Attitudes of Europeans*), which is a very important indicator that more education and information must be provided by the EU.

In May 2011, the European Commission adopted the *EU Biodiversity Strategy to 2020* to halt the loss of biodiversity and improve the state of the species, habitats and ecosystems, as well as the services they provide, in Europe. Overall, the 2020 target is: "Halting the loss of biodiversity and the degradation of ecosystem services in the EU by 2020 and restoring them, in so far as feasible, while stepping up the EU contribution to averting global biodiversity loss" (European Commission, *The EU Biodiversity Strategy to 2020* 6). Six major targets address the main pressures on the natural world and establish policy foundations for EU-level actions. In collaboration with member states, one of these proposed EU-level actions will be the development and launch of a major communications campaign about biodiversity and *Natura 2000*.

Better information about biodiversity and its links to *Natura 2000*, among other EU priorities, is important for building public interest and support for the *EU Biodiversity Strategy to 2020*. This statement is not meant to imply that nothing was done prior to the development of the Strategy; several educational initiatives were underway, and some are currently being conducted in European countries. It is worth mentioning that events such as the UN International Year of Biodiversity 2010 (IYB) provided the impulse, and a good example from Central Europe was the huge campaign to increase awareness of biodiversity in Poland and to celebrate IYB 2010. This nationwide campaign, entitled *Many Faces of Biodiversity*, was initiated and managed by the Centre for Environmental Studies and Sustainable Development (UCBS) at the University of Warsaw, an inter-faculty, interdisciplinary academic unit that performs research on education and communication related to various areas of sustainable development. The Centre operates on the principle that spectacular successes related to the protection of biodiversity cannot be expected until the conservation message is broadly spread to various social groups that are diverse in age, education level and profession. Each group represents a specific audience with specific communication expectations, so a successful campaign should encompass a wide range of potential supporters, such as students from all the faculties of Warsaw University, fine artists, journalists, teachers, school children, and the general public. This was the principle underlying a multimedia educational campaign aimed at raising public awareness of biological diversity under the *Many Faces of Biodiversity* theme. The profile of the campaign was raised by the patronage of the Minister of the Environment and the Secretary General of the Polish Committee for UNESCO.

Various synergies were designed to increase awareness. First, a competition launched among Fine Arts students was aimed at promoting the International Year of Biodiversity with slogans and design posters. Two among 40 projects across Poland were printed as posters and 4,000 copies were distributed to municipal institutions, schools, town libraries, teacher training centres, etc. Young artists are a very influential group of actors in social media, and their emotional message on the importance of biodiversity in our life can be key to attracting a broader audience through informal education. Second, several exhibitions of award-winning posters were combined with popular lectures on biodiversity for visitors and the media. Third, a lecture series on *Many Faces of Biodiversity* was offered to students from all faculties as well as journalists, teachers and the public. Leading scientists and practitioners explored a wide range of topics related to biodiversity at the National Conference *Let's Talk about Biodiversity*, which brought together more than 120 academic and media representatives to discuss more effective ways to communicate and popularise biodiversity. Training programmes and workshops for more than 100 teachers and environmental educators supported pedagogic preparation for biodiversity education. Fourth, two books completed the synergy of methods in the Biodiversity Campaign. In 2010 Batorczak and Kalinowska published *Na spotkanie różnorodności biologicznej* (*Let's Meet Biodiversity*), a training manual aimed to help teachers carry out lessons outside the classroom, and *Różnorodność biologiczna w wielu odsłonach* (*Many Faces of Biodiversity*), a collection of interdisciplinary lectures and conference papers intended for students, teachers and journalists.

An evaluation of the groups directly targeted by the above-mentioned activities confirmed positive effects regarding familiarity with biodiversity and *Natura 2000*. However, no effect was observed at the national level. The Flash Barometer survey even indicates decreasing familiarity from 2007 to 2013.

What conclusions can be drawn from the Centre's campaign? Should it be regarded as an example of good practice for other European countries? According to the results of a survey conducted in Poland (Burger), television is the main source of information about the environment for the majority of the public (more than 70 per cent); only a minority derive information about the environment from sources other than television. Similar results were reported in the USA (Hannigan). However, this does not mean that educational and informational activities other than television should be neglected; such activities can be sound if the flow of information is directed at the most influential groups in society. However, the above results become problematic if educational activities are restricted to a short period, such as the IYB. Because these concerns are shared globally, the United Nations announced the 2011–2020 Decade on Biodiversity, so the IYB 2010 campaign promoted by the University Centre for Environmental Studies and Sustainable Development can be useful for planning long-term activities during the United Nations Decade on Biodiversity. The Centre promotes the principles of environmental protection and biodiversity conservation along with models for sustainable production, consumption and living, and it also plays an active role in the ongoing UN Decade on Biodiversity (2011–2020). The title *Let's talk about Biodiversity*, which points to the need for dialogue while also implying the need for individual approaches to different recipient groups, has been adopted for a programme of celebrations organised by the Centre and associated with the entire Decade on Biodiversity.

Additionally, the long-term action programme for the Decade on Biodiversity entails a wide range of activities: in 2013 the *Forests Mitigate Climate* exhibition was organised with the Information Centre of the State Forests in the context of the Global Landscapes Forum under UNFCCC COP19 and was convened in Warsaw; in 2014 the *Ideal City – Sustainable City* conference focused on spatial planning and urban green space as remedies for climate change in cities; in 2015 the *Biodiversity at a High Level* exhibition of artistic photography by well-known photographer-scientists was associated with the International Biodiversity Day; in 2016 the *Greening the University Campus* exhibition of posters offered examples of environment-friendly innovation at the higher-education establishments in Poland and other countries.

In the context of the Decade on Biodiversity, the Centre has launched programmes for students, teachers, urban planners and business circles. A comprehensive programme on biodiversity and sustainable development addressed to senior citizens is particularly innovative. In partnership with the NGO *Ziemia i Ludzie*, the Centre launched *Zielona Wiedza dla Uniwersytetów Trzeciego Wieku* (*Green Knowledge for the Universities of the Third Age*), a programme for the 60+ group. The Centre now pays significant attention to the education of senior citizens by preparing teaching materials for numerous Universities of the Third Age across Poland. The *Uniwersytety Trzeciego Wieku* forms a network of more than 80 Polish educational institutions engaging in educational and cultural activities for senior citizens, several hundred thousand of which participate in courses as mature students each year. Currently in place for an entire year, the aforesaid *Green Knowledge* Programme includes university lectures as well as field trips, all of which are based on *Biological Diversity – We Are All in this Together*, a handbook published by Kalinowska and Batorczak in 2013. Divided into two parts, the first targets senior citizens and addresses biodiversity and associated benefits, while the second indirectly targets children by advising grandparents on how they may best develop a feeling of responsibility for the environment among their grandchildren, such as by observing nature together within a citizen science framework.

Demographically in both Poland and throughout Europe, senior citizens are a growing social group whose awareness of biodiversity issues can be very important. The guiding principle of the UN Decade on Biodiversity framework is that the youngest are trained so that they might act to preserve nature in the future and the elderly can act in the present, as they possess experience and traditional knowledge about biological diversity. The reason why the Centre attaches such importance to educating Senior Citizens is that demographically this is the most rapidly growing age group, and it is significant in terms of its lifestyle and means of utilising biological diversity. These are people with experience who can act as custodians of traditional knowledge, and intergenerational solidarity is essential if biodiversity is to be preserved.

Educational initiatives of several Central European countries have converged into the BEAGLE (Biodiversity Education and Awareness to Grow a Living Environment) project, which aims to increase the level of familiarity with biodiversity. This online project focusing on the quality of learning outside the classroom is co-ordinated by the Centre for Environmental Studies and Sustainable Development. Formed by institutions from Poland, Slovakia, Hungary, Germany, the United Kingdom and Norway, it has set up the Pan-European Biodiversity Observation Project (BOP), based on monitoring the phenology of trees across Europe. Three hundred schools have participated in BEAGLE observations.

Despite the many previous educational initiatives aimed at various target groups, there is still a great need to seek new educational methods and, especially, to increase the involvement of local and electronic media in raising social awareness of biodiversity issues in Europe.

VI Education and communication in the convention on biological diversity

Despite major efforts by many countries on different continents, the *Global Biodiversity Outlook 4* (Convention on Biological Diversity) report on the implementation of the Strategic Plan for Biodiversity 2011–2020 is not encouraging. Many indicators of threats to biodiversity presented in the report will remain at current levels until at least 2020 and the values of some will rise. At the same time, resources for preservation will continue to decline.

Since over 70 per cent of the loss of biological diversity is due to agriculture and food production, specialists argue that if the rate of destruction is to be kept at bay, it will be necessary not only to strive for sustainability in production and consumption but also to encourage new lifestyle trends. If the above changes are to be made a reality, it will be necessary to steadily increase the awareness of sustainable development in every circle which can influence the state of ecosystems and makes use of ecosystem services. These circles include decision-makers, producers and consumers around the world, i.e. the entire global community.

In fact, the Convention on Biodiversity (CBD) attaches great importance to the role of social awareness, as Article 13 is devoted to education and communication with the public. To ensure that recommendations of this kind are followed, successive plenary sessions of signatory countries to the CBD were convened as Conferences of the Parties and used as opportunities to initiate, develop and introduce various practical solutions. Not least among these is an international initiative under the CBD entitled *Communication, Education, and Public Awareness* or CEPA/CBD (Kalinowska, *Article 13. Towards Public Participation*; Convention on Biological Diversity, *CBD-Biodiversity*).

To achieve better results within the CEPA Initiative framework, the CBD Secretariat, whose seat is in Montreal, was called upon by the COP to establish an international team of experts to be known as the Informal Advisory Committee for CEPA (CEPA-IAC), which acts to support the CBD with respect to Article 13. My participation in the work of IAC as its Representative for Poland and Central Europe allows me to directly present the main thrusts of the Committee's work as well as examples of educational initiatives in different CBD areas, both in different countries and internationally. The Committee is also in charge of ensuring the emergence of a Global Strategy for Communication and Education in the name of Biological Diversity. Its goal is to establish a theoretical and practical action framework for state institutions and organisations active in the community as well as educational and scientific institutions.

The results of international surveys commissioned by the Secretariat indicate that practitioners are expecting education in the name of the CBD to do more to activate people on behalf of the environment, consequently changing their lifestyles as opposed to merely increasing their level of knowledge. Of additional importance is a need to account for matters that are particularly interesting to such diverse groups as farmers, young people or businesses. Matching the ways of presenting content related to biodiversity to the needs of different audiences is a matter of great discussion worldwide and certainly in Poland. Modern biodiversity education must also consider the achievements of psychology and behavioural studies to adjust its message to the capabilities and needs of particular target groups in the context of the different Aichi Biodiversity Targets in *Living in Harmony with Nature to Transform the World*, a ten-year framework for action by all countries and stakeholders to save biodiversity and enhance its benefits for people.

Awareness of biodiversity in society can be influenced by communication, whereas the state of public knowledge and awareness will depend on the achievement of the Strategic Plan for Biodiversity 2011–2022 and the Aichi Biodiversity Targets. One of the tasks of IAC-CEPA Members is to exchange experiences in communicating these Targets in relation to the key problems of biodiversity preservation from the points of view of their countries or organisations. The presentation of programmes and good practices requires very diverse forms of action with an emphasis on concrete practical effects, not only in the African and Asian cases but also where Europe is concerned. Here, we find that the most effective educational initiatives are those that offer practical experience in the field as opposed to the static conveyance of knowledge about nature. Such practical experience should be organised in combination with the far-reaching promotion and dissemination of educational publications that clarify how individuals can take practical steps to protect biodiversity.

In most countries, social media are increasingly becoming the universal means for communicating biodiversity-related content, and the CBD Secretariat also recommends the full use of such platforms, noting a steady increase in activity on Twitter, Facebook and Instagram. However, while appreciating the use of social media, experts point to certain limitations, not least of which is accessibility by older age groups; this shows how necessary it is for different communication techniques and technologies to be combined. Emphasis is laid also on the need to recognise specifics when communicating issues related to biodiversity, such as whether recipients are male or female. A differentiated approach is seen as imperative when the cultural specifics of many countries are taken into account; for example, different genders may play different roles and thus have different ways of impacting biodiversity.

Currently, a change in the approach to biodiversity awareness involves moving away from the threats towards the positive actions that must be taken in support of the sustainable use of natural resources. This approach is exemplified by the major international campaign of the IUCN Commission on Education and Communication, *Nature for All*, which represents a continuation of *Love, Not Loss* and is based on awakening positive feelings for nature as opposed to dire warnings about dramatic losses. More than 50 major strategic partners are committed to this campaign, including Google. The need to fully appreciate the role that senior citizens play in intergenerational education is also a point of emphasis.

Additionally, interest in this approach was raised by the results of the European *University Educators for Sustainable Development* (*UE4SD*) programme, which has brought together 54 institutions of higher education across Europe including the University of Warsaw (Kalinowska and Batorczak, "Universities for Sustainability"). The materials drawn up within the *UE4SD* framework are designed to train academic staff to introduce sustainable development principles into their curricula and many experts consider these activities as worth recommending for tertiary-level activity in the context of the UN Decade on Biodiversity.

VII Expectations about the UN Decade on Biodiversity

Among its tasks, the IAC-CBD advises the Secretariat in such areas as the global organisation of the Decade on Biodiversity (DBD), and the experts have agreed that a coalition of the various DBD players would allow for better co-ordination as well as promotion of less-known aspects of the CBD such as the protection of genetic resources and the Access and Benefit-Sharing system. Also of interest is the opportunity to connect sustainable development, quality of life and wellbeing of local communities, and biodiversity conservation. Participants in the discussion have underscored the necessity of adjusting what was being conveyed to specific groups of recipients and the importance of seeking ways to reach these people; charismatic leaders or celebrities could be involved in this process. Additionally, biodiversity campaigns targeted at different sectors of public life or business will be implemented. However, from the point of view of biodiversity education in Europe, it is worth stressing: first, the need to make better use, by raising public awareness, of the years 2016–2020, the second half of the Decade on Biodiversity, given that most countries, even in Europe, have failed to make full use of such opportunities in the first half of the DBD; second, the promotion of the topics selected by the CBD Secretariat for the International Day of Biological Diversity each year (22 May), not least because these also serve as the themes for each year during the Decade on Biodiversity; third, the involvement throughout the Decade on Biodiversity of a variety of different partners, including those from the world of business, as well as national co-ordination to inspire DBD-related activities as was achieved in many countries during the 2003–2014 UNESCO Decade of Education for Sustainable Development; fourth, the translation of the CBD documentation and materials into different languages in a manner that takes full account of the cultural context; fifth, greater use of the entire social media spectrum to promote biodiversity issues while including other forms of communicating with the many and varied societal groups.

A conclusion expressed by all participants is the conviction that, while the current results are unsatisfactory, the most important way of reducing rates of biodiversity loss continues to be an accelerated educational process that raises awareness. This idea has clear support in Europe thanks to social studies of awareness (Kalinowska, "Biodiversity Loss and Public Opinion"). The CEPA-IAC experts also widely agree that knowledge acquisition should be deemphasised and that greater focus should be placed on developing capacities to take action locally, on sustainable lifestyles, and on the interrelationship between the state of biodiversity and human wellbeing.

VIII Conclusions

Like the rest of our planet, Europe faces a biodiversity crisis. European habitats face strong pressure from agricultural intensification, urban sprawl, infrastructure development, land abandonment, acidification and eutrophication. Many species are directly affected by overexploitation, persecution and impacts from alien species as well as climate change, which will be an increasingly serious threat in the future. Although the state of biodiversity and the public perception of the problem differ from country to country, there are some commonalities due to similar histories. Austria and Germany are old EU members that belonged to the West politically, while Poland, the Czech Republic, Slovakia, Hungary and Slovenia are new members that joined the EU in 2004. Citizens of Austria and Germany are better informed about biodiversity than citizens from new member countries with communist historical backgrounds. During the accession process, new EU countries concentrated on building an environmental infrastructure and implementing the EU environmental policy, so the "brown issues" dominated education and the interest of the media. The topic of biodiversity was related to the problems of environmental pollution and waste management; even the establishment of the *Natura 2000* sites was aimed at the implementation of EU directives rather than biodiversity. However, the media focus on the *Natura 2000* Network resulted in a better dissemination of information to citizens and the level of awareness due to *Natura 2000* in Central Europe is relatively high compared to other EU countries. In most Central European countries, the proportion of respondents who have heard of *Natura 2000* and know what it is matches or exceeds the EU average of 11 per cent. Additionally, there has been a steady improvement in awareness in all Central European countries from 2007 to date, compared to the percentage of all EU citizens who have heard about *Natura 2000*: it is much higher than the average for all 27 EU countries. *Natura 2000* is still quite new in all the Central European states, which became EU member states in 2004. However, the network constantly attracts media attention and involvement by environmental NGOs, which has impacted the sphere of public environmental consciousness. What is common among all European countries is a strong need for better information about the importance of biodiversity. The citizens' drive towards information should persuade the media as well as scientific and educational institutions to prepare and present examples of good practice.

The establishment of the United Nations Decade on Biodiversity by the United Nation General Assembly represents an opportunity to link national awareness-raising activities to broader international processes to increase the visibility and create traction for such actions. In the opinion of the CBD Secretariat, a very important action towards this target could be to assess the current level of biodiversity awareness in order to identify gaps as well as those groups whose awareness of biodiversity values is more important for conservation in the country. The information from such an assessment could then be used to identify and prioritise methods of communication and education.

Bibliography

Batorczak, Anna, and Anna Kalinowska. *Na spotkanie różnorodności biologicznej. Szkolne obserwacje drzew. Poradnik Nauczyciela* [*Let's Meet Biodiversity. Observation of Trees by school children. Teacher's Manual*]. Uniwersyteckie Centrum Badań nad Środowiskiem, 2010.

Bołtromiuk, Andrzej, and Tadeusz Burger. *Polacy w zwierciadle ekologicznym. Raport z badań nad świadomością ekologiczną Polaków w 2008 r.* [*Poles in an Environmental mirror. Report from the Studies on the Environmental awareness of Poles 2008*]. Instytut na rzecz Ekorozwoju, 2008.

Burger, Tadeusz. *Świadomość ekologiczna społeczeństwa polskiego* [*Environmental Awareness of the Polish Society*]. Instytut Gospodarki Przestrzennej i Mieszkalnictwa, 2005.

Convention on Biological Diversity. *CBD-Biodiversity and the 2030 Agenda for Sustainable Development.* July 2016, Secretariat of the Convention on Biological Diversity, www.cbd.int/development/doc/sdg-jul2016-flyer.pdf. Accessed 10 September 2016.

Convention on Biological Diversity. *Global Biodiversity Outlook 4. A Mid-term Assessment of Progress towards the implementation of the Strategic Plan for Biodiversity 2011–2020.* Secretariat of the Convention on Biological Diversity, 2014.

Dias, Braulio Ferreira de Souza. "Mainstreaming Biodiversity: Ensuring Sustainable Development". *Understanding Synergies and Mainstreaming among the Biodiversity Related Conventions. A Special Contributory Volume by Key Biodiversity Convention Secretariats and Scientific Bodies,* UN Environment, 2016, pp. 29–33.

Euro + Med PlantBase, Dipartimento di Scienze ambientali e Biodiversità ed Orto botanico, Università degli Studi di Palermo, 2005, www.emplantbase.org. Accessed 10 September 2016.

European Commission, *Special Eurobarometer 421. The European Year for Development – Citizen's Views on Development, Cooperation and Aid.* 2015, http://ec.europa.eu/public_opinion/archives/ebs/ebs_421_en.pdf. Accessed 2 January 2017.

European Commission. *Attitudes of Europeans towards the Issue of Biodiversity. Analytical report – Wave 3. Flash Eurobarometer 379.* European Commission, TNS Political and Social, 2013.

European Commission. *The EU Biodiversity Strategy to 2020.* Publications Office of the European Union, 2011.

European Commission. *Attitudes of Europeans toward the Issue of Biodiversity. Analytical Report – Wave 2. Flash Eurobarometer 290.* European Commission, The Gallup Organisation, 2010.

European Commission. *Attitudes of Europeans toward the Issue of Biodiversity. Analytical Report – Wave 1. Flash Eurobarometer 219.* European Commission, The Gallup Organisation, 2007.

European Commission. *LIFE and Endangered Plants. Conserving Europe's Threatened Flora.* Publications Office of the European Communities, 2007.

European Environmental Agency. *European Environment: State and Outlook 2010.* Luxembourg: Office for Official Publications of the European Union, 2010.

European Environmental Agency. *Halting the Loss of Biodiversity by 2010: Proposal for a First Set of Indicators to Monitor Progress in Europe.* Technical Report no. 11. Luxembourg: Office for Official Publications of the European Union, 2007.

European Environmental Agency. *State of Nature in the EU.* Technical Report no. 2. Luxembourg: Office for Official Publications of the European Union, 2007.

Fauna Europaea. www.faunaeur.org/. Accessed 10 September 2016.

Hannigan, John A. *Environmental Sociology*. Routledge, 1997.

Harrison, Paula A. and the RUBICODE consortium. *Conservation of Biodiversity and Ecosystem services in Europe: From Threat to Action*. PENSOFT, 2009.

Informal Committee on Communication, Education and Public Awareness, www.cbd.int/ doc/meeting=CEPAIAC-2016-01. Accessed 1 October 2016.

International Union for Conservation of Nature. *IUCN List of Threatened Species*. IUCN, 2010.

Kalinowska, Anna. "Biodiversity Loss and Public Opinion. What is the Situation in Central Europe?" *Current Challenges of Central Europe: Society and Environment*, edited by Jan Vávra, Miloslav Lapka, and Eva Cudlínová, Univerzita Karlova v Praze, 2014, pp. 68–82.

Kalinowska, Anna. *Różnorodność biologiczna w wielu odsłonach* [*Many Faces of Biodiversity*]. Warszawa: Uniwersyteckie Centrum Badań nad Środowiskiem, 2010.

Kalinowska, Anna. *Artykuł 13. W poszukiwaniu społecznego wsparcia dla w zarządzaniu Konwencją o różnorodności biologicznej. Polska praktyka na tle doświadczeń światowych* [*Article 13. Towards Public Participation in the Management of the Convention on Biodiversity. Practice in Poland against the Background of the Global Trends*]. Warszawa: Agencja Wydawnicza A. Grzegorczyk, 2008.

Kalinowska, Anna, and Anna Batorczak. "Universities for Sustainability – New Challenges from the Perspective of the University of Warsaw". *Environmental and Socio-Economic Studies*, vol. 3, no. 1, 2015, pp. 26–34.

Kalinowska, Anna, and Anna Batorczak. *Różnorodność biologiczna to także my. Zielona Wiedza dla Uniwersytetów Trzeciego Wieku* [*Biodiversity – We are All Together in This. 'Green Knowledge' for the Third Age's Universities*]. Warszawa: Fundacja Ziemia i Ludzie, 2013.

Kottelat, Maurice, and Jörg Freyhof. *Handbook of European Freshwater Fishes*. Publications Kottelat, 2007.

Ministerstwo Środowiska. *Badanie świadomości i zachowań ekologicznych mieszkańców Polski. Raport TNS Polska dla Ministerstwa Środowiska* [*Research on Environmental Awareness and Behaviours of Polish Citizens. Report TNS Poland for the Ministry of Environment*]. Warszawa: TNS, Ministerstwo Środowiska, 2012.

Natura 2000 Web: www.ec.europa.eu/environment/nature/natura2000/index_en.htm. Accessed 1 December 2016.

Pietrzyk-Kaszyńska, Agata, Joanna Cent, Małgorzata Grodzińska-Jurczak, and Magdalena Szymańska. "Factors Influencing Perception of Protected Areas – The Case of Natura 2000 in Polish Carpathian Communities". *Journal for Nature Conservation*, vol. 20, no. 5, 2012, pp. 284–292.

Pimm, Stuart, Maria Alice S. Alves, Eric Chivian, and Aaron Bernstein. "What is Biodiversity?" *Sustaining Life. How Human Health Depends on Biodiversity*, edited by Eric Chivian and Aaron Bernstein, Oxford University Press, 2008, pp. 3–27.

Reichholf, Josef H. *The Demise of Diversity. Loss and Extinction*. Haus Publishing, 2009.

State of Nature in the EU Commission Report: http://ec.europa.eu/environment/nature/ index_en.htm. Accessed 2 January 2017.

Strategic Plan for Biodiversity 2011–2020 and the Aichi Targets www.cbd.int/doc/strate gic0plan/2011-2020/Aichi-Targets-EN.pdf. Accessed 2 January 2017.

Temple, Helen, and Andrew Terry. *The Status and Distribution of European Mammals. IUCN Red List of Threatened Species – Regional Assessment*. Luxemburg: Office for Official Publications of the European Union, 2007.

TNS, Special Eurobarometer 436: *Attitudes of Europeans towards Biodiversity Survey Conducted by TNS Opinion and Social at the Request of the Directorate General for Environment.* 2015. http://ec.europa.eu/COMMFrontOffice/PublicOpinion/index/Result DocumentKy/68148. Accessed 20 June 2016.

United Nations. *Convention on Biological Diversity*, 1992, www.cbd.int/doc/legal/cbd-en. pdf. Accessed 20 June 2016.

United Nations Decade on Biodiversity, www.cbd.int/2011-2020. Accessed 30 March 2017.

van Swaay, Chris, Annabelle Cuttelod, Sue Collins, Dirk Maes, Miguel López Munguiro, Martina Šašić, Joseph Settele, Rudi Verovnik, Theo Verstrael, Martin Warren, Martin Wiemers, and Irma Wynoff. *European Red List of Butterflies.* Published by IUCN (International Union for Conservation of Nature) and Butterfly Conservation Europe in collaboration with the European Union. Publications Office of the European Union, 2010.

Wilson, Edward O. "Foreword". *Sustaining Life. How Human Health Depends on Bio-diversity*, edited by Eric Chivian and Aaron Bernstein, Oxford University Press, 2008, pp. VII–VIII.

II.5 Is the current global role of English sustainable?

Richard Chapman

I Introduction

While the current status of English as a world language might seem obvious and with little need for elucidation, a closer examination reveals quite the opposite. Its multi-faceted character and worldwide reach render the claims that are almost habitually made about English as often excessive, poorly substantiated and rarely detailed enough to provide us with anything but the most generalised of pictures. At various times in the recent past English has been claimed to be the "liberal" tongue, the efficient language, the language of empowerment (Dunton-Downer, "Preface"), the language of science, study and debate, the language of law and international diplomacy, the language of popular entertainment and the arts, and the future (or even current) lingua franca. The list provided by Montgomery (35–36) includes 36 highly significant fields where English is claimed to be the dominant code, the implication being that without it, no one can have a realistic hope of operating successfully in these areas on an international level. In a vaguer, but perhaps still more powerful way, English has been spoken of as "the international language of problem-solving" (Walker), crystallising a common thread running through these descriptions that endows English with a peculiar power, perhaps rooted in its intimate historical relationship with science,[1] technology and post-war economics, to be both the language that initiates or brings development and change, and the language that provides commentary and analysis of these processes.

So, the claims associated with the rise of English are significant and impossible to ignore in any debate about sustainability, worldwide policy or geopolitical development, and they are, at the same time, extremely grandiose. English is not only one of the most widely spoken languages on the planet, but it is the most influential, dictating politics and economics on the one hand, and intellectual and scientific growth on the other. It has become the world's "default tongue" (Ostler; Dunton-Downer xv). It is hardly surprising to find that the picture is a good deal more complex and this brings into question any prognostications about its future.

II Measuring global English

Nowhere is the picture more confused than when it comes to quantification. Despite many attempts at calculation (with varying degrees of rigour and reliability) we have very little idea at all of just how many people speak English in the world today. And these inaccuracies matter because they explicitly or implicitly underpin many of the assumptions and claims made about the language. Even Kachru's well-known three-circle model (Kachru 356, presented clearly in Jenkins, *Global Englishes* 14–15, and Montgomery 34–35) can be highly deceptive if not treated with care. The figures given are estimated *total* populations in each of the areas and countries listed and not indications of the actual numbers of English speakers involved. This caveat may seem rather superfluous for any attentive reading of Kachru's work, but already we are faced with an indication of the fundamental complexity at hand: the estimated two billion inhabitants of the "expanding circle" are clearly not all supposed to be users of the language, but what is often neglected in discussions of the Kachru model is that even the "Inner Circle" is highly problematic from a quantification point of view. We have an estimate of between 340 to 380 million inhabitants of "classically" English-speaking states (e.g. the United Kingdom, the United States, Australia, etc.). But even here the count becomes difficult: how many of the *c.*60 million inhabitants of the British Isles or the 250 million of the United States are speakers of English? Not all, by any means. Not only the migrations of the recent period of crisis (perhaps caused by the subprime economic collapse of 2008 and the effects of the so-called "Arab Spring"), but also the presence of relatively closed communities within the British or American populations make any linguistic description of even historically stable countries complex and at risk of error. In *English as a Global Language* Crystal gives figures of 1.5 million British and 25.6 million American citizens that use English as an L2 and not as their mother tongue, but this hardly suffices. Largely the analysis is geographical, and the figures we find depend mostly on census returns.

Jennifer Jenkins critiques the Kachru model clearly and effectively in *Global Englishes* (13–17), perhaps most importantly of all drawing attention to the fact that the original model was the basis of an argument that was more geopolitical than strictly linguistic. The linguistic populations concerned are highly varied and only fit into a three-tier diagram if our descriptive intentions are limited. But to describe the overall development and spread of English in a manageable fashion risks excessive simplification and misinterpretation. Kachru has defended the model, essentially claiming that it does what it is supposed to do and should not be over-interpreted, but this merely highlights our point: it is notoriously hard to count users of a language and with English in its current forms this is particularly so. Simply assessing the numerical extent of English is thus a tendentious business.[2]

Jenkins correctly emphasises the danger of ignoring fundamental complexities of use: for example, in the "Outer Circle" there are myriad realities of English speaking abilities and habits (*Global Englishes* 16). In India, social classes enjoy different access to English and use the language in very different ways, and their varieties of English reflect their first languages, while in Singapore a much greater proportion of the population speak English and it operates in a more homogenous linguistic context. Both of these countries are to be found in the same "Outer Circle" category. And all this when we are still supposed to be attempting a simple headcount of English speakers. The truth is, of course, that it is impossible to quantify something before we have identified and categorised it unequivocally, and English presents severe problems in this regard.

The most commonly employed solution is to rely on census data and make arithmetic calculations. While these may be of some limited use if employed with caution and caveats, they risk throwing analyses of the present and future status of English well off the mark. Authors invariably use *Ethnologue: Languages of the World* (Simons and Fennig) as a source of data, or rely directly on "the two Davids", Crystal (*English as a Global Language*) and Graddol (*English Next*), in order to have some estimate as to numbers, but here we must bear in mind not only their own warnings as to the accuracy of their figures, but also the *historical* dimension: all figures pertaining to numbers of speakers are, by definition, out of date, and this is hugely significant if we are proposing a dynamic picture of the growth and spread of English.[3] Census data are already out of date when used by authors (censuses are usually carried out once every ten years at best), and this process of ageing goes on as publications are used and cited. Data that are ten or 15 years old may seem perfectly acceptable at first glance, but when we look at "the rise of English" from the 1980s to the early years of this millennium and accept the supposedly dramatic speed of spread, then data even five years old might pose problems. Indeed, Montgomery cites the growth of scientific publications in English that moved from 70 per cent in the 1980s to as much as 90 per cent in the 1990s "in some fields" (11) to emphasise a fast-developing situation.

The *quality* of the data is also in question as linguistic information provided by a census is invariably self-reported with no other control mechanisms to make the numbers more reliable. When we consider the descriptions of the kinds of speakers we are attempting to quantify (native or non-native? L1, L2 or other? etc.) and the uses they claim (or we assume) they make of the language, we are again faced with complexity that can be simplified only at some peril. However, if we wish to describe the current state of affairs, and construct hypotheses about the development of English, we have to take our courage in our hands and use the figures we have at our disposal. Kachru's three-circle model and the work of Crystal (*English as a Global Language*) and Graddol (*The Future of English?*; *English Next*) do give us some basis for observing and understanding the emerging or emergent roles of English, as long as they are treated with the requisite cautions and provisos.

III The roles of English

To appreciate the real role of English requires exactly this kind of awareness of complexity and tentativeness of approach. Studies dealing with the spread of

English need not only to posit some kind of "ballpark figure" for speakers, but also to try and give a measure of the influence of the language. Kachru's model immediately emphasised this aspect by referring to *norm-providing*, *norm-developing* and *norm-dependent* circles of English use, corresponding to the inner, outer and expanding circles he proposed. Critics might question the precision of "fit" of geographical entities and sociolinguistic notions, but the issue is at least laid bare: we are not simply dealing with a numbers game when we consider the chances of survival or development of English as a global language and as the effective depository of various global discourses. Influence is naturally very hard to assess, but certain aspects of global language use are worthy of consideration. Ammon suggests various forms of classification that take into account factors other than raw speaker numbers as ways to understand the relative influence and roles of world languages. Perhaps the most striking table lists languages based on the total GDP of their speakers (what he refers to as "economic strength"). Again, the caveats as to the reliability and age of the data apply, but we can suggest that if English is clearly ahead in this table (it almost triples its nearest rival, Japanese, and has extended its lead between 1987 and 2005), then economic power may well be on its side, consolidating if not defining its global role. We might, however, also be tempted to suggest with Ammon that "the poor predictability of economic events … are reasons to be cautious about predictions concerning language status and function" (119).

Another simple but potentially significant indicator might be the geographic spread of the English language (Ammon 112, Table 4.5): present in all six continents and in 50 countries, the next closest language in this regard being French (five continents, 29 countries). The emergent language thus has a truly worldwide role in the most literal sense and this may be more important for its permanence than economic power, adding to an aura of omnipotence coming from its ubiquity. This association with wealth, power and effective presence in virtually all communities around the globe will become important when we consider international discourses below.

IV What language and whose language?

But what language are we talking about exactly? English might have global reach and a global role, but is it the language of Shakespeare? The Queen's English? Or General American? One of the most serious problems with the quantification of English and assessments of its importance is an assumption of homogeneity. Native speakers of English have been mentioned as reaching between 360 and 380 million in total, but this figure has only relative importance when we consider the genuine differences between American, British, Australian, Canadian and Indian varieties, to name just the most numerous. Again, the picture we are faced with is complex: there are more non-native than native speakers of English, but there are also varieties within the native-speaker version coexisting and presumably influencing one another. The relationship of native and non-native forms of English will also be subject to greater and subtler variation than has been assumed as the norm-providing inner circle does not produce consistent norms itself.

And this begs the question of the native speaker herself. What constitutes a *native speaker*? Davies, among others, has underlined the difficulties associated with the term, and MacKenzie summarises the issue succinctly (6–9). Are native speakers born, and so guaranteed to remain "native" for their whole lives, or can they risk losing that mystical birthright by prolonged absence from the mother country? Does a native have special rights as regards the language? Is it possible to acquire a language to a level that is on a par with the native speaker? Is the cultural component of a language only available to a born-and-bred speaker?[4] Do non-natives sometimes know the language better than their native peers? Pinner analyses this confusion of issues briefly and underlines the "uselessness" of the term while, quite rightly, admitting its continued currency (44–47). In an age of international travel, migration and dislocation, it is reasonable to imagine that the border between native and non-native will only get fuzzier. For us, it is enough to stress that the interface between the English of the United States or Great Britain on the one hand, and international varieties on the other, is far from unproblematic and will most probably have an effect on the future development of the language.

The native/non-native debate has many facets pertaining to the way English is taught, the norms it may have, and the wider issue of "ownership". But it is in regard to international forms of English that the native speaker becomes a point of contention and perhaps pivotal in the role of English and its sustainability. The most confident assertions as to the international role of English project it as an international language that is potentially quite different from the home-grown varieties of America or Britain. EIL (English as an International Language) or ELF (English as a Lingua Franca) are conceptualisations that attempt to explain the new role English has gained in the last 30 years or so. Although some might insist on differences between these two, they are similar as regards the international role of English, indeed, so much so that Ostler equates them (276). Essentially, they are similar in their claim to accommodate second-language speakers of English with a tongue that is not culturally specific or bound to one people's history, but instead offers an equal playing field to all users, enabling faster, easier communication and reducing misunderstanding and (to some extent) the effort of learning. The debate around ELF has been intense, and so it is perhaps useful to concentrate on this in order to examine the future sustainability of the role of English.

V A lingua franca?

Proponents of English as a lingua franca are increasingly numerous (e.g. Jenkins *English as a Lingua Franca* and *Global Englishes*; Seidlhofer; MacKenzie, who offers an interesting and challenging critique). It is customary now to refer to the "lingua franca movement" and an annual international conference devoted to ELF had its eleventh edition in the summer of 2017. While there are strands

strands within the concept of ELF, the basic claims are clear: English is a language spoken more by non-native than native speakers. This means ownership should pass to this majority population, resulting in English being endo- rather than exo-normative with grammar, syntax and vocabulary growing naturally as products of use and mutual understanding by second-language (L2) speakers and not imposed by exo-normative behaviour from the traditional heartlands of English. There is an inevitability to this change of arbitration in the development of English precisely because of the weight of the majority (what we might term the "democratic argument"). There is a moral obligation on the part of native speakers to modify their language, avoiding the use of excessively idiomatic forms that are obscure and not transparent to non-natives, and an expectation that natives will relearn English by becoming capable of communication in the lingua franca as well as their local mother-tongue variety. There is also insistence on the reduction in importance (and power) of traditional varieties; a laissez-faire attitude to error (perhaps more correctly termed *variation*?) and stress on the fundamental importance of understanding and being understood, along with an expectation of tolerance of the other's attempts to communicate (the so-called "let-it-pass principle"), even to the extent of accepting code-switching within an ostensibly International English or lingua franca situation.

The debate about lingua franca is surprisingly acute and touches on issues that will directly affect the sustainability of English. Perhaps it is not too much to say that the way English develops, either as an international, largely culture-free lingua franca or as a group of varieties with varying degrees of authority, prestige and power, will determine its future sustainability. If the idealised lingua franca model is not unrealistic and so the demotic variety used by around two billion people for simple, clear and largely effortless communication takes hold and becomes an established norm (or group of norms), then we can see a potentially stable future for English and one that could then contribute on a long-term basis to international understanding, and to discourses regarding climate change and other environmental issues, world economic and scientific policies or education and artistic expression. If instead we can see that the lingua franca model is unrealistic or flawed, then the sustainability of English is very much in doubt, depending as it would on colonial residues and economic power, risking accusations of imperialism (Phillipson, *Linguistic Imperialism* and *Linguistic Imperialism Continued*) and so being subject to the vicissitudes of geopolitical developments, attaining a world role only to lose it when the historical conditions change.

In language, prediction is of little value and we do not know whether ELF will in some way prevail and take on the global role etched out for it, encouraging and facilitating communication. What we can do, however, is understand the present better and analyse the features of ELF that render it likely or unlikely to fulfil its promise. There are tensions in the ELF model proposed by Jennifer Jenkins and other theorists: the conception of error as variation or creativity, while perfectly legitimate from a principled point of view, may not possess adequate social reality. Error is, of course, a socially negotiated (or imposed) feature of language and the hope that the traditional English-speaking nations will forgo their exo-normative role is perhaps optimistic. The democratic argument that the greater number of non-native speakers will be able to impose new norms, or greater flexibility upon international English is attractive, but by no means demonstrated. Jenkins quotes Brumfit on the effects of use in determining the development of language (*English as a Lingua Franca* 5), but the strong version of the argument is essentially hers: there is defensiveness in native speaker normative claims "at a time when the numerical balance ... is shifting so dramatically in favour of non-native speakers" (*English as a Lingua Franca* 44). But there is little hard evidence in linguistics that indicates the fundamental or decisive importance of numbers in defining language change. What seems natural and somehow ethically right is hardly scientifically proven. Instead, we could construct an equally convincing argument for the role of elites in promoting or effecting alterations in a language, even if this is done unconsciously. The desire to engage economically with the more powerful, or to emulate their linguistic practices would likely affect the speech of the mass of users. Jenkins herself mentions Wolff's study on non-reciprocal intelligibility (67), which underlines the importance of power and economics in linguistic behaviour. Yet again, the picture is probably much more complex, and we should refrain from assigning a pivotal role in linguistic changes to one simple factor.

Connected with this is the strongly-held notion that users of a lingua franca are "all in it together" and so will be supportive of each other's efforts to communicate. The "let-it-pass principle" is an aspect of this and Jenkins' Lingua Franca Core (a very brief summary is presented in *Global Englishes* 91), which essentially attempts to reduce phonological difficulty in English by eliminating superfluous and difficult-to-produce sounds and ensuring mutual intelligibility through certain phonetic and prosodic features, is another. The emphasis on the primacy of comprehension is clear, and, at first glance, seems reasonable.

The co-operative idea brings us to a further challenge for a lingua franca, however. Ostler states that a lingua franca has to eliminate cultural content in order to really be effective: "denial of any cultural overlay to a language is one mark of an established lingua franca" (14). But this de-cultured language has real communicative potential only for limited interactions. If the conversation is largely transactional, then the lingua franca model can hold. If, on the other hand, the communicative purpose is more complex, we might question the efficacy of this even playing field that tries to impose on no one. Language has more than an information-bearing function (cf. the three functions, *informational*, *social* and *textual*, in Halliday). Language is used for social interaction, presenting and maintaining identity and reaffirming hierarchies and social structures. A lingua franca that has been "purified" of its historically accreted peculiarities and implicit cultural messages will be potentially very effective for simple, clear-cut communicative situations but much less so for subtle social interaction. The risk is that we invest in English as a lingua franca, hoping to facilitate international trade, debate and understanding but in doing so we lose some of the features that have made English so popular today. "Neutral language does not exist" (Ricoeur 162). Ridding English of its colonial and imperialist past obviously seems highly attractive, but if this entails emptying it of a substantial part of its pragmatic capabilities, then we might lose more than we gain.

Put simply, the question is whether a lingua franca can support complex (even hostile) linguistic interaction (e.g. negotiation, political argument, philosophical analysis, etc.) and whether it can express the nuanced ideas and feelings of its users. Can you produce a Hollywood film in a lingua franca? A pop song? A video game? These are all artefacts that depend on the fuzzy, culturally layered aspects of language and often they deliberately do *not* help the listener or participant with tolerance of difference in form or a keen dedication to understand the other. Precisely the contrary: grammatical variation of form may be permitted, but is often replete with meaning (as a social signal of identity and belonging, or an intensifying marker), rather than being a morphological curiosity to be accepted a priori. As sociolinguistics has repeatedly discovered,[5] free variation is rarely free or without social significance, and deep linguistic knowledge and cultural awareness are required to interpret it. With a lingua franca there is always the risk that a message will be perfectly understood referentially, but its social and pragmatic content will be lost. It is perhaps redundant to say that this "secondary meaning" is often far more important.

The English as a lingua franca movement has not ignored the issue of culture-based linguistic competence; instead it has consciously brushed it aside as being of little relevance to the project at hand. Indeed, de-culturating English can be seen as an implicit aim of the lingua franca movement, probably in response to critiques of the imperialistic element in the spread of English in the post-war period (e.g. Phillipson's concept of *Linguistic Imperialism*). Jenkins, quoting Phillipson, provides in tabular form a clear dichotomy between what is called the "Global English Paradigm" (referring to the English exported and imposed by the United States and the United Kingdom, which is exo-normative, assimilationist, accepting British and American linguistic norms and viewing the native speaker as the model for learning) and the "World Englishes Paradigm" which supports diversity, respects local linguistic norms and has the good ESL user as target (*English as a Lingua Franca* 19).

The political content of this approach may be very attractive and convincing, but for us the question is whether this choice will enable the lingua franca to survive and prosper or whether it leaves a fatal flaw in the system: is a cross-national, culturally open linguistic core adequate for the needs of the international community and does it have the capacity to be sustainable? There are two aspects of this question that require elucidation: the pragmatic efficacy of any international language and the contextual elements that will play a part in its continued use.

VI Pragmatic considerations

Perhaps the most under-developed area in the English as a lingua franca debate is that of pragmatics. We have already mentioned the transactional nature of many lingua franca interactions (as reported by Jenkins, Seidlhofer and many others). Essentially the avowed aim is to communicate, to get the message across as clearly and painlessly as possible, with the usually unspoken assumption that this will result in reduced effort and cost on the part of the lingua franca learner in acquiring the global ability to communicate. This is vital for a lingua franca: it has to represent a good deal to its users or they will simply go elsewhere (Ostler 286).

But wo/man cannot live by trade alone. In reducing the intricate, characterising elements that may be tricky to learn and seemingly of little surrender value, the risk is to lose sight of how a language works. Not just words as tools to say something, but how we say it, the effects our contributions can have. And these pragmatic signals are not just there for the speaker to utilise, they exist for the listener or reader to exploit to aid understanding or to influence perception. Even the clearest, most basic exchange of information has its pragmatic aspect (expression or reaffirmation of status or power on one side, and need, recognition or appreciation on the part of the respondent). Pragmatics here are much more complex than researchers seem to appreciate. Besides conscious intention on the part of the speaker/writer there may be implicit or even unconscious aims underpinning an utterance. In the same way, reception may be influenced on many levels, with interpretation affecting perception that can be based on all sorts of assumptions, previous knowledge and prejudices. Garrett's work on language attitudes underlines unequivocally the importance of accent and hidden scales or indexes of personal preferences in decoding or interpreting messages. Miller puts it succinctly: "Being understood in the L2 is not sufficient, 'sounding right' is also important" (291). We can never ignore the complexity and layeredness of these linguistic interactions. Identity is presented, asserted or reaffirmed, and social norms are enacted and so re-established. The question of course is whether a lingua franca can do this and do it well enough and in enough diverse contexts to be perceived as useful by a critical mass of users to be sustained in future. If instead it is perceived as offering "reduced linguistic capital" (Prodromou 250, following Canagarajah), then it will either dwindle, be reduced to very limited contexts or play the eternal second fiddle to US and UK varieties.

The risk is that ELF becomes a low-density or low context project in its efforts to facilitate international and intercultural communication. Language that is highly informational and geared towards relatively simple, shared goals avoids cultural curiosities, rituals and formulae and so runs the risk of imposing low-context cultural norms.[6] This in turn makes the language itself highly dependent upon external context in order to be communicatively successful. While this is in no way a weakness of a particular language system, and is perfectly reasonable in a lingua franca, it does naturally render the language particularly vulnerable to cultural changes, as we shall see later. It is also worth noting that the attempt to achieve language neutrality, while praiseworthy, is in itself barely sustainable: a language is rarely if ever neutral and each interaction that occurs will contribute to the creation of narratives that will inevitably take this neutrality away.[7] A more hostile observer might also comment on the danger that any push for an international role for English merely falls into the trap of mouthing neoliberal platitudes that are ultimately false: the claim is often made that English is somehow free of political bias and liberating (Dunton-Downer, "Introduction"), or de facto the language of modern science (Montgomery, especially Chapter 3). The lingua franca movement has striven hard to avoid becoming a mouthpiece for an imperialist or post-imperialist agenda (indeed, in abandoning native speaker norms it aims at exactly the opposite perhaps), but this effort will be in vain if a low-culture interactional pattern is adopted, as this fits closely with neo-liberal ideas.

It is perhaps unfortunate that most of the research effort into the development of ELF has been focused at the lexico-grammatical level, because the issues we have just mentioned are to be found at the level of discourse. Discussions of pragmatics or the holistic view of interactions (using conversation analysis and methods from linguistic anthropology) are essentially issues of discourse. It is here that ELF will prove to be sustainable or will fall away as something not aspired to by the estimated one and half a billion people who are learning it at present (Montgomery 6). And discourse is where, of course, reality bites into our linguistic understanding. Analyses of pragmatic choices and interpretations are really social and political constructs and depend on our appreciation of discourse as an interaction of language and context. It is at the discourse level that we can assess whether a particular language is "working" for a community (Blommaert, *Sociolinguistics, passim*). It does not matter what I or you or Barbara Seidlhofer think about an instance of deviation from NS norms, whether it is an example of linguistic creativity, an acceptable variation or a crass error that compromises comprehension. Obviously, this is a subjective judgement. What matters is the whole discourse (immediate context, the power relations, the society as a whole), the variation it is posited in, and the effects it has on participants. When we bear in mind the move in Critical Discourse Analysis towards an understanding of the internationalisation of shared discourses (Fairclough), their fundamental import-ance as a way of dealing with the complexity of language use that we have men-tioned repeatedly in this chapter is even clearer.

Discourse has history (indeed it may be said to be a product of the cumulative effects of innumerable historical events) and this history provides the narratives that create our understanding and limit our linguistic choices (Blommaert, *Discourse*, Chapter 6). "People speak *from* a particular point in history, and they always speak *on* history" (Blommaert, *Discourse* 126, italics in the original). A lingua franca may try to excise itself from this burden of accumulated experi-ence but it is unlikely to succeed, despite the best efforts of its champions, because the narratives that are so influential are often implicit and unconscious, and probably different for people around the globe. "Group identity is based on important narratives and the language in which they are told" (Edwards 254). Repertoires are vital elements of discourse construction and they are by defini-tion intricate, subjective and localised. Now, ELF is supposed categorically not to be a homogenous new form of English imposed on all just because it is easier (Jenkins, *English as a Lingua Franca* 20), but if not then it needs to negotiate a detailed and variegated terrain. Can a language sustain itself as a promise "to be all things to all men" when the danger is of miscomprehension in the interaction of localised forms (see Pennycook for localised specificities of use)? How can competing narratives and social conventions, implicit repertoires and different varieties really communicate across the globe (or in low-context situations such as tweets and blogs)? A sustainable version of ELF will have to resolve this conundrum.

VII A universal language?

However, the doubts expressed regarding the ELF project risk doing it an injustice. To really understand a lingua franca, we do need to take note of the contexts of its use and the potential pitfalls in interactions, but we also have to be constantly aware of the monolingual fallacy in all its various manifestations. If we assume some form of the "universal character" envisioned by John Wilkins (mentioned in Montgomery 21), which is entirely neutral and favouring no particular group, linking scientific progress to the spread of peace and shared values, then, for all our idealism, we are in truth falling into a modernist trap. Having a single universal tongue would only be superficially attractive. A single global language would represent a threat to our linguistic patrimony (the statistics for language death are always startling, if again they are notoriously hard to verify), reducing the 6,000–7,000 dialects we now recognise[8] to irrelevance and extinction. This would also mean the loss of precisely that *historical* feature of language mentioned earlier: some experience is only codified in a specific language and may be lost when the last native speaker disappears. The blithe assumption that everything can be expressed equally well in every tongue is also highly improbable. A fully developed language can express any idea (any gaps in lexis, for example, do not represent a limitation for a language as it is relatively easy to steer around them for communicative purposes), but this is not to say that all of these ideas can be communicated with the same level of rhetorical skill, nuance and pragmatic effect. Some narratives that add meaning to linguistic choices might cease to be available. In a more obvious way, we could lose certain idiomatic expressions that have no equivalents in other languages but are highly effective, both referentially and rhetorically.[9] It is unlikely that a one-size-fits-all solution to international communication will be beneficial or sustainable. Quite the contrary. It is more probable that the imposition, be it through economic advantage or political power, of a single dialect would merely result in resentment or at least a feeling of inadequacy: not only of the non-native speaker struggling with her/his inevitable language deficit, but also an inadequacy of the dialect itself in being able to express realities in specific contexts. There would be economic gains of scale and savings in translation costs if there were to be one monolithic lingua franca, but the losses would be inestimable.

The ELF movement is not monolingual in this way, in contrast to some other global English projects. From the outset, it was multi-lingual in encouraging code-switching even during ostensibly English conversations and this has developed into the encouragement of translanguaging: the active employment of varied linguistic resources even in the completion of a single task. Developed in classrooms in Wales (an interesting case of a country that is at once at the very centre of Kachru's Inner Circle and at the same time living with a linguistic landscape that involves dominant NS English and a revived local tongue that is replete with historical, cultural and political meaning), translanguaging demands active employment of varied linguistic resources and fits with a bilingual or multilingual approach: for any language policy to be truly sustainable it should encourage high levels of *lingual* ability, not near-native competence in one dialect that happens to be on top at present.

This lingual skill, involving a high level of linguistic awareness and a broad range of linguistic experience, entailing, as it presumably will, a developed capacity for translation and subtle awareness of cultural norms, might also reset the balance in the native/non-native debate. Indeed, Montgomery goes so far as to say, "The real casualty from the spread of global English may well be the native speaker himself" (65). A monolingual, highly accentuated speaker from the United Kingdom may be seriously disadvantaged at work or socially[10] if forms of communication become more interlingual and depend on trained dexterity in more than one code.

Thus, the hegemony of the English language, for all its numbers and influence, may not be sustainable at all. Graddol is modest in his vision of the long-term growth in English (*The Future of English?* and especially *English Next*) and Ostler provocatively entitled his 2010 work on the subject *The Last Lingua Franca*, predicting the demise of English as the globally dominant tongue and suggesting that current systems of communication have rendered the need for a universal dialect passé. Instead, it is not really a language we should be hoping to preserve but *a linguistic situation*. A world with shared discourses that are not exclusive of communities or currents of thought and allow bottom-up discussion rather than imposed norms from an economic centre. These shared discourses will contain linguistic repertoires that will empower speakers to express their identities and develop them; the language resources will be broad and open, enabling richer communication and recognising varieties with creative, other morphologies and syntax. "We produce language as a result of our local practices" (Pennycook 41). Rather than the inauthenticity of language students studying texts generated in the United Kingdom, downloaded and brought to a classroom in another continent and thus utterly divorced from any but the most generalised context (Pinner, Chapter 4), this sustainable version of language, which might be recognisably English at least in percentage terms, will be anchored in contexts that are both virtual and concrete, connecting the world with the local and particular.

Context was mentioned earlier in connection with lingua franca: they are inevitably highly context-dependent because less can be encoded into their lexico-grammatical forms. The context today is rapidly changing: Brexit in Europe presents the possibility that the European Union will no longer have English as one of its official languages. Yet English will still very probably be used heavily as the lingua franca of a great deal of Union business. Native speakers will be largely absent (only the Irish will be native speakers at many European meetings, and they naturally have a more nuanced linguistic identity) and so an endo-normative dialect of English really could develop. At the same time, we are experiencing a period of migration flows that will transform the linguistic context. Presidential policy in America is at least promised to be more isolationist and less engaged in foreign policy. For any language to sustain itself in this changing context it must maintain its appeal by remaining useful and so worth the investment, and flexible enough to be able to give voice to peoples' concerns.

The hegemony English has enjoyed is still powerful: witness the repeated research evidence showing students' desire to learn British or American English in preference to other less culturally-weighted forms (e.g. Jenkins, *English as a Lingua Franca*, Chapter 6), but a combination of war, environmental change and migration might render this situation untenable. Even language policies are not all going in English's favour: China has recently cut the importance of English in the infamous *Gaokao* examination from 25 per cent to 20 per cent and increased the weighting of Chinese (Pinner 51), and although not without controversy, this measure was largely popular.

If English is to continue in its central role in global affairs it will need to disassociate itself from the discourses of neoliberalism and "traditional" growth-based economics. Indeed, English will have to develop the discourses of *degrowth*. Predictions are of little value, but if English continues to be closely associated with ideas of freedom, efficiency and capitalist ideals it might lose exactly this ineffable link to the future that has made it the language to be emulated by so many around the globe. In a period of revived nationalism and international (e.g. environmental) threats, the language that succeeds will be the one that comfortably expresses specific identities and realities in an internationally meaningful way. A tall order we might think, but if we recognise the language as emergent (and so changing continually, as of its very nature) and highly context-dependent, then there is no reason why a form of English should not continue to play a significant global role well into the future. Sustainability is certainly an emergent concept, having constantly to adapt to changing, environmental and political and economic issues. It will require an emerging language, or portfolio of languages, to sustain it.

Notes

1 Montgomery includes an entire chapter on the current status of English as the language of science and charts the rapid rise of publication in English as compared to German, French etc.
2 Perhaps the most reliable generally agreed figures come from Crystal (*English as a Global Language*) and Graddol (*The Future of English?* and *English Next*) and are reported by Montgomery succinctly (27–28): between 360–380 million native speakers. About the same figure use it as a second language (which implies a certain level of institutionalised importance); and between 800–850 million using it as a foreign language, with very varying levels of skill, presumably.

3 Both Graddol and Crystal published works in the 1990s that were effectively updated within ten years precisely because the spread of English had been so rapid.

4 The debate about native speakers especially as models and teachers for learners of English has been significant and informative. See in particular Medgyes; Phillipson, *Linguistic Imperialism*; Holliday (for critiques).

5 Labov's famous research into the *rhortic r* in New York, or pronunciation variations in Martha's Vineyard are classic examples: see Spolsky for a brief and clear account of free variation in linguistics.

6 The concept of low- and high-context cultures originates with Durkheim: see Jeffrey C. Alexander 112.

7 See Pennycook 41: "grammar is not a set of norms that we adhere to or break, but rather the repeated sedimentation of form as the result of ongoing discourse".

8 *Ethnologue* (Simons and Fennig) counts 6,909 different languages, but I would urge caution: it is notoriously difficult to cleave one dialect from another (even in Europe, which is well studied and familiar to linguistic enquiry, there are differing opinions as to status). Of this figure, around 95 per cent are indigenous languages with between 1 and 10,000 speakers.

9 For idiomaticity in regard to ELF see Prodromou: he rightly stresses not only the information-bearing effectiveness of idioms, but also their social character in bonding user and recipient.

10 For example, a recent article on the BBC website: "You Need to Go Back to School to Learn English" (Morrison).

Bibliography

Alexander, Jeffrey C. *The Civil Sphere*. Oxford University Press, 2006.

Ammon, Ulrich. "World Languages: Trends and Futures". *The Handbook of Language and Globalisation*, edited by Nikolas Coupland, Oxford, Wiley-Blackwell, 2013, pp. 101–122.

Blommaert, Jan. *The Sociolinguistics of Globalisation*. Cambridge University Press, 2010.

Blommaert, Jan. *Discourse*. Cambridge University Press, 2005.

Brumfit, Christopher. *Individual Freedom in Language Teaching*. Oxford University Press 2001.

Canagarajah, A. Suresh. *A Geopolitics of Academic Writing*. Pittsburgh University Press, 2004.

Coupland, Nikolas. *The Handbook of Language and Globalisation*. Wiley-Blackwell, 2013.

Crystal, David. *English as a Global Language*. 2nd edn, Cambridge University Press, 2003.

Davies, Alan. "The Native Speaker in Applied Linguistics". *The Handbook of Applied Linguistics*, edited by Alan Davies and Catherine Elder, Blackwell, 2006, pp. 431–450.

Dunton-Downer, Leslie. The *English is Coming!* Touchstone/Simon and Schuster, 2010.

Duranti, Alessandro. *Linguistic Anthropology*. Cambridge University Press, 1997.

Edwards, John. *Language and Identity*. Cambridge University Press, 2009.

Fairclough, Norman. *Critical Discourse Analysis*. 2nd edn, Longman/Pearson Education, 2010.

Garrett, Peter. *Attitudes to Language*. Cambridge University Press, 2010.

Graddol, David. *English Next*. The British Council, 2006.

Graddol, David. *The Future of English?* The British Council, 1997.

Halliday, M. A. K., and Christian Matthiessen. *An Introduction to Functional Grammar*. Hodder Education, 2004.

Holliday, Adrian. "The Role of Culture in English Language Education: Key Challenges". *Language and Intercultural Communication*, vol. 9, no. 3, 2009, pp. 144–155.

Jenkins, Jennifer. *Global Englishes. A Resource Book for Students*. Routledge, 2015.

Jenkins, Jennifer. *English as a Lingua Franca: Attitude and Identity*. Oxford University Press, 2007.

Kachru, Braj B. *The Other Tongue. English Across Cultures*. University of Illinois Press, 1992.

MacKenzie, Ian. *English as a Lingua Franca: Theorising and Teaching English*. Routledge, 2014.

McArthur, Tom. *The English Languages*. Cambridge University Press, 1998.

Medgyes, Péter. *The Non-native Speaker Teacher*. Macmillan, 1994.

Miller, Jennifer. "Identity and Language Use: The Politics of Speaking ESL in Schools". *Negotiation of Identities in Multilingual Contexts*, edited by Aneta Pavlenko, Multilingual Matters, 2004, pp. 290–315.

Montgomery, Scott L. *Does Science Need a Global Language?* University of Chicago Press, 2013.

Morrison, Lennox. "You Need to Go Back to School to Learn English". *BBC Capital*, 16 December 2016, www.bbc.com/capital/story/20161215-you-need-to-go-back-to-school-to-relearn-english. Accessed 23 March 2017.

Ostler, Nicholas. *The Last Lingua Franca. English until the Return of Babel*. Allen Lane, 2010.

Pennycook, Alastair. *Language as a Local Practice*. Routledge, 2010.

Phillipson, Robert. *Linguistic Imperialism Continued*. Routledge, 2009.

Phillipson, Robert. *Linguistic Imperialism*. Oxford University Press, 1992.

Pinner, Richard S. *Reconceptualising Authenticity for English as a Global Language*. Multilingual Matters, 2016.

Prodromou, Luke. *English as a Lingua Franca*. Continuum, 2008.

Ricoeur, Paul. *The Rule of Metaphor. The Creation of Meaning in Language*. Routledge, 2003.

Seidlhofer, Barbara. *Understanding English as a Lingua Franca*. Oxford University Press, 2011.

Simons, Gary F., and Charles D. Fennig, editors. *Ethnologue: Languages of the World*. 20th edn, SIL International, 2017, www.ethnologue.com. Accessed 23 March 2017.

Spolsky, Bernard. *Sociolinguistics*. Oxford University Press, 1998.

Walker, Jay. "The World's English Mania". *TED Talks*, www.ted.com/talks/jay_walker_on_the_world_s_english_mania?language=en. Accessed 15 December 2016.

Wilkins, John. *An Essay towards a Real Character, and a Philosophical Language*. Samuel Gellibrand, 1668.

Wolff, Hans. "Intelligibility and Inter-ethnic Attitudes". *Anthropological Linguistics*, vol. 1, no. 3, 1959, pp. 34–41.

Part III

Sustainable wellbeing via habitat and citizenship

Prologue

Paola Spinozzi and Massimiliano Mazzanti

Sustainable wellbeing eludes the definition of quantifiable targets and takes shape in diverse forms of social development embedded in idiosyncratic local features. Levels and types of co-operation, forms of funding and sharing public goods, interconnections and complementarities between markets and the state, the institutional role of technologies and cultures and its support through policies, political views on knowledge and other forms of social and human capital are factors, among others, which define a given "region". One-size-fits-all approaches would fail to represent the rich and variegated map of "well beings" arising from different contexts, invoking human, natural, and man-made components, thriving on tangible and intangible capital. An empirical map of diverse cases from the bottom up is key to representing a complex scenario.

The theoretical, historical and institutional policy approaches to sustainable wellbeing in Part I and Part II are here thoroughly explored through case studies encompassing South America, Africa, Europe and Asia. The aim is to assess conceptualisation with regard to geopolitical criteria, by selecting specific areas and examining them through the lens of cultural geography, sociology, urban planning, and architecture. In particular, the impact of urbanisation on rivers is examined with a focus on the Belém River and Curitiba in Paraná, Brazil; the effect of urban planning and public policies on life quality and wellbeing is studied in Medellin, Colombia; the management of seismic events in Italy exemplifies how the preservation of cultural heritage can be attached to diverse national, regional and local agendas; why and how cultural policies should be supported by political programmes is related to a developing African city like Saint-Louis du Sénégal; the issues arising from the preservation of cultural memory and the development of cultural tourism are assessed by exploring the Mapuche people from the Araucanía Region in Chile and traditional landscape design in Japan; how a developing economy integrates global sustainability approaches in the development of local economies and sectors is analysed in the case of Vietnam.

The case studies illustrate how sustainability and wellbeing are pursued by an adaptive approach characterised by interactions with the global perspective on the one hand and local idiosyncratic strategies on the other hand. Top-down international views on sustainable strategies are thus presented along with bottom-up approaches that preserve and develop a national/regional response to sustainability and wellbeing. While highlighting many critical points to address along the paths towards a sustainable wellbeing, the diversified case studies offer insights on how to incorporate global perspectives while maintaining local identities. This synergy is fruitful and necessary to implement feasible policies that achieve efficiency, effectiveness, and social acceptance.

Most cities in Brazil have grown in an environmental policy vacuum, resulting in the vast clearing of green areas and the transformation of urban rivers into open sewer lines and garbage dumps. "The Impact of Settlements on Urban River Basins and the Case of the Belém River in Curitiba, Brazil" by Gilda Cassilha, Marta Gabardo, Sylvia Leitão and Zulma Schussel examines this phenomenon in Curitiba, the capital of the state of Paraná, where the built environment started to be controlled according to environmental criteria introduced in the last decades of the twentieth century. The relationships between urbanisation and urban rivers have been studied with a specific focus on urban and environmental legislation, economic assessments and perception of vulnerability.

Urban and social transformation in Medellín has been possible thanks to major investments in public infrastructure and civic architecture. In "Creative Social Innovation and Social Urbanism: The Case of Medellin" Ana Elena Builes Vélez and María Florencia Guidobono examine how public projects founded on the social urbanism model and involving public policies and strategies have provided better connections throughout the city as well as its urban and social tissue. Moreover, over the last decade inclusive and socially equitable spaces have been opened up for the communities living under the poverty line and areas affected by unresolved social and environmental issues.

A major challenge in post-disaster planning is to achieve a balance between ordinary processes of rebuilding and long-term visions for revitalising communities and their wellbeing. Gianfranco Franz's "Long-term Visions and Ordinary Management: Post-Earthquake Reconstruction in the Italian Region of Emilia" examines the difficulties arising from reconstruction processes and policies enacted after the earthquake that struck the valley of the River Po in the Italian region of Emilia in May 2012. The results, limits and uncertainties of rebuilding offer ample ground for investigating connections and gaps between post-disaster reconstruction and wellbeing. While traditionally urban planning is concerned with forecasting and with the spatial and functional growth of the city, post-disaster planning is aimed at finding immediate answers to the emergency phase and fast, cost-effective solutions for reconstructing cities, facilities and infrastructures. Processes of material and physical reconstruction can gradually secure a new life to cities and towns (as evidenced by several cases, among which San Francisco, Messina, Tokyo), but for the communities it is arduous to regain the level of wellbeing and safety achieved before the event. How to rebuild by stimulating the sense of belonging and maintaining the cultural heritage must be regarded as one of the major tasks pursued by post-disaster planning. "In a Prescient Mode. (Un)Sustainable Societies in the Post/Apocalyptic Genre" by Paola Spinozzi shows how, by evoking post-apocalyptic environments, utopian writers can represent sustainable or unsustainable future societies and explore diverse forms of wellbeing, ranging from hyper-technological societies geared towards efficiency and material comfort to post-industrial communities thriving on the integration between the urban space and the countryside.

In "Urban Life and Climate Change at the Core of Political Dialogue: A Focus on Saint-Louis du Sénégal" sustainable wellbeing is connected to civil society, local communities and, more recently, to an active citizenship at the neighbourhood level. Adrien Coly, Fatimatou Sall, Mohamed Diatta and Chérif Samsédine Sarr describe a social movement led by sports and cultural associations through the slogan Set Setal (Make clean). Saint-Louis relies on the network of these associations and their relevance within the community in order to establish District Councils, which function as meeting places for the organisations, the officials elected and the delegates designated by the administration. This structure extends public action to citizens and facilitates the management of climate change issues. Saint-Louis stands out as an emblematic example of a sustainable city created via the direct action of citizens, the involvement of individuals and their impact on urban metabolism. Decentralisation invites further reflections as does climate change in relation to African and global policies, especially after the UN Climate Change Conference held in Paris in 2015.

"Is Ethno-tourism a Strategy of Sustainable Wellbeing? A Focus on Mapuche Entrepreneurs" is based on semi-structural interviews to Mapuche people providing tourism services in different areas of the Araucanía Region in Chile. Gonzalo Valdivieso, Andrés Ried and Sofía Rojo study why the Mapuche people started to provide tourism services; what characterises their motivations and conceptions of tourism as well as their relationships with visitors; whether this is a long-term sustainable activity that can improve their life quality and preserve their culture. Motivations are economic but also connected to the preservation of their culture, the ecosystem, and sense of place. Moreover, these activities give them the opportunity to value and share their knowledge within their own communities and with visitors. Interviews also show that this is a complementary activity which allows them to preserve their traditional farming labours and be connected with the land, in accordance with their worldview and philosophy.

After the tsunami disasters in Southeast Asia, 2004, and in North-eastern Japan, 2011, landscape design and environment control have become crucial global issues. A contemporary dilemma seems to exist between the application of technologies globally developed to subdue Nature, or to coexist with Nature in ways that bring to mind the skills of a good sailor who knows how to take advantage of the tide and wind. In "Japanese Castle Towns as Models for Contemporary Urban Planning" Shigeru Satoh explains how this dilemma may be overcome by blending modern technology and traditional techniques such as those applied to the traditional landscape design in around 300 Japanese castle towns from the Edo period. The Chinese theory of Feng shui, literally meaning wind-water, has been developed all over Asia in Vietnam, Japan, Korea, and Taiwan. In Japan the principle of Shang shui, literally meaning mountain-water, acquires a broader significance, as landscape design theories present specific local characteristics and share a philosophy of harmony. While respecting traditional principles of urban design, communities strengthen their public image and citizens are actively involved in collaborative projects of sustainable regeneration.

The trade-off between short-term economic growth and long-term sustainability makes it challenging for countries to commit to their public policies on sustainable development (SD). "Vietnam's Pathway towards Sustainability: Stories Half-told" illustrates how the fast growing, resource-based economy must face critical environmental obstacles in Vietnam and make use of limited resources to address such concerns. The country's participation as a signatory member of the conventions established in the Earth Summit on Environment and Development has remained active over the last 25 years. Policies and institutions have been rectified to regulate and support sustainable development in Vietnam ever since. However, the enforcement of such policies can still be significantly improved. Nhai Pham and Dan Tong examine the relationship between economics and the environment. Particularly, they highlight the costly trade-off between rapid economic growth and severe environment pollution as well as the excessive disturbance to the environment due to flawed development plan(s). Despite strong commitment, backed up by an extensive legal and institutional framework, a gap between policies and practices still remains in Vietnam's pathway towards sustainability.

Policies are not enforced in a vacuum and their success certainly depends on the proper integration of theories and forms of disciplinary thinking. How concepts and frameworks may work in reality becomes clear in the case studies. Failures can occur, but a robust interdisciplinary approach does reduce the risk of taking unsustainable paths that cannot be reversed. The possible limits of self-enclosed working hypotheses are overcome through interdisciplinary approaches, which can manage a trial and error process better than narrow approaches, given their "portfolio diversification" and intrinsic complementarity of components. Among other factors, the various case studies highlight the centrality of intangible assets behind the achievement of wellbeing. Education, social capital, social cohesion, trust are as fragile as they are necessary to support the complex challenges posed by development and sustainability. Technologies do not produce innovations without intangibles based on human capital. The emergence of modes of dialogue and socialisation lead citizens to identify and share sustainable goals. A case by case mapping of institutional, cultural and economic contexts and experiences exemplifies forms of present wellbeing and points to sources of future wellbeing.

III.1 The impact of settlements on urban river basins and the case of the Belém River in Curitiba, Brazil

Gilda Amaral Cassilha,
Marta Maria Bertan Sella Gabardo,
Sylvia Ramos Leitão, and
Zulma das Graças Lucena Schussel

I Introduction

In the twenty-first century, the sustainable development of cities has been a recurring theme in urban studies, which have generally addressed the topic from a multidisciplinary perspective. Urban rivers have also been the subject of numerous studies and continue to be of importance, as new theories have expanded or altered existing diagnoses, providing a better understanding of the relationship between rivers and cities.

Criticising economic growth because it produces prosperity is not significant; rather, the question that deserves reflection is what is understood by "prosperity" (James *et al*. xiii). Questions about sustainability at a time of crisis brought about by unsustainable pressures on the planet suggest there is a need to rethink the fundamental concepts behind sustainability. It is in this context, according to the authors, that the Circles of Sustainability approach allows "sustainability" to be understood in relation to other conditions of human existence, such as resilience, habitability, adaptation, innovation and reconciliation, and to be reconstructed in these relations.

According to Morin, transdisciplinarity is a way of dealing with complex situations. A transdisciplinary approach presupposes that ideas, methods and competencies used in a particular discipline can be applied to another discipline: a new context encompasses different perspectives and, more significantly, the inter-relation between them (5–10). It is against this background that the present study seeks to assess the impact of settlements on urban river basins by selecting the Belém River basin in Curitiba as a case study for exploring the complexity of Brazilian urbanisation.

Like all major Brazilian cities, Curitiba, the capital of the state of Paraná, is facing problems associated with the unsustainability of the built environment caused by land-surface impermeabilisation and the straightening and channelling of its rivers, as well as irregular settlements and slums on floodplains.

International recognition of a sustainable city, which is part of the planning process in Curitiba, requires a rethinking of various definitions of urban sustainability. According to one such definition, defended by Thierry Paquot, sustainable urban development is not an ideal model to be reached – as many authors maintain – but rather a development process that should be flexible, in order to incorporate both the complexity of existing social relations in urban spaces and aspects related to the environmental conditions in these spaces (15–25). Complementing this definition, Ferreira observes that progress in a sustainable society is measured by the quality of life, which comprises health, longevity, psychological maturity, education, a clean environment, community spirit and creative leisure (102).

These definitions – which introduce the concept of planning for the future and a broader approach to urban sustainability – indicate that the issue of sustainability in cities in the southern hemisphere is directly related to the elimination of urban poverty, the capacity for governance and the socioenvironmental role of urban planning. Each of these factors poses various quantitative challenges to public authorities, including housing, sanitation, urban mobility, public transport, management of urban waste, health and education facilities. A particularly important challenge is the growth of substandard settlements in urban areas that has characterised the expansion of these cities.

According to the Instituto Brasileiro de Geografia e Estatística – IBGE, in Brazil there were 6,329 substandard settlements – illegal occupations of land, public or private, characterised by a pattern of urbanisation outside the current standards and by the precariousness of essential public services. These settlements, known as favelas, were concentrated in 323 municipalities and corresponded to 11,425,644 inhabitants (6 per cent of the Brazilian population). Contradictorily, the largest concentration of favelas is in the major cities of the country, where the Human Development Index is considered very high or high, ranging from 0.847 (Florianópolis) to 0.721 (Maceió), being 0.805 in São Paulo, 0.799 in Rio de Janeiro and 0.823 in Curitiba.

This pattern of Brazilian urbanisation has gained speed from the 1950s onwards. Initially driven by an exodus of the lower-income population from the countryside, it has later been fuelled by migration from small towns to large cities. This growth has led to the spontaneous urbanisation of areas that used to be unsuitable from an environmental point of view. The many negative impacts such urbanisation has caused include pollution of rivers, changes in natural drainage patterns and the extinction of native flora and fauna. In the 1960s and 1970s, a legal conservation framework was set up in response to this situation, most notably Federal Laws No. 4.771/65, which established the Forestry Code, and No. 6.766/79, which established the subdivision of urban land.

In the 1980s a new phase in urban management began in Brazil as a result of a series of international conferences on the environment and the intervention of international funding agencies such as the World Bank and the Inter-American Development Bank. Environmental issues had to be taken into consideration in all projects, and environmental impact assessments were mandatory for projects which required funding.

As far as water resources were concerned, even though specific legislation establishing the National Water Resource Management System (Federal Law No. 9.433/1997) was introduced, the legacy of the past remained. In particular, serious drainage problems during periods of high rainfall (floods) caused by settlements on the floodplains of urban rivers had not been solved by structural engineering.

II The circles of sustainability and the case of the Belém River in Curitiba

The method used to assess the impact of urban settlement on the Belém River basin in Curitiba draws upon the Circles of Sustainability, according to which relations in the social dimension are analysed in four conceptual domains: economic, ecological, political and cultural (James *et al.* 10–12).

The economic domain encompasses issues such as the production, exchange, consumption, organisation and distribution of goods and services as well as the value criteria associated with these relations. While economics as a social sub-discipline deals with aspects of the economic domain, traditional economic analysis usually concentrates exclusively on quantitative issues, assessing production and distribution costs as well as opportunities in consumer markets.

Even though the natural environment extends beyond human experience (despite our increasing scientific capacity for reconstructing elements of nature), the ecological domain is taken to be the intersection between the social and natural realms, including the dimension of human engagement with and within nature.

The political domain is defined in terms of practices related to authorisation, legitimation and regulation, where these parameters extend beyond their conventional meanings in politics. Sustainability indicators in the political domain can include conventional indicators related to the State and citizenship, such as citizen participation in decision-making processes about the future of cities.

The cultural domain is defined in terms of practices, discourses and material expressions which, over time, express continuities and discontinuities, points in common and differences in meaning. While the currently prevailing concept of "cultural" refers to popular art or culture, in the Circles of Sustainability approach it is defined more broadly to emphasise standardised expressions of social meaning that extend the concept beyond the aesthetic domain to include community integration.

James stresses that the Circles of Sustainability approach aims to be flexible, modular and systematic. Drawing an analogy with a tool shed, he notes that parts (tools) can be added to or moved around for use in different locales, i.e. the approach can be used as the basis for integrated planning in a city or an urban settlement.

The hydrography of Curitiba is characterised by small rivers that feed into the Iguaçu River, the largest in Paraná. The Iguaçu River starts in metropolitan Curitiba and ends in the Paraná River in Foz do Iguaçu, bordering on Brazil,

Paraguay and Argentina. The Belém River basin has been chosen for two reasons. It starts and finishes in Curitiba, extending over 21 km within the municipality; its drainage basin is completely urban, covering some 87.85 km² in the central region of the city and several of the more densely populated neighbourhoods, or approximately 20 per cent of the total area of the municipality (432 km²). The sub-basin of the Pinheirinho River and its tributary, the Guaíra River, have been chosen for this study of the impact on the Belém River basin because, while they share characteristics with the other sub-basins, Valetão slum, the oldest slum in Curitiba now known as Vila Parolin, is located on their banks.

III The analysis

The following sections describe the results, based on an analysis of the settlement process in the Belém River basin and the unsustainability of this settlement from an ecological, political, economic and cultural perspective.

III.1 The settlement process

The settlement process in Curitiba is deeply connected to the River Belém basin as well as to the sub-basin of the Pinheirinho River and its tributary, the Guaíra River. Since the first settlements in the area where the city of Curitiba now stands, various indigenous populations, including the Guarani, Kaingang and Xokleng, have depended on the rivers for their survival, all the while maintaining a respectful relationship with them. With the passing of time, the area has gradually undergone urbanisation, producing increasingly adverse impacts on the river channels.

In the seventeenth century, the "village", as it was then, was a circle of merely 12 km in diameter, and the Belém River was the main source of water for the population, although it was taken for granted. According to Santos, for Rafael Pires Pardinho, a magistrate of the Portuguese court in 1721, the situation was such that the government was forced to introduce a law obliging the population to clean the river, as the population was using it to dispose of their waste, causing disease and flooding (Santos 42).

Since then, Curitiba has grown significantly, and between 1940 and 1980 the population increased from 127,000 to 1,025,000. The first slum in the city, Favela do Valetão, dates from 1953 and is situated on the banks of the Guaíra River by the railroad in the Parolin neighbourhood. One of the most traditional neighbourhoods in the city, Parolin grew from a community of Italian immigrants in the nineteenth century originally known as the Dantas settlement. The inhabitants of Curitiba saw the increasing poverty in Parolin as synonymous with promiscuity and disease and a danger to society in general, as poverty was associated with frequent floods. The first drainage infrastructure for the Iguaçu, Belém, Ivo, Pinheirinho, Guaíra, Juvevê, Atuba and Areãozinho Rivers was installed in 1963 and included works to straighten, dredge, clean and channel these rivers (IPPUC, 1991, 13–24). In the same year, the first municipal housing department in Brazil, COHAB-CT (Curitiba Social Housing Company), was set up (Trindade 70–73).

In October 1964, in the aftermath of the military coup in March of the same year, a nationwide tender process was launched by the City of Curitiba for the Curitiba Preliminary Urban Plan. The plan was based on a physical, economic and social diagnosis and took into account the first emerging slums in the city, which were close to the railroad and the Guaíra River, as well as at various places along the Belém River. After the Master Plan was approved in 1966, Vila Nossa Senhora da Luz, the first social housing facility in the city, was built. This consisted of 2,500 houses built by COHAB-CT for residents of the Valetão slum in Parolin and Vila Pinto in Prado Velho, on the banks of the Belém River (Leitão 62–63).

The rehousing of slum dwellers reflected the hygienist mentality of the time, which dictated that poorer areas had to be kept at a distance from wealthier neighbourhoods. According to Ribeiro and Cardoso, urban issues went hand in hand with social reform, for which the therapy adopted was "changer la ville pour changer la vie" (77). Because Vila Nossa Senhora da Luz was built 10 km from the original slums in an area without any infrastructure or services, the population that had been rehoused gradually returned to the old slums on the river banks. This led to the Parolin neighbourhood becoming segregated into two distinct areas: "Upper Parolin", a more select neighbourhood on the higher ground, and "Lower Parolin", near the river and close to the slum.

In the 1980s the slum was taken over by organised crime controlled by drug dealers, blighting the lives of the 6,000 people who lived there, more than 45 per cent of the population of the neighbourhood.

In 2006 COHAB-CT carried out a census in the slum so that a property regularisation project could be started. The findings indicated that the slum, which extended over 240,000 m^2, was home to 11,000 people. Of the 1,507 families, 41 per cent lived in wooden sub-dwellings, 70 per cent had access to basic sanitation facilities and 57 per cent used illegal connections to the electricity supply. Just over two-thirds (68 per cent) of the residents were 30 years of age or older, only 63 per cent had completed primary school and 25.7 per cent had formal employment. In all, 70 per cent of youths in the slum were not attending school and 92.5 per cent of families had a household income of at most three times the minimum monthly wage. A total of 42.1 per cent of residents earned between US$90 and 180 a month in today's values, placing them in the extreme poverty bracket, and 15 per cent stated that they were unofficial trash collectors (carrinheiros).

It was assumed that when the land regularisation project was implemented, the majority of the residents of Lower Parolin, around 830 families, would remain there, while 677 families that were at risk or living in unhealthy conditions would be rehoused in other areas available in the neighbourhood, including Upper Parolin. On the one hand, the fact that families would be staying in the occupied areas, as advocated in the City Statute, Federal Law 12.257/2001, generated discontent among the residents of Upper Parolin. On the other hand, the project was awarded the Brazilian Association of COHABs Seal of Merit

Award in 2010 for two main reasons. First, it ensured the involvement of the local community at all stages of the implementation. Second, it provided for environmental rehabilitation of the banks of the Guaíra River and urbanisation of 25 alleys some 1.5 m wide, used as hideouts by criminals and drug dealers. An immediate, but short-lived consequence of reurbanisation was a substantial (57 per cent) drop in the homicide rate in the period up to 2008, after which, however, it started to rise again (Fernandes).

Because of the increasing violence in Parolin, a Safe Paraná Unit (Unidade Paraná Seguro, UPS) was set up there in 2012. The murder rate in the region was 80 per 100,000 inhabitants, more than twice the figure for the city as a whole (39.8) and more than three times the national figure (26.2). The UPSs were based on the Police Pacification Units (Unidades de Polícia Pacificadora, UPPs) in Rio de Janeiro, which sought to occupy areas dominated by crime and return them to residents, as the region was always a strategic point for the distribution of crack and cocaine throughout the city. At first, the slum was occupied around the clock by some 300 military and civilian police. After this initial period, 150 police remained for 30 days, and finally a contingent of 30 police stayed to ensure the public's safety and implement community actions (Peres and Tavares).

Today, ten years after the beginning of the property regularisation project, only 295 of the original 677 families have actually been resettled. The resettling in Upper Parolin of poorer residents from Lower Parolin generated discontent among the residents of the more affluent neighbourhood, first because the value of their properties dropped, second because they were obliged to live alongside with the new neighbours. The cultural shock was heightened by the new residents' habits: storing recyclable waste in their backyards, making kebabs on the sidewalk, listening to loud music, letting their children fly kites in the street and bringing their relatives to live with them. Nevertheless, the residents of Lower Parolin believe that social interaction is still possible as the slum community is organised and has been there for more than 50 years.

III.2 *The ecological domain: vulnerability and risks*

According to Marcovitch, "environmental risks, together with economic and sociopolitical risks, are issues of particular importance in this, the first century of the third millennium" (140). This assertion can easily be observed in the direct relationship between ever-greater urbanisation and the increasing tension between society and nature, with the attendant negative socioenvironmental consequences, such as floods caused by settlements on the floodplains of urban rivers.

Similarly, urban form and density can have social and environmental consequences, an example being high-density informal settlements in risk areas, which are socially and environmentally very vulnerable to climate extremes. According to Marândola, the responses of human settlements affected by these events depend on two factors: the vulnerability and the resilience of the natural, technological and social environment (29–53). For Cutter, vulnerability can be understood as a biophysical risk as well as a social response in a specific geographic area or domain. This can be the geographic space, where the vulnerable people and places are, or a social space, where people are more vulnerable (529–539).

The inclusion of environmental variables in urban planning led in turn to the inclusion of new alternatives for assessing land use and cover, most notably management by watershed. This helps identify areas unsuitable for occupation and areas that are already occupied but require environmental and urban rehabilitation. In this sense, the Curitiba Municipal Master Plan plays an important role by creating environmental protection zones, as certain areas are environmentally very vulnerable.

From 1989 onwards the change in the Forest Code that defined locations at the bottoms of valleys as Permanent Conservation Areas (APPs) had an undesired effect in some cities as it led to the emergence of slums in these areas, which were no longer of interest to the real-estate sector. Many of these slums were highly vulnerable both socially and environmentally: in addition to the risks of landslides and floods to which the residents were exposed, they caused environmental problems that were very difficult to solve.

This situation worsened throughout the 1990s until the early twenty-first century, when the National Housing Agency was set up in the Ministry of Cities. This new agency has worked towards improving squatter settlements by opening up streets, installing urban drainage systems and resettling families living in risk areas. Yet, the impact of urbanisation on rivers can be observed in the increased amount and flow rate of surface runoff caused by land-surface impermeabilisation, the construction of artificial channels (gutters and drainage networks) and channelling of watercourses.

Federal Law No. 12.651, dated 25 May 2012, changed the Forest Code with regard to urban areas, metropolitan regions and conurbations: the width of the strips along the banks of natural watercourses demarcating the flood passage strip should be determined by Master Plans and Land Use Laws according to the recommendations of the State and Municipal Environmental Councils. Hence, it is the responsibility of the municipal authorities to draw up plans for recovering or revitalising strips along river banks and incorporating these into the surrounding urban environment. The City of Curitiba has taken a responsible approach to this issue and introduced specific legislation, although with only limited success insofar as established settlements are concerned. The population density is very high along some stretches of the Belém River, and in several cases, areas that are environmentally vulnerable overlap with others where the population is socially vulnerable. These areas present squatter settlements, which by their very nature are home to low-income families and show greater susceptibility to floods caused by their topographic characteristics, vegetation and level of land-surface impermeabilisation.

III.3   The political domain: failure to comply with urban and environmental legislation

In general, there are many squatter settlements in the sub-basins of the Belém River. These neither comply with the parameters for land use stipulated in the municipal legislation nor satisfy the environmental requirements for permanent conservation strips along watercourses specified in the national legislation. Furthermore, the rivers have been channelled and the banks have become highly impermeabilised to allow roads to be built. In addition to this, disregard for urban land use legislation has led to degradation of the landscape and urban environment in the various neighbourhoods through which the river passes.

Daily practices have a cumulative positive or negative effect on the city, degrading the natural environment and urban landscape and in turn producing negative impacts. The most significant political and legal impacts in the sub-basins of the Pinheirinho and Guaíra Rivers are a result of the relationship between areas where the land cover is impervious and others where it is still pervious. *In loco* research, images and a survey of land-division records in Curitiba City Hall were used to determine the percentage of land that complies with the land permeability requirements laid down in the federal legislation on the subdivision of urban land (Federal Law No. 6.766/79) or the municipal legislation on zoning in Curitiba (Municipal Law No. 9.800/2000). The research outcome showed that the Permanent Conservation Areas provided for in the legislation should extend, in the case of the rivers studied here, at least 30 m from each bank.

Measurements were taken in two areas: one in the sub-basin of the Pinheirinho River, corresponding to a block next to the river in the Lindóia neighbourhood, and the other in the sub-basin of the Guaíra River, covering three blocks on the banks of the river in the Parolin and Guaíra neighbourhoods. According to the Land Use and Cover Zoning Law, both areas belong to Residential Zone 3 (ZR-3), in which single-family dwellings and single-family row houses (a maximum of 3 per lot) are allowed, with a minimum lot size of 360 m² and a minimum percentage of pervious surface of 25 per cent.

III.3.1   Area 1 – The Pinheirinho River Basin

Analysis of Figure III.1.1 shows that the block in Area 1 in the Pinheirinho River sub-basin covers 4,630 m² and that the minimum 25 per cent pervious surface coverage requirement would correspond to 1,158 m². The actual percentage of pervious surface, however, is less than 10 per cent. The APP extends over on average 8 m although according to the law regulating the Curitiba Master Plan it should be at least 30 m wide.

III.3.2   Area 2 – The Guaíra River Sub-basin

Analysis of Area 2 in the Guaíra River sub-basin, shown in Figure III.1.2, reveals that the minimum percentage pervious cover required is very low, indeed

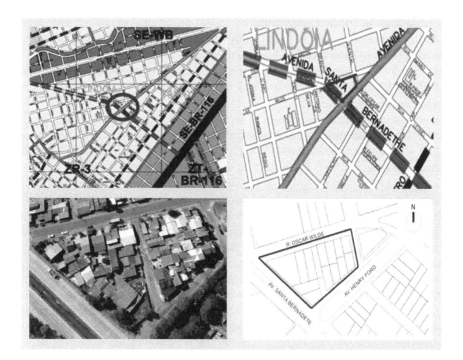

Figure III.1.1 The Pinheirinho River Basin.

Figure III.1.2 The Guaíra River Sub-basin.

almost zero. In Block 1, amounting to 15,300 m², 25 per cent of the total area should be pervious, i.e. 3,825 m²; however, the actual percentage of pervious surface observed was less than 8 per cent. Block 2 covers an area of 9,300 m², of which 2,325 m² should be pervious, yet the percentage of pervious surface measured was less than 5 per cent. In Block 3 the situation is even more critical: the block covers 18,750 m², and 25 per cent of this area, or 4,688 m², should be pervious, but the percentage of pervious area actually measured was less than 2 per cent. The APP specified in the Municipal Master Plan (a minimum of 30 m) was 12 m wide on average in Blocks 1 and 2 and dropped to 6 m in Block 3. In both areas, the number of lots is greater than the number registered in the City Hall.

When residents were asked about the low percentage of pervious surface, they showed a complete lack of information about environmental laws and restrictions on settlements in the APPs along urban streams and rivers. Neither the local community nor the public authorities has set a good example in terms of occupation and urbanisation along the banks of these rivers. No inspections are carried out or procedures implemented to prevent these areas from being occupied. Thus, the local population ends up building on the strip protected by law to allow riparian vegetation to grow on it. This type of squatter settlement, grown out of environmental and urban laws, causes pollution and limits the absorption of rainwater, drastically reducing real-estate prices and leading to (sometimes catastrophic) consequences for the local population during the rainy season.

III.4 The economic domain: real-estate prices

The area of influence of the Pinheirinho and Guaíra Rivers extends from the Lindóia and Fanny neighbourhoods, which are divided by the Pinheirinho River, to Guaíra and Parolin, which are crossed by the Guaíra River. Located approximately 5 km from the city centre, the area should have some of the highest real-estate prices in Curitiba. However, Parolin in fact has some of the greatest social contrasts in the city, reflected in the great variety of housing, ranging from mansions to slum dwellings.

Villaça offers theoretical tools for understanding the results of this empirical analysis of real-estate dynamics in the area of influence of the Pinheirinho and Guaíra Rivers. According to the author, the value of the use of an intra-urban structure is the value of that location as part of a social settlement (72). Based on Henri Lefèbvre's theory of use value and exchange value of urban land, Villaça develops a theoretical and conceptual framework for the *location*, *value* and *price* of urban land, here used to interpret the findings of this study. The real-estate supply in the area of influence has been evaluated in three stages. Initially, a spatial analysis of the Parolin, Guaíra, Lindóia and Fanny neighbourhoods was carried out to decode the real-estate dynamics. Based on this, the prices of the 54 properties for sale between August and September 2014 in the main real-estate web sites for Curitiba were recorded and mapped by property type (plot of land, detached house, duplex/row house and apartment). The figures were updated in

December 2016. The factors affecting real-estate prices were then analysed, including location and the relation with the socioeconomic dynamics of the study area. The analysis considers development rights, urban mobility, the presence of major amenities such as hypermarkets and shopping malls and, last, proximity to the Pinheirinho and Guaíra Rivers and the Valetão slum.

Location and urban mobility are undoubtedly the most important factors affecting real-estate prices in the neighbourhood. Being within 500 m of the hypermarket and shopping mall increases the prices of plots of land and apartments by on average 10 per cent and 25 per cent, respectively, compared with other locations in the neighbourhood. Also of importance is the fact that the buses serving the city centre run along the main thoroughfares.

In Upper Parolin, the area of influence of the hypermarket, which increases prices, overlaps with the area of influence of the slum, which has an opposite effect on prices. Both areas extend over five blocks, and the effect of one almost cancels the effect of the other. The average price of a detached house or duplex/ row house in Upper Parolin is US$1,350.00/$m^2$ while the average price of a plot of land is US$600.00/$m^2$. In contrast, within two blocks of the slum, in Lower Parolin, prices for similar properties drop to US$600.00/$m^2$ and US$330.00/$m^2$, respectively, some 45 per cent and 55 per cent less.

Proximity to the Linha Verde, the newest transport corridor which extends over 22 km in a north-south direction, tends to erase the influence of proximity to the Pinheirinho River, which runs alongside Henry Ford Avenue, as the average price of a house within 3 blocks of the Linha Verde is between US$1,350.00/$m^2$ and 1,550.00/m^2.

Development rights also have an influence on real-estate prices, particularly near Wenceslau Braz Avenue, which runs alongside the channelled Pinheirinho River. This avenue connects the Linha Verde to the bus terminal in the Portão neighbourhood and has the characteristics of a linear park with sports and leisure facilities. Because Wenceslau Braz Avenue is considered as a Special High-density Axis Sector under municipal legislation, development rights gained from conservation of historical and architectural heritage in the city can be applied to plots on the avenue. The presence of the river in this case does not depress real-estate prices. What affects prices is proximity to the Valetão slum on the banks of the Guaíra River: while a plot of land on the block next to the river commands a price in the region of US$310.00/$m^2$, a plot 500 m from the river costs on average US$440.00/$m^2$, a difference of approximately 30 per cent.

III.5   The cultural domain: quality of life and perception of wellbeing

Residents were surveyed to find out how they perceive and appropriate the area they live in, how they relate to their neighbourhood and the natural environment and what their community references are. In the article "How Many Qualitative Interviews Is Enough?" Baker and Edwards discuss the minimum sample size for qualitative research by referring to some authoritative authors: according to Charmaz a small number of interviews may be sufficient in the field of applied research. When one wants to discern common opinions and experiences in a community with homogeneous characteristics, in most cases 12 interviews are sufficient for a qualitative research (21–22). According to Adler and Adler the sample for qualitative research can vary from 12 to 60 interviews, but when dealing with interviewees in illegal or vulnerable situations, they advise the use of the minimum level, that is, 12 interviews (10).

Therefore, 12 interviews were carried out, using parts of the Social Life Questionnaire as a script. This tool, deriving from the Circles of Sustainability method, is used to convey a picture of the subjective attitudes of community members towards the sustainability, liveability and resilience of their communities. In this way, the problems faced by the local community could be identified by using a participative, relational methodology that was easily understood by the respondents.

The interviewees represent the current state of the sub-basins in terms of settlements and vulnerability, with a focus on the slum and surrounding areas, i.e. the same areas used for the political and economic domains. Block 1, i.e. Valetão slum, which is now known as Vila Parolin, shows the worst situation as regards habitability; in Block 2, which is also in the Parolin neighbourhood, habitability is greater, but sustainability is not; and in Block 3, in the Lindóia neighbourhood, the dwellings are of a higher standard and the built densities greater.

The age of respondents in Block 1 varied between 25 and 40 years, confirming official municipal data about the population in Vila Parolin, according to which 67 per cent are between 15 and 25 years of age; more than 90 per cent live on up to three minimum monthly wages (equivalent to US$960.00); around 40 per cent live in wooden sub-dwellings (IBGE, *Censo 2010*). Area 1 was treated in greater detail, as this area has traditionally been more vulnerable owing to poor social and living conditions. In Blocks 2 and 3 respondents were mainly between 50 and 78 years of age.

The questions about residents' perceptions produced surprising as well as predictable results. The first surprise was that residents are satisfied or very satisfied to be part of the community they live in. Only one person in Block 2 said they were dissatisfied. Respondents in Blocks 2 and 3 said their financial situation was comfortable, while half of those in Area 1 said they were happy with their income. However, this clashed with observations *in loco*, as the signs of poverty were clear and the difficulties people face in order to achieve a minimal standard of living were evident in their long-suffering expressions.

Only two of the interviewees had lived in their neighbourhood for less than 20 years, and the remainder had lived in the same place for between 24 and 47 years. In other words, the latter were witness to the history of the settlement and decided to stay despite the problematic conditions because they had family ties or ties with neighbours and the community itself.

Another unexpected finding was that, with one exception, respondents said they felt comfortable living with people from different social classes, showing that social differences do not lead to social exclusion among members of the community nowadays. The researcher carrying out the survey, however, had a different perception, as violence and lack of security in the area are well known; if people feel threatened or are being watched by drug dealers, they avoid telling the truth about community life and their own lives.

When asked about their own health and that of other members of their family, residents of Blocks 2 and 3 said they considered themselves healthy and felt that they had enough money to take care of their family. They said they go to the health centres whenever they need to. In contrast, although residents of Block 1 said they were healthy, half of them said that they cannot afford to spend money on their family's health. This reflects a worrying situation, as family health is directly related to all other aspects of the population's wellbeing.

According to their answers, respondents interact with the community mainly at religious events, at the workplace, and in the company of other family members. Although half the interviewees said that there are leisure and recreational spaces such as parks, squares and community centres in their neighbourhood, only those in Block 2 said that these spaces were near their homes.

Less than half of the respondents said they felt they had some say in decisions affecting the community, but nearly all said they believed the decisions taken by community leaders reflect the interests of the majority. Only three of them believe that government laws and decisions are good for their community, but respondents generally believe that specialists can solve the environmental problems.

Respondents regard distribution of wealth as a source of exclusion affecting their quality of life, and nine of them agree that their level of consumption is not compatible with an environmentally sustainable future. Respondents expressed opposite perceptions of their individual relationship with the environment: six reported they felt connected to the rivers, while the other six did not see the natural landscape around them as a reference in their lives.

When asked if the settlements had been affected by natural disasters, two residents in Block 1 said their homes had been destroyed in fires but made no mention of the constant flooding in this area. In Block 3, which is on ground sufficiently high to be unaffected by floods, two residents said their houses had in fact been flooded. In Block 2, residents did not mention any type of natural accident.

Even though environmental rehabilitation measures have been implemented in Block 1, which is in the slum, it is still the most socioenvironmentally vulnerable area. The urban morphology, characterised by a web of narrow lanes and bends, favours the activities of drug dealers and places residents in danger.

IV Coping with an unsustainable way of life

The Circles of Sustainability has proved very useful to this analysis of settlements in the Belém River basin. A transdisciplinary analysis of sustainability in the economic, ecological, political and cultural domains has revealed how society's actions affect the environment. The effects are: failure to comply with environmental, urban and land-division legislation; systematic floods and their effects on the wellbeing of the local population; poor quality of the environment; depressed real-estate prices in areas next to the rivers.

For Soja the geographically unequal development results from socio-spatial dialectics shaping the relations of production, therefore class conflicts are simultaneously social and spatial (99). On the other hand, Gottdiener states that planning, even when inefficient, is a control mechanism of class conflict, an ideological mask that seduces the working class to believe that state intervention in the environment promotes indeed the representation of their interests in society (136). Here it is necessary to highlight an ambivalent attitude: residents attribute unsustainability as a way of life to both nature itself and the virtual absence of the State. Living on the edges of the rivers and society, they have nevertheless accepted their lot and created a sense of belonging to their neighbourhood that gives them a feeling of wellbeing despite the socioenvironmental discomfort they continue to experience year after year.

Bibliography

Baker, Sarah E., and Rosalind Edwards. "How Many Qualitative Interviews Is Enough?" Discussion Paper. National Centre for Research Methods, 2012. eprints.ncrm.ac.uk/2273/. Accessed 20 January 2017.

COHAB Curitiba. "Projeto de Urbanização da Vila Parolin receberá prêmio". www.cohabct.com.br/conteudo.aspx?conteudo=128. Accessed 10 November 2016.

Cutter, Susan L. "Vulnerability to Environmental Hazards". *Progress in Human Geography*, vol. 20, no. 4, 1996, pp. 529–539, inti-phg.sagepub.com. Accessed 5 December 2016.

Fernandes, José C. "Luta de Classes no Parolin". *Gazeta do Povo*, 2 April 2011, www.gazetadopovo.com.br/vidaecidadania/conteudo.phtml?id=1112369. Accessed 10 November 2016.

Ferreira, Leila C. *A Questão Ambiental: sustentabilidade e políticas públicas no Brasil*. Boitempo, 2003.

Fortunato, Rafaela A. "Subsídios à prevenção e controle das inundações urbanas: bacia hidrográfica do rio Belém, Município de Curitiba, PR". MS Thesis, Universidade Federal do Paraná, 2006.

Gottdiener, Mark. *A produção social do espaço urbano*. Translated by Geraldo Gerson de Souza. 2nd edn, EDUSP, 1990.

Global Compact Cities Programme. "Appendix 1. Assessing the Sustainability of Cities". citiesprogramme.com/aboutus/our-approach/circles-of-sustainability. Accessed 5 December 2016.

Instituto Brasileiro de Geografia e Estatística (IBGE). *Censo 2010*. www.ibge.gov.br. Accessed 5 December 2016.

Instituto de Pesquisa e Planejamento Urbano de Curitiba. *Memória da Curitiba Urbana* (Depoimentos, 7). IPPUC, 1991.

James, Paul, with Liam Magee, Andy Scerri and Manfred Steger. *Urban Sustainability in Theory and Practice: Circles of Sustainability*. Routledge, 2015.

Leitão, Sylvia Ramos. "O discurso do planejamento urbano em Curitiba: um enigma entre a prática e a cidade real". Dissertation, MS Thesis, Universidade de São Paulo, 2002.

Marândola, Eduardo J. R., and Daniel J. Hogan. "Vulnerabilidades e riscos: entre Geografia e Demografia". *Revista Brasileira de Estudos de População*, vol. 22, no. 1, 2005, pp. 29–53.

Marcovitch, Jacques. *Para mudar o futuro: mudanças climáticas, políticas públicas e estratégias empresariais.* Edusp, 2006.

Morin, Edgar. "Sur l'Interdisciplinarité". *L'autre Forum: le Journal des Professeurs de l'Université de Montréal*, vol. 7, no. 3, 2003, pp. 5–10.

Paquot, Thierry. "Économie, écologie et démocratie". *L'Écologie Urbaine?*, edited by François Séguret and Henri-Pierre Jeudy, Éditions de La Villette, 2000, pp. 15–25.

Peres, Aline, and Osny Tavares. "Parolin recebe a 2ª. UPS do Paraná". *Gazeta do Povo*, 3 May 2012, www.gazetadopovo.com.br/pazemvozemedo/conteudo.phtml?id=1250966. Accessed 10 November 2016.

Prefeitura Municipal de Curitiba. *Lei nº 9.800/2000. Dispõe sobre o Zoneamento, Uso e Ocupação do Solo no município de Curitiba e dá outras providências*, multimidia. curitiba.pr.gov.br/2010/00084664.pdf. Accessed 15 December 2016.

Ribeiro, Luiz Cesar de Queiroz, and Adauto Lucio Cardoso. "Da cidade à nação: gênese e evolução do urbanismo no Brasil". *Cidade, Povo e Nação*, edited by Luiz Cesar de Queiroz Ribeiro and Robert Pechman, Civilização Brasileira, 1996, pp. 53–80.

Santos, Antonio C. "Provimentos do Ouvidor Pardinho para Curitiba e Paranaguá (1721)". *Revista Monumenta*, vol. 3, no. 10, 2000, pp. 27–80, archive.org/search. php?query=pardinho. Accessed 10 November 2016.

Senado Federal. *Código Florestal: Lei Federal nº 4.771, de 15 de setembro de 1965, alterada pela Lei Federal nº 12.651 de 25 de maio de 2012*, www.senado.gov.br. Accessed 15 December 2016.

Senado Federal. *Estatuto da Cidade*: *Lei Federal nº 10.257/2001. Regulamenta os arts. 182 e 183 da Constituição Federal estabelece diretrizes gerais da política urbana e dá outras providências*, www.senado.gov.br. Accessed 15 December 2016.

Senado Federal. *Lei de Parcelamento do Solo Urbano: Lei Federal nº 6.766/79, alterada pela Lei Federal nº 9.785/1999*, www.senado.gov.br. Accessed 15 December 2016.

Senado Federal. *Lei n.º 9.433/1997. Institui a Política Nacional de Recursos Hídricos, cria o Sistema Nacional de Gerenciamento de Recursos Hídricos, regulamenta o Inciso XIX do Art. 21 da Constituição Federal*, www.senado.gov.br. Accessed 15 December 2016.

Senado Federal. *Política Nacional do Meio Ambiente: Lei Federal nº 6.938/1981*, www. senado.gov.br. Accessed 15 December 2016.

Soja, Edward W. *Geografias pós-modernas: a reafirmação da teoria social crítica.* Translated by Vera Ribeiro. 2nd edn, Jorge Zahar, 1993.

Trindade, Etelvina M. *Cidade, homem, natureza: uma história das políticas ambientais de Curitiba.* Unilivre, 1997.

Villaça, Flávio. *Espaço intra-urbano no Brasil.* Studio Nobel – Fapesp-Lincoln Institute, 1998.

III.2 Creative social innovation and urbanism

The case of Medellín

Ana Elena Builes Vélez and
María Florencia Guidobono

I Introduction

Latin American countries have been searching for development alternatives, as half of the population lives under the poverty line (Pereira Guimaraes). Local and regional governments must devise ways of creating and improving competitive capabilities and of transforming local productive systems. These two aspects should be included in urban planning policies, specifically in the development of a culture that integrates them.

Economic analyses rarely consider the full environmental and human costs as a whole, and this deficiency has led to detrimental and even life-threatening consequences. Some tasks should be fulfilled for inclusive development, such as closing the gap between territorial inequalities. Economic and social disparities are expressed in the geography of development. Land is the greatest wealth of every country because it is the space in which citizens develop relationships and ideas. Latin American cities, such as Medellín, are not only large conglomerates of people affected by poverty, violence, social inequality, and environmental and urban deterioration; they are also characterised by miscegenation and cultural complexity, which can become catalysts for creative solutions.

II Conceptual approaches and definitions

In 1987 the United Nations stated that sustainability would involve development that "meets the needs of the present without compromising the ability of the future generations to meet their own needs" (World Commission on Environment and Development 43). Throughout the decades, many conceptual modifications have been suggested to consider and include multiple variables: social aims, time period, psychological and physical needs, and therefore human-oriented needs. Sustainability does not focus on an isolated activity; rather, it provides an ongoing process that accounts for how various systems influence each other. The goal is then to involve as many perspectives as possible.

Wellbeing concerns optimal experiences and functioning. According to Ryan and Deci, "How we define wellbeing influences our practices of government, teaching, therapy, parenting, and preaching, as all such endeavours aim to change humans for the better, and this requires some vision of what 'the better' is" (142). This vision is important for sustainability processes. The theories on wellbeing can be grouped into two categories: (1) the hedonic approach is based on the idea that wellbeing involves the pursuit of pleasure and the avoidance of pain. The most commonly applied theory in this approach is called subjective wellbeing, which measures life satisfaction; (2) the eudaimonic approach presupposes that not all pleasures or positive emotions lead to wellbeing. Instead, the focus is on self-realisation and self-development, so that wellbeing is conceptualised as the extent to which an individual is fully functioning. The most commonly applied theory founded on this approach is psychological wellbeing.

Sustainable wellbeing intersects wellbeing research and sustainability processes to strengthen the sustainability of wellbeing approaches and to clarify the final aims of sustainability processes. Moreover, wellbeing research should shed light on the needs to be satisfied, promote a holistic and balanced perspective, and foster discussions on values and constraints in order to generate a more inclusive and longer-lasting wellbeing. Sustainable wellbeing should highlight that an individual cannot satisfy his/her own needs only, as the need to adapt to the demands of other human beings and nature also exists. It involves being in tune with the environment and the city. Planning cities and urban developments need to consider sustainability processes and wellbeing research by using interdisciplinary approaches and tools aimed toward inclusive and sustainable cities.

In 2006 the government of Medellín and urban planners developed a programme called *Medellín cómo Vamos MCV* (*Medellín, how are we doing*),[1] which analysed the quality of life by quantifying the citizens' perception of wellbeing. The Municipal Administration used this report to develop strategies leading to a continuous improvement of wellbeing in the city. Once the local government finished its mandate, the MCV programme produced a report in which data on the evolution of wellbeing and the quality of life had been collected over a three-year period. The report focused on multidimensional and multisectoral aspects, which enabled the programme to incorporate transversal analyses of equity, poverty, demographic changes, education, healthcare, employment, security, and an urban habitat dimension that includes housing, public services, environment, transportation, and public space. Two other dimensions, governance and citizenship, have also been considered: they include public policies, finances, competitiveness, and citizenship responsibility.

The report uses a definition of quality of life that includes an objective dimension but also a subjective one, and both dimensions consider the integral wellbeing of people. These concepts are analysed in the Citizen Perception Survey conducted by the MCV programme annually. The survey shows that Medellín is the city with the largest per capita social investment in Colombia. It also indicates that the city has grown significantly over the last ten years in terms of the most important aims of public policies: gaps in the quality of life in urban and rural territories have been reduced, the levels of poverty and extreme poverty have decreased, and sustainable processes in urban transformations have improved. In this way, the strengthening of the community is presented as a form of support to social processes and not as derived from commercial logic.

In "Reflexiones estratégicas sobre la innovación en el campo social" the Comisión Económica para América Latina y el Caribe – CEPAL defines innovation as a social process (23). Triggered by an original change in the presentation of a service or the production of a good, social innovation can achieve positive results with regards to one or more poverty situations, marginality, discrimination, exclusion, or social risk, and can be replicated or reproduced. Clearly, reducing the concept only to its economic dimension would be inadequate, because social, political, and environmental dimensions are also essential.

When the territory is included in the debate, equity, justice, and sustainability are involved along with productivity or economic growth. One of the basic elements is the development of innovation, and therefore social cohesion is fundamental. Creating competitive conditions based on innovation and creativity requires collective actions. This approach aims to develop a growth and development strategy in which the activation of different territorial players leads to an integrated personal vision encompassing the economic, social, political, environmental, and cultural aspects. The concept of territory emphasises how cultural dynamics and creativity become relevant vectors of the inclusion and construction of the social tissue in different social groups.

III Urban development through creative social innovation

Urban sustainability implies structural changes within institutions as well as guidelines of social behaviour. Sustainable cities offer liveable, safe, and fair socialisation places that preserve their cultural and environmental characteristics and allow for the development of every human being without compromising the environment of future generations. They must provide elements to create a more equitable, egalitarian, and democratic access to the natural or socially generated wealth. They should develop institutional, educational, and moral ways of developing a mentality and a social sensibility toward nature as a value in itself.

Rethinking public space in cities has become a fundamental element in reconstructing and transforming Latin American cities through creative social innovation procedures, mobilising social and economic processes to regain wellbeing, and becoming a more socially and environmentally sustainable city. These creative processes, which are driven by government and some non-government institutions in Medellín, have fostered an enormous transformation.

Inhabiting the world in an urban way has generated a physical, environmental, and social degradation in most Latin American cities, and the transformations have resulted in social spaces that hinder the development of social interaction, sense of citizenship, and territory cohesion. Certain characteristics of space and territory are necessary to the growth of individuals and communities, in which wellbeing is not a feeling but a sustainable status because of a reconstructed, reorganised, and re-established community. In order to achieve this urban and social development in

Medellín, it is fundamental to work alongside the community directly affected by the transformation by using creativity as the most important asset for launching innovating ideas instead of just having them. Cities where creative social innovation takes place are those where new administrative, institutional, and structural paradigms are established. Cities where citizens build a network and contribute to urban planning enhance wellbeing and a sense of community.

Urban development through creative social innovation leads to creative sustainable cities that are resilient and can react to urban and social conflicts by using creative solutions but also communication strategies that help citizens understand the risks and opportunities in their own territories. This new way of understanding, planning, and constructing the city has led to the development of communities characterised by sustainable wellbeing and equal social relations.

City researchers, urban planners, community leaders, and scholars agree that Medellín has become an outstanding example of urban reformation, social reconstruction, and sustainable wellbeing. Social innovation, resilience, sustainability, wellbeing, and creativity have been successfully pursued through innovative public policies and a new urban planning model. Social urbanism in Medellín began in 2003 with different municipal government actions involving parts of the city that had been neglected in urban development projects for more than a decade. It can be defined as a territorial intervention model that includes physical transformation, social transformation, institutional management, and community participation. These elements have promoted territorial equity, sustainability, and wellbeing in the city, previously characterised by violence, non-government or institutional legitimacy, loss of territorial control, deficit of security and healthcare systems, discrimination and social exclusion of multiple groups.

Medellín's model of social urbanism works through acts of governing through educational actions; it implies that social innovation and social transformation can only happen when people are educated and considered. Seven facets have been identified: city of knowledge, city for the family, city to be travelled, city for coexistence, green city, thoughtful city, and city for the citizens. These multiple facets have been the pillars of every urban transformation or city project in Medellín in the last ten years. Different strategies allowing planners, architects and designers to work with their interdisciplinary groups have been devised to accomplish their goals in terms of wellbeing, quality of life, and sustainability. Five specific strategies have been established: connecting a divided city, reading the territory, developing urban projects, promoting all-inclusive interventions, encouraging collaborative and participative work with the citizens. Several urban interventions planned and executed in Medellín in previous years have empowered the citizens, connected the city, and improved the perception of life quality.

Expanding on the social urbanism model, the Municipal Administration has designed an enhanced strategy called *Proyectos Urbanos Integrales* (PUI) (*Integral Urban Projects*) as an overall programme integrating all aims and strategies devised in the model. The PUI is defined as a set of tools for urban intervention to be used simultaneously within a designated area. Housing, schools, library parks (*Parque Biblioteca*), linear parks, kindergartens, and mobility corridors are part of the PUI projects.

IV Transformation through creative social innovation or social urbanism

The core aim of Medellín's Municipal Administration in adopting the social urbanism model has been to achieve a substantial social and environmental impact on communities with low income and in a marginality situation throughout the execution of integral interventions involving sustainability, wellbeing, and social equity. The result is the creation of multiple projects based on high-quality design and materials, which have generated a sense of satisfaction and pride in the citizens of those communities and provided reconciliation spaces. These transformations include the environmental recuperation of the Moravia sanitary landfill through human development, coexistence, and integration.

Moravia, the north-eastern part of the city right on the border of the Medellín River, was a shantytown on a small artificial mountain that resulted from the first sanitary landfill. This zone received solid wastes from the city from 1972 to 1984. Under the concept of an open-air rubbish tip, an artificially created promontory rose up to 40 m in a 7.5 ha extension. In 2001, more than 15,000 people camped in this place among tonnes of waste and multiple infestations, escaping from violence and hoping to live by recycling. For three decades, the population had a strong relationship with the hill, recycling plastic materials, cans, and waste cardboard to build their homes. A total of 2,300 slum houses were built precariously on the improvised land.

The conditions of the population were uncomfortable, as each inhabitant could only have $0.37 m^2$ of free space without potable water and access to public services. In addition, the population was exposed to geotechnical and chemical risks brought about by toxic liquid waste, which came from the rubbish mountain and threatened the stability of the terrain.

In 2008 the UNESCO Chair on Sustainability in collaboration with the Pontifical University of Cataluña launched an initiative aimed at transforming degraded urban areas and recovering Moravia, where the main municipal waste dump was located. The socio-economic and environmental conditions of Moravia and its surroundings were improved by strengthening the participation of the community in the urban transformation and by lowering the risks of disease through appropriate technologies. Every goal pursued through public, private, and communitarian partnerships linked to the Moravia programme and its area of influence was led by the Ministry of Social Development. The ministry adopted the social urbanism model to improve the environmental, socio-cultural, and economic conditions of the inhabitants.

The UNESCO Chair on Sustainability focused on the environmental recovery of Moravia and the involvement of the community to make this process sustainable. Activities generating sustainable wellbeing included environmental restoration and decontamination actions, socio-environmental recovery, promotion of citizens' participation, urban recovery, landscaping intervention, and communication strategies. These activities have made it possible to re-establish the social tissue, decontaminate and recover the area, empower the community with the help of public and private agents, and develop long-term social and economic activities.

The initiative in Moravia exemplifies the transformation of degraded urban areas through social creative innovation, which improves the quality of life from various perspectives, including urban integration, environmental restoration, economic reactivation, and social participation. The most empowering activities in Moravia involve the group called Communitarian Gardens. Through these activities, the population has taken part in the recovery, maintenance and transformation of the space through training in planting techniques, multiplication of species, and composting. Not only does the active participation of the community maintain and manage the regenerated spaces, it also strengthens the cohesion and social stability of marginalised groups.

This innovative initiative addresses specific socio-environmental issues by including different social, academic, and government actors. Since the beginning, it has been presented as a participatory process toward transformation. The specific aspects of the initiatives to be held and all that is needed to ensure sustainable projects are not clearly defined in any economic or public policy in Colombia. This is why the community has worked in order to form associations and non-profit organisations that co-ordinate and gather different groups as well as legitimise their informal economy activities. Citizenship initiative and creativity have arisen and proved capable of self-convocation and of reactivating and autonomously organising the labour capacities and production means available. They can also efficiently fulfil legitimate necessities ignored by the market. These initiatives will help strengthen the market and boost an economic system based on social reconstruction and conducive to sustainable and social wellbeing for those communities. This overall structural option implies a qualitative change in the economy, which will encourage society to generate more democracy and more resources, ultimately favouring corporate sectors and the State's economic viability and competitiveness.

The Moravia programme is not the only successful integral transformation project conducted in Medellín. Several important and interesting examples have been used in multiple academic, social, and political gatherings to demonstrate the importance of collaborating, co-working, and building alternatives for the community, urban planners, and policies in order to transform and improve the sustainability, wellbeing, and life quality of the communities.

V Ways of inhabiting and urban migrations in Medellín caused by recent transformations

Ways of inhabiting must be understood as the system of relations that an individual establishes with the place inhabited, and this system includes practices, representations, and meanings related to the space. This connection with the living space an individual has created is called territoriality and is a fundamental issue in inhabiting. Although anyone can inhabit a place, inhabiting also mentally implies a network of places knitted by the lives of the people who inhabit it (Lindon).

In modern societies, inhabiting has ceased to provide security to the people, as it is viewed as a place where one does not belong but can live for a limited period. This crisis seems linked to space mobility, and it forces displacement within different urban spaces.

The ruptures that occur in the inhabited spaces and the existing relations between them and the inhabitants are not necessarily due to everyday mobility but rather to the high residential mobility through people's life. In cases like Medellín, the high residential mobility is more often than not a consequence of violence in peripheral communes, inhabited by people also displaced by violence from other regions in the country.

This situation shows an attitude of indifference toward the inhabited place and indicates a dysfunctional inhabitation. In communes 1, 8, and 13 in Medellín this indifference is manifested in almost all cases by the feeling of contempt towards the inhabited place, voiced through the common expression "I will not live here forever" or "We came here because there was no other option".[2]

This manner of inhabiting results from continuous forced displacements that occur in the city, where families are forced to move from one commune to another. In these cases, people are not interested in establishing a connection with the territory where they were forced to move. The new living space is conceived of as a simple location that represents no kind of familiar root, as the root is where people establish home and knitted social relations and connections. The absence of a memory about this new place and the lack of belief in a future in that same place reflect the condition of migrants. This condition worsens when migrants are forced to move once again from one place to another in the city owing to violence or urban transformation projects, which is the case in Medellín.

According to Mexican urban scholars Duhau and Giglia, a number of social logics regulate the relationships between people and territory that tend to organise community behaviour in any urban space. Therefore, if the proposed transformations are not structured or organised after studying and understanding the territory and the communities where a project will be built, irregular human settlements will develop and these projects will transform the ways of inhabiting the territory. Human displacements caused by violence are some of the most problematic phenomena in Latin American countries, particularly in Medellín, because they greatly affect the life quality of the displaced community but also of the receiving community.

In Medellín, communes 1 and 13 are characterised by communities living in extreme poverty. In 2009 the Unidad Permanente de Derechos Humanos (Permanent Unity for Human Rights) reported that 163 families suffered from forced internal displacement in the city, and 582 people had to migrate from one commune to another (Personeria de Medellín 17). The report also stated that the number of people forced to move inside the city had grown in relation to the displacement caused by violence in other departments of the country. These displacements have caused major spatial segregation, known as the residential separation of a group from a large population (Ortiz y Aravena). These spatial phenomena show incomplete and complex connections in growing communities and reflect the social differences and inequalities suffered by migrants in Medellín. These migrants are forced to live in degraded communes, where their presence limits the possibilities of urban renovation and local investment.

Intra-urban migrations in Medellín occur not only because of uncertainty, threats from illegal groups, poor quality of life, or unsustainability. In some cases, they are also caused by displacements forced by urban transformation, large housing projects, and integral urban renovation planned and/or executed in the city in order to unify it by improving the quality of life and sustainable conditions in those communes (Secretaría de Bienestar Social, 2010). In addition, the price increment of properties settled around the transformations has led the inhabitants of these zones to a new displacement in other communes in the city.

The mapping and categorisation of creative social and sustainable innovation initiatives implemented in Medellín using the social urbanism model, which was conducted by an interdisciplinary research group from different universities,[3] began with the analysis of how territories had been inhabited before an urban project was executed and which social issues were addressed. In some cases, the communities exhibited a lack of morphological context analysis and any understanding of everyday life. Consequently, the interventions are somehow forgotten and not received by the community. In the worst cases, they cause intra-urban migration.

Two projects, namely, the Moravia Cultural Centre and the España Library Park, which are part of different programmes within the social urbanism framework, have been examined because their results indicate that both positive and negative effects are involved in the communities. Without planning and after the Moravia project was implemented, the community transformed into a real neighbourhood, where wellbeing and life quality developed and a great territorial identity was enhanced. The good results of the Moravia programme prompted the restructuring of the entire place, not only the mountain, into an area of social and urban reconvention and reconstruction through a cultural programme that offers different cultural and education activities for families in the community. The territory where the centre was built was occupied by families that had lived there since their migration from other territories in the country. Thus, the government had to expropriate the houses of 150 families and relocate them within the area while attempting to preserve the sense of belonging they had formed with that territory. When the building was opened, the government made a great effort to make the community understand the importance of the centre in terms of social reconstruction and wellbeing.

The España Library Park is located in the Santo Domingo Savio neighbourhood in commune 1, founded in 1964 when a family settled in the north-eastern hills of the city. This commune has been one of the most violent, unsustainable, and socially unstable groups since its foundation. Before the construction of the library park, all the housing and buildings in this territory were improvised and modest, as the families had been displaced from other territories in the country owing to violence. The hills were considered by the government as high-risk areas and shantytowns.

The Municipal Administration expropriated the houses and forced families to move to other communes. The social conflict that generated the project involved different actors and perspectives and caused unusual ways of inhabiting the commune. The project has modified the existing relations among territoriality, inhabiting, and the subjects who lived in that territory (Builes Vélez).

VI Creative social innovation for the development of sustainable communities

This analysis shows that the city should be built not only by the Municipal Administration, architects, designers, and urban planners but also by the people who inhabit it. These people must actively participate in the city's construction and constant transformation, as they are the ones who give life to the city. The need for an urban transformation that responds to market needs or to the government on duty is also evident. The transformation of society implies innovation through actions, behaviours, speeches, schemes, and ways of thinking and feeling, as achieved through the Moravia Programme, among others.

When analysing life in the communes that receive migrants, it transpires that progress is linked to the constant transformation of the infrastructure, which must be connected to the needs of the subjects who inhabit the territory. A city in fragments, without networks and social relationships, is unsustainable. Urban and social transformation thrives on creative social innovation, because the relationships between people and the inhabited spaces are crucial. The networks created by the people who already inhabit a territory are essential to understanding the varieties of space as well as the ways of inhabiting them.

Notes

1 For the institutional presentation of the *Medellín cómo Vamos* programme see www. medellincomovamos.org/download/presentacion-institucional-medellin-como-vamos/. Accessed 13 March 2017.
2 The interview conducted by Plano-Sur, an interdisciplinary group that examines wellbeing and human rights in the communes of Medellín by using storytelling and sociological methods, is not described in detail in their final report.
3 The Research Project entitled *Analysis of the impact caused by the urban transformations in Medellín, Willemstad, Córdoba, and City of Mexico. Multiple case studies* has been conducted by Universidad Pontificia Bolivariana, Universidad Católica de Córdoba, Argentina, and Universidad de San Buenaventura, Medellín.

Bibliography

Builes Vélez, Ana Elena. "Nuevas maneras de habitar la ciudad como resultado de las continuas migraciones intraurbanas y las transformaciones sociales y culturales". *Arquetipo*, no. 9, 2014, pp. 12–18.

Comisión Económica para América Latina y el Caribe – CEPAL. "Reflexiones estratégicas sobre la innovación en el campo social". *Claves de la innovación social en América Latina y el Caribe*. CEPAL, 2008, pp. 21–36.

Duhau, Emilio, and Angela Giglia. *Las reglas del desorden: habitar la metropoli*. Siglo XXI, 2008.

Empresa de Desarrollo Urbano. *Proyectos Urbanos Integrales – PUI*. Alcaldía de Medellín, 2006.

Lindon, Alicia. "El mito de la casa propia y las formas de habitar". *Revista Electrónica de Geografía y Ciencias Sociales*, vol. IX, no. 194, 2005, www.ub.edu/geocrit/sn/sn-194-20.htm. Accessed 14 March 2017.

Ortiz Véliz, Jorge, and Evelyn A. Aravena. "Migraciones intraurbanas y nuevas periferias en Santiago de Chile: Efectos en la sociogeografía de la ciudad". *GeoFocus. Revista Internacional de Ciencia y Tecnología de la Información Geográfica*, no. 2, 2002, pp. 49–60.

Pereira Guimaraes, Roberto. *Tierra de sombras: desafíos de la sustentabilidad y del desarrollo territorial y local ante la globalización corporativa*. United Nations Publications, 2003.

Personeria de Medellín. "Informe Ejecutivo de Derechos Humanos". XIX Semana de los Derechos Humanos. Tejiendo Alternativas y Resistencias, 1 December 2009, Personería de Medellín, 2009.

Porter, Michael E., and Mark R. Kramer. "Creating a Shared Value". *Harvard Business Review*, 89, January–February 2011, pp. 62–77.

Ryan, Richard M., and Edward L. Deci. "On Happiness and Human Potencials: A Review of Reseach on Hedonic and Eudaimonic Well-Being". *Annual Review of Psychology*, vol. 52, 2001, pp. 142–161.

Secretaría de Bienestar Social. *Análisis del contexto y la dinámica del desplazamiento forzado intraurbano en la ciudad de Medellín*. Alcaldía de Medellín, 2010.

World Commission on Environment and Development. *Our Common Future*. Oxford University Press, 1987.

III.3 Long-term visions and ordinary management

Post-earthquake reconstruction in the Italian region of Emilia

Gianfranco Franz

I Introduction

Five years after the earthquake that struck the Po River Valley in Emilia-Romagna on 20 and 29 May 2012, the reconstruction process has now entered its middle stage. Considering three stages: (1) emergency; (2) most urgent reconstruction, regarding schools, public buildings and facilities, etc.; (3) repair of minor damages, it is clear that work for the recovery of cultural heritage in the area affected by the seismic impact has recently started, while the recovery of private housing, factories and farms is ongoing.

The Regional Government and the techno-structure of the Special Commissioner for reconstruction have supported step by step the many municipalities affected by the earthquake, sharing their efforts in particular with regards to the execution of: (1) Reconstruction Plans; (2) Plans for Public Works; (3) the drafting of the *Programma Speciale d'Area*, a sort of plan/programme for the whole impacted area; and, finally, the definition of (4) the *Piano Organico*, especially designed for the revitalisation of ancient city centres. In immediate terms, a tremendous job of predisposition and preparedness to the reconstruction in the coming years has been done.

II Effects of the earthquake and early results of reconstruction

Two strong shocks in May 2012 caused the displacement of 45,000 out of 760,000 residents, subsequently reduced to 14,400 at the end of 2013 (11,900 with rent contributions and 2,500 accommodated in temporary housing containers) and 4,645 by May 2015, 71 per cent less than in June 2012.

Damaged in various ways were 33,000 homes; 1,540 public buildings and facilities; 293 churches. After the earthquake 40,752 out of 270,000 workers employed in industry – more than 15 per cent – were suspended from the workplace. By the end of 2013 this was reduced to only 250 workers suspended. 3,748 companies registered serious damages, for a total of 13,000 productive buildings damaged. 14,000 agriculture and livestock farms (18.7 per cent of the entire regional agro-industry) were affected by significant damages, with a loss of €2.4 billion during the 6 months after the seismic event.

It has been established that the Emilian earthquake was the first in the history of Italian disasters hitting a region widely urbanised and heavily industrialised, in which about 2 per cent of the national GDP is produced, generating €12.2 billion in hi-tech exports. In April 2015, 7,369 projects for the reconstruction of private homes were submitted to the Municipalities; 5,066 have received public aid, for a total of one billion €89 million, 50 per cent of which already disbursed.

By the end of 2015, 25 out of the 33 municipalities in the impacted area accomplished or nearly accomplished the phase of private reconstruction. Out of 12,351, 4,069 damaged private houses were rebuilt or repaired, allowing more than 17,621 people to return home.

No doubt the most important result, five years after the earthquake, is the construction of schools, town halls, gymnasiums, libraries, auditoriums and churches as temporary buildings, for a total of €425,922,996.28. Funds amounting to a total of 563 million were granted by the European Union Solidarity Fund: it is the highest amount ever allocated by the EU after a natural disaster.

Within the span of two months, the Committee for Reconstruction supervised the rebuilding as well as the *ex novo* construction of 37 temporary and 32 prefabricated schools (now largely dismantled) with a capacity of 18,000 students, for a total of €86.5 million. During the first six months, the Committee funded the construction of 26 school gyms, for a total of €33 million, nine city halls and three temporary, prefabricated structures for a total of €50.5 million. A number of temporary churches, the Pico della Mirandola's new public library and a temporary auditorium were also built. All these temporary buildings were realised via international tenders, paying close attention to Class A energy efficiency rating and to anti-seismic criteria. In addition to these achievements, during the first 18 months another 320 schools, 99 hospitals and health centres which had been damaged in various ways but did not require demolishing, were restored and returned to operation.

In the long history of Italian post-earthquake recovery processes this was the first time that the emergency phase was handled with such efficiency and effectiveness: procedures for international tenders were strictly followed and new buildings were realised according to the highest level of energy efficiency and anti-seismic criteria.

III Planning innovations

Some innovative planning and management tools in support of a future economic recovery have recently been adopted, particularly those that follow.

1 Programma speciale d'area *(special programme for the area)*

With regards to 16 municipalities, the programme allowed a significant integration between public funding and private investment. Unfortunately, what appears as inadequate is: (1) the level of integration between many different public financial resources allocated to manage various policies; (2) the level of integration between national resources and resources from the European Structural Funds. Paradoxically, this was the intended goal of the Special Programme.

There is a strong feeling that the contents of the Special Programme were defined all too quickly in order to meet the target of allocating financial resources already available to projects submitted by the Municipalities. Although this result is certainly reasonable, it does not allow for evaluating the success of long-term strategic planning. In fact, within the short span of three years, communities and institutions were able to successfully carry out:

- the emergency phase;
- the phase of recovery from small and medium impacts;
- an update of the whole regional legislative framework;
- an innovating system for monitoring all building practices and ensuring the quality control of projects: see MUDE and SFINGE digital platforms;
- coherent tools for reconstruction, urban redevelopment, and economic and social regeneration;
- a Plan for Public Works and for the recovery of the most important monuments.

In conclusion, there is an obvious risk that the Special Programme can be reduced to a mere shopping list drawn to satisfy the demands of each Municipality.

2 Piano organico *(operational plan)*

Among these innovations, the Operational Plan stands out as a relevant tool. However, it would have been more correct to call it an Operational Programme rather than an Operational Plan, because the concept of programme is totally different from the concept of plan. Plans are static, while programmes are dynamic and pursue the specific aim of integrating and managing land use decisions by means of (1) financial programmes; (2) available budgets; (3) redevelopment strategies for economic regeneration. In any case, the Operational Plan is intrinsically connected to the *Piano della Ricostruzione* (Reconstruction Plan) and specifically aimed at analysing and implementing objectives and choices for redevelopment in space and time. Because urban planning tools focus on the spatial dimension and are thus static, it is particularly important to associate them with dynamic tools, which consider time and the availability of money.

Four years after the earthquake, 24 out of 28 municipalities have relied on an Operational Plan and a number of such plans have been attached to specific revitalisation strategies. In contrast, others are focused on projects aimed at obtaining public funds for reconstruction. Even though a survey of case studies is not yet available, it is clear that the effectiveness of these tools will be relevant for the future redevelopment of this region. In fact, for the first time in Italy, thanks to the influence and the political pressure exerted by the President of the Regional Government, Vasco Errani, on the National Government, the post-disaster recovery process can, and hopefully will continue to, rely on Operational Plans. Such plans do have great potentialities, as they intersect land use plans, which are static and thus unable to represent and convey a long-term vision, and tools for managing public and private funding, which are unable to spatially represent strategies for future development, in spite of the huge investment they can allocate (Isola and Zanelli).

3 Zone franche urbane *(urban tax-free zones)*

Urban Tax-free Zones are the latest innovation introduced at the national level thanks to the experience of the recovery process activated in the Emilian impact zone. Thanks to another solicitation coming from those who have boots on the ground in the impacted area, in 2015 the National Government implemented a specific law, importing Urban Tax-free Zones for the revitalisation of neighbourhoods in social and economic decline from the French context. The Urban Tax-free Zone is specifically a management tool allowing for financial support and tax incentives to sustain the recovery of small business, trade, and craft or micro-enterprises within smaller historic centres. In fact, before May 2012 in many of the villages and small towns struck by the earthquake more than 1,700 small businesses were active. Many of them are now located in temporary zones and buildings. The strategic importance of this aim is really a priority, because without a resumption of at least a part of the many small businesses active before the seismic disaster, and without a desirable number of start-ups created by young entrepreneurs, it will be very difficult to achieve the social and economic regeneration of the damaged ancient city centres.

Urban Tax-free Zones and Operational Plans are very important innovations, specifically because for the first time in Italy the recovery process has relied on parallel tools, both for physical and spatial planning, managing resource allocation by driving the focus of the reconstruction managers towards the social and economic revitalisation, and in this way avoiding the risk that the reconstruction may end up as a mere technical and engineering exercise.

IV Successes and limits

Despite its relative brevity, this descriptive preamble is necessary to synthesise the situation following the earthquake and what has been achieved over the first five years. A simple illustration of data and figures, however incomplete, helps gain an insight into the level of efficiency and effectiveness achieved by the Regional Government of Emilia-Romagna and the institutional system implemented for the reconstruction, consisting of the techno-structure of the Special Commissioner, the Regional Directorate of the Ministry for Cultural Heritage, and the Municipalities.

However, there are limitations and risks, some of which still high, resulting from uncertainty about the future and the possibility of unpredictable delays, which could have serious consequences and even lead to failure.

V Dynamic stalemate, over-sized equipment, and the dilemma of sustainability

Recovery poses a problem of long-term management and, paradoxically, the best results from the reconstruction, at least as regards the stages of first emergency and first repairs, pose a dilemma in terms of sustainability and, in particular, financial sustainability. Today, many small municipalities that suffered through the earthquake in 2012 find themselves in the paradoxical condition of a dynamic stalemate.

The reconstruction and recovery of places that were heavily damaged is a process that takes a long time, roughly between ten and 15 years. During the first phase of reconstruction (usually the first two years) many activities often take place in an atmosphere of organised frenzy, coinciding with the first six months of the emergency phase. Towns are full of construction sites with cranes, scaffoldings, men at work, and factories resuming their activities. The most affected cities, towns or villages are often visited by political authorities, while local institutions or community-based organisations promote a variety of activities. During the second phase of the recovery process (between two and five years) work progresses more slowly, and strategic projects for new infrastructures, factories and clusters of factories are defined and undertaken, to respond to the post-disaster depression.

Despite these activities, communities often struggle to perceive how or whether reconstruction is progressing. Indeed, it is precisely in the transition between the first and the second phase that everything appears to be suspended in a stalemate phase. The post-earthquake shock is felt for a long time and the many reconstruction activities can only partially help the communities feel that they are returning to the normality of their pre-disaster everyday lives. Meanwhile, public strategies for recovery, followed by private companies involved in the process, proceed relentlessly. It is possible to define this specific condition as a sort of dynamic stalemate.

The moment of transition between the two phases is the most dangerous in terms of adequate public policies and efforts. Often with the goal of healing the wounds of the community, local authorities promote oversized projects, trying to compensate people for their losses. These projects are usually disproportionate, if compared to the real needs of the community, in the present and even more through the decades, in view of the fact that social preferences and objectives are likely to change.

In Emilia, various experts in charge of managing the recovery processes have discussed the future financial risks for municipal budgets once the reconstruction is complete. Small towns of 15,000, 10,000 or even less than 5,000 inhabitants are now equipped with beautiful new schools, new and functional municipal offices, auditoriums for 300 people, assets that will pose a cost of management and maintenance for the next 50 years. This cost must be added to the costs of all the pre-existing buildings – schools, libraries, municipal offices, gyms – that have been and will continue to be largely repaired, becoming more efficient and safe than before the earthquake. Where will these municipalities find the financial resources, the energy and the critical mass of users to sustain new, revitalised, and improved facilities? In fact, these communities are equipped today with new and sustainable buildings which will contribute to the unsustainable financial management of the local governments.

VI Between reconstruction and new development

In the reorganisation of a territorial/sub-regional area, as well as of a city or a small town anywhere in the world, there constantly emerges a particularly complex problem: the relationship between the management of the emergency phase on the one hand and the ability to rebuild, pursuing a different vision of the future, on the other (Lindell). In fact, long-term social impacts usually tend to be minimised during the recovery phase, but more importantly, opportunities for radical transformations through long-term visions are often overlooked, especially in terms of reorganisation, redevelopment, and new development of a city, or a network of small towns and villages, or even a region.

Possibly, the most famous case of reconstruction promoting a radical change compared to the pre-disaster condition is San Francisco (California), after the Loma Prieta earthquake in 1989 (Cervero). In 1986, San Francisco voters rejected the first proposal to remove the Embarcadero Freeway. Then, in 1989, after the damages caused by the 7.1 magnitude Loma Prieta earthquake, the Embarcadero Freeway was closed, and the city opted for removing the infrastructure. After the freeway was removed, in 1991, real estate values in adjacent neighbourhoods went up by 300 per cent, and the whole San Francisco waterfront experienced an urban "renaissance", soon becoming a sightseeing place for people and tourists.

In Italy, the most interesting case is certainly represented by the process of recovery and reconstruction after the 1976 earthquake in Friuli, the north-eastern region of Italy. This case exemplifies, for the first time on the Italian peninsula, reconstruction based on the "where it was, as it was" model regarding physical buildings and monuments. At the same time reconstruction radically transformed the social and economic organisation of the entire impacted area, changing its economy and its dynamics for the next 40 years. The region, known as a poorly developed area before 1976, after the reconstruction was able to capitalise on the small and medium enterprise development model that was characterising the growth of Italian industrial districts in that period, changing the economy of many rural regions. This is a successful model that continued to function from the 1970s to the end of the twentieth century.

The model of material reconstruction and social and economic recovery of the impacted area in Friuli has paradoxically showed that old villages and monuments can be rebuilt "almost as they were" as well as fulfil modern safety standards, while social relations, the territorial organisation, and the economic system may undergo substantial, and not necessarily positive, changes.

A different result was achieved after the earthquake in Umbria and Marche, two regions in Central Italy, in 1997. This earthquake has been defined as the "earthquake of cultural heritage", because it struck mostly many small medieval settlements, the small city of Foligno, and Assisi, one of the world capitals of religious tourism. In this case, the "where it was, as it was" recovery method, based on the previous Friuli experience, was planned, managed and achieved, even at the cost of huge financial investments, but the final social and economic outcomes could not be foreseen. Today, in fact, many of the smaller towns and ancient villages are no longer populated by the original inhabitants, who moved to the nearby towns. The medieval villages have become, at the end of the recovery process, holiday resorts and vacation houses, bought by affluent people from Rome. No member of the committee for the recovery of Umbria and Marche could have foretold or wished for this result. At the beginning of the recovery process nobody among them could have chosen to achieve this result, because no one would have been able to politically support this kind of objective. It simply happened, and it happened because the reconstruction focusing on "where it was, as it was" takes a long time (and huge amounts of money), as a result of: (1) the recovery of ancient materials from the debris; (2) their classification and reorganisation; (3) their contemporary reuse along with new materials and new techniques, while preserving the morphology and the aesthetic quality of the ancient buildings. Nowadays this model, which seemed to be successful from a technical point of view, has been totally reconsidered and redefined, owing to an earthquake swarm that struck a much wider area between 2016 and 2017, causing approximately €23 billion of damage and the destruction of numerous monuments already restored after the 1997 earthquake.

The "where it was, as it was" reconstruction method, nowadays regularly practised by Italian technicians, requires a period of approximately ten years, often extending to 15 years, in order to complete the process of reconstruction. From a spatial, morphological, historical and artistic point of view the outcome of reconstruction is always very successful. However, from the social point of view the outcome of this long process is quite the opposite.

As a matter of fact, during this period the traditional communities previously living in the impacted areas have undergone a process of radical transformation. In fact, in such situations elderly people usually have to be relocated, and only few of them are strongly determined to keep living in the places where they spent most of their lives. Many others never return to their homes, while some grow sick and die during the first 24 months after the disaster. Families and younger residents move to other cities, while their children grow up elsewhere. The identity and cultural connection with these places are broken. After ten years spent elsewhere and with a family and a professional life fully reorganised, for many people it becomes easier to sell or rent their former houses as vacation residences, thereby actually supplementing their income and family assets.

This outcome, understandable at a rational level, cannot be programmed by the experts and explained to people, at least during the first phases of post-disaster recovery. Communities emotionally and materially affected by the disaster could never accept predictive planning that proposes uprooting from their original places. For many people, however, uprooting happens anyway in a relatively short span of time, usually ten years.

In Friuli after the 1976 earthquake there was the ability to manage the physical and spatial reconstruction, while also achieving a broader social and economic revitalisation. In Umbria after 1997 this same result was not fully achieved. In 1976 the adoption of the new "where it was, as it was" model was immediately connected to the objective of economic regeneration of the region affected by the earthquake and linked to the model of emerging Italian industrial districts, even though it was alien to that territory. However, in Umbria the recovery process after the 1997 earthquake did not have the power to achieve an economic transformation.

The conspicuous effort to recover the cultural heritage and many small medieval villages – clearly a plausible effort – led to the concentration of public and private resources on physical reconstruction, resulting in insufficient attention being devoted to the aspects of economic diversification and regeneration. In brief, from a technical point of view, it has not resulted in a long-term strategic planning capable of transforming the ongoing dynamics of the area affected by the disaster. Indeed, looking at this case as a non-resident, it is possible to consider the touristification of the ancient villages as a positive new force that has alleviated economic stagnation.

The different outcomes of these two Italian experiences dating between 1976 and 1997 call for a broader examination. New processes of development could still be triggered with relative ease, based on exogenous models, in an age which was socially, economically and industrially still not particularly complex, if compared to the enormous difficulty of activating, nowadays, a process of economic revitalisation based on non-pre-existing paradigms.

As Scira Menoni has observed in a recent article published in the special issue of *Urbanistica* on Italian post-earthquake reconstructions, despite the high efficiency and excellence of results that Italy has reached in the reconstruction of ancient centres and monuments destroyed or damaged by earthquakes, especially in Umbria and Marche in 1997, the culture of post-disaster recovery has not been able to combine the technical operations and tools of engineering, architecture and urban planning with the social sciences, especially sectors such as sociology and political economy. A decade-long commitment to this specific research field allows Menoni to explain how European research, after the 5th Framework Programme, has not been able to combine the technical aspects of reconstruction with cultural, social and economic components, focusing, almost without exception, on technical contributions by engineering.

In Italy, the most relevant difficulties reducing the ability to "use" the disaster in order to determine a new model of development – or strengthen an existing one – is due to the fact that in almost every Italian seismic event most of the damages and recovery costs are connected with two specific national characters:

1   The relevance of cultural heritage, often not complying with the anti-seismic safety standards;

2   The particularly high percentage of home-owners and the strong and historical preference of private savings toward investments in real estate.

These two issues tend to characterise post-disaster reconstruction in Italy in terms of identity and belonging and high social value assigned to the property. As Michael K. Lindell wrote in 2013, for most people the main goal of the disaster recovery is restoring housing and factories, facilities and infrastructures, which they hope will return to the situation prior to the event. For Lindell, the main problem in this case is that "restoring the community to its previous condition will also reproduce its previous hazard vulnerability" (815).

With regards to Italy and the specific topic of post-earthquake recovery, it would be wrong to maintain that previous conditions of vulnerability are automatically reproduced owing to the "where it was, as it was" method, but it is also true that the acritical employment of such method prevents from establishing after each disaster how to reconstruct according to up-to-date criteria and safety standards.

Reconstruction strategies in Italy are almost always too much geared towards restoring the conditions and quality of the damaged properties, including cultural heritage. This approach, of course, determines a lower attention, at least in the early years of the recovery process, to the choice of the model of development to adopt for the future of the affected territory.

As Webb *et al.* have pointed out, businesses with a larger regional, national or international market are able to recover their buildings and their activities more rapidly than the smaller economic activities and firms that are only locally based. Often large companies can count on insurances, on an easier fund-raising process, and on experts and skilled professionals, while small businesses face greater obstacles in restoring their pre-disaster levels of operations.

This is exactly what happened to the bio-medical cluster located in the area affected by the earthquake in Emilia, consisting of numerous small and medium-sized companies, largely owned by multinational corporations which could immediately react to the disaster, while most of the problems actually reside in the revitalisation of local small businesses (Regione Emilia-Romagna, *L'Emilia dopo il sisma*).

It is nonetheless true that for the first time in the history of Italian earthquakes the national government, following strong pressures by the regional government, has started to fund activities other than the ones existing before the disaster and aimed at making the existing cluster more complex and functional. It is the case of the €50 million allocated to research activities focusing on the productive potentials of the areas hit by the disaster, and of the European Structural Funds 2007/2013 and 2014/2020 used to support the expansion and upgrading of small and medium-sized enterprises, with investments oriented towards product and process innovation (ERVET).

VII Conclusions

The recovery of Emilia has also addressed for the first time in Italy the issue of the recovery of a highly industrialised economic system. Before 2012, in fact, most of the Italian earthquakes had struck predominantly rural and mountainous areas. No expert was aware of the heterogeneous forms of property of industrial buildings (ownership, rent, leasing, etc.). As property is a key factor for receiving financial contributions, procedures for the allocation of public contributions have been significantly delayed. Multinational corporations can wait and have waited, but many small and medium-sized enterprises have faced great difficulties caused by the discrepancy between general legislation and the heterogeneity of the context. The law prescribed that damage should be refunded to the company, not to the owner of the building. Since many damaged buildings were not owned but rented by the companies, it has been necessary to change the law on the basis of this specific and previously unknown context in order to be able to refund the landlords (Regione Emilia-Romagna, *Terremoto, la ricostruzione*).

However, a real cultural deficit emerging in the first five years of the recovery process has been the lack of key strategic elements such as the political ability to reorganise a widespread area lacking critical mass and the scarcity of legal and planning tools aimed at improving the region of the seismic crater with the aid of modern infrastructures.

Up to now, a real process of participation and inclusion in the main choices has been missing. Only four years after the earthquake, the Democratic Party, which presides over a large part of local authorities, started to discuss the future structure of governance and government with the objective of reducing the number of local entities by merging many small municipalities. This is a correct choice in terms of efficiency and effectiveness as well as long-term sustainability. However, it is also a controversial topic involving the citizens' sense of local identity and traditions. A broader participation and the immediate involvement of many local communities is highly advisable and the definition of a tactical strategy in order to achieve the final objective is essential. The Democratic Party has now come to envision a large municipality of 70,000 inhabitants in place of a dozen small municipalities. The goal is even more correct in theoretical terms, but it is also challenging, as is the creation of two new municipalities by uniting neighbours with closely related historical and territorial identities and by involving citizens in the design of the new institution of government.

A second mistake was made by the Regional Government in relation to the construction of a new highway. The area hit by the earthquake has long needed a more modern road, able to connect the villages that still depend on a network of roads dating to the Middle Ages. Yet, the Regional Government has defined other political priorities in terms of investments and infrastructures, and the demographic weight of the sub-region has been too low. In 2010 the Regional Government prepared a first draft for a new fast road connecting most of the urban settlements, but with a lower environmental impact compared to a highway. After the earthquake, the higher levels of government have suddenly decided to entrust the construction of a highway to a private group. This new highway is oversized compared to the area, the environmental impact is huge, and the cost will be very high, because it must be built in compliance with the anti-seismic criteria. For these reasons, some communities and some local groups have strongly protested against the highway project.

These two examples show the deficiency of a strategic vision and expose the inability of democratic institutions (and their technocratic structures) to involve the inhabitants in the policy process. And this is one of the most serious obstacles preventing government institutions from achieving goals of sustainability. All the communities affected by a disaster should have the opportunity, during the first phases of the reconstruction, to think collectively about how they want to react and rebuild and how they see their future in a timespan of at least 20 or 30 years. Only a broader vision of the future developed in the early stages of reconstruction allows for specific actions toward smart long-term development (EPC).

Whenever, after a disaster, institutions and communities fail to stimulate long-term thinking, especially at the regional level, they lose a tremendous opportunity for strategic reorganisation, both intangible, regarding governance, new forms of institutional organisation, new management entities and actors as public/private agencies, and tangible, related to infrastructure, advanced technology poles, productive clusters and existing production and supply chains.

Nowadays crucial issues are the organisation of resilient communities, adaptation to climate change, reduction of vulnerabilities and environmental impacts, preservation of the economic and cultural wellbeing of local communities. Thirty years after the concept of sustainable development was first defined, these issues should be regarded as the new frontier of sustainability, especially when it is necessary to activate a post disaster recovery process.

Bibliography

Cervero, Robert. "Freeway Deconstruction and Urban Regeneration in the United States". Paper delivered at the *International Symposium for the 1st Anniversary of the Cheong-gyecheon Restoration*, Seoul, Korea, October 2006, www.uctc.net/research/papers/763.pdf. Accessed 25 March 2017.

Environmental Planning Collaborative (EPC). *Participatory Planning Guide for Post-Disaster Reconstruction*. EPC, 2004, http://eird.org/cd/recovery-planning/docs/10-additional-resources/TCGI-Disaster-Guide.pdf. Accessed 25 March 2017.

ERVET. "La ricostruzione post-sisma: contesto economico e misure di intervento". *Rapporto 2014 sull'Economia Regionale*. Unioncamere Emilia-Romagna, Bologna, 2014.

Isola Marcella, and Michele Zanelli. "La prospettiva dei Piani Organici per la rigenerazione dei centri storici colpiti dal sisma". *Inforum*, Regione Emilia-Romagna, no. 48, 2015, pp. 13–36.

Lindell, Michael K. "Recovery and Reconstruction after Disaster". *Encyclopedia of Natural Hazards*, edited by Peter T. Bobrowsky, Springer, 2013, pp. 812–822.

Menoni, Scira. "Urbanistica e rischio sismico: appunti per uno stato dell'arte a livello internazionale". *Urbanistica*, no. 154, 2014, pp. 74–80.

Regione Emilia-Romagna. *L'Emilia dopo il sisma. Report su quattro anni di ricostruzione.* Agenzia di informazione e comunicazione della Giunta regionale dell'Emilia-Romagna, 2016, www.regione.emilia-romagna.it/terremoto/numeri/report-4-anni-ricostruzione-1. Accessed 25 March 2017.

Regione Emilia-Romagna. *Terremoto, la ricostruzione*, www.regione.emilia-romagna.it/ terremoto/nove-mesi-dal-sisma/la-ricostruzione. Accessed 25 March 2016.

Seattle Department of Transportation. *Case Studies. Lessons Learned. Freeway Removal.* DOT, 2008.

Webb, Gary R., Kathleen J. Tierney, and James M. Dahlhamer. "Business and Disasters: Empirical Patterns and Unanswered Questions". *Natural Hazards Review*, vol. 1, no. 2, 2000, pp. 83–90.

III.4 Urban life and climate change at the core of political dialogue

A focus on Saint-Louis du Sénégal

Adrien Coly, Fatimatou Sall,
Mohamed B. C. C. Diatta, and
Chérif Samsédine Sarr

I Introduction

The commitment of Sports and Culture Associations is a particularly recent expression of urban governance in Senegal. Long marginalised in the spheres of decision-making or relegated as localised practices with no impact on the functioning of the city, these structures are increasingly needed not only to promote participatory democracy but also to manage public policies.

Civic participation via communication occurs at the "palaver tree", which is a traditional institutional gathering place in the shade of trees where citizens express themselves freely about life in society, the problems of the village, or the city, the policies to adopt and the future in general. Elaborating on Habermas' *Theory of Communicative Action* and *The Structural Transformation of the Public Sphere*, Fweley Diangitukwa has defined the palaver tree as "the most ancient form of governance and democracy" and stressed the importance of arenas of debate, exchanges and construction of an urban identity that addresses the peculiarities of the living environment in Africa.

This form of participation tends to upset the traditional pattern of decision-making throughout the municipality and contributes to the reconfiguration of modes of public action with the establishment of new structures in which people voice their concern about their living conditions. In Senegal, the idea of strengthening community participation beyond elective or administrative representation emerged in 1994. The city of Saint-Louis, through its network of associations, created the "District Councils", in which grassroots organisations, elected officials and delegates from neighbourhoods interact. How does participation foster wellbeing? What are the characteristics of this relationship in the context of climate change?

The MACTOR analysis (Method Actors, Objectives, Strength Ratios) proposed by Godet illuminates the dynamics of alliances and potential conflicts between actors and raises questions about the development of relations involving different actors. Applied to the context of a city that wants to be resilient to climate change such as Saint-Louis, this method allows us to study strategic interactions and the direction of actions resulting from co-operative play between the various components of decision-making mainly in the context of climate change.

II Flooding and climate change in Saint-Louis

Saint-Louis is a city exposed to multiple risks linked to climate, the presence of the river and the sea, and the proximity of the water table (Figure III.4.1). This context favours the existence of districts structurally vulnerable to flooding, erosion and/or salinisation (Weet *et al.*).

Saint-Louis is considered one of the city's most vulnerable to climate change in Africa, and particularly in Senegal, where it accounts for half of the damage and estimated economic losses of the country. The total discounted cost of river

Figure III.4.1 Risk situation in Saint-Louis.
Source: original by the author.

flooding in 2080 would amount to 818 billion CFA francs, the equivalent of 13 per cent of the GDP in 2010. Estimates of flooding in Saint-Louis indicate a low of 80 cm and a moderate level of 203 cm (Sall). Without protection, 80 per cent of the territory will be submerged by 2080. Since colonial times, in the context of urban development, energy has been increasingly focused on raw variables with the establishment of dikes, docks, and banks. If river flooding seems controlled, sea and rain submersion and water rising still threaten populations. The latter are organised at the institutional level, creating new synergies at the local level to bring structural solutions under the control of associative movements at the neighbourhood level.

III The organisation of Saint-Louis neighbourhoods

In Senegal, the idea of associations has long prevailed, particularly in rural areas (68–08 Act of 26 March 1968). Decree No. 76–0040 of 16 January 1976, laying down special obligations to which associations for popular education and Sports and Cultural Associations are subjected, and the 8437 Law of 17 April 1984 were adopted to promote microeconomic activities and foster Economic Interest Groups (GIE). Such legislation has led to a proliferation of associations and popular movements in the city. These structures are designed for self-care, especially in the fields of environment, living environment, socio-cultural mobilisation, health, education, women's promotion and the economy.

A very strong community involvement in the district's associative dynamic is therefore noticed (Table III.4.1). Associations are usually thematic. Economically, these basic community organisations represent specific Economic Interest Groups and aim to improve the income of their members.

The Sports and Culture Associations are the institutions that foster the greatest adherence by enhancing the identity of their neighbourhoods. These associations enjoy a good social integration in their neighbourhood and pre-existing social structures, because they generate mobilisation of energy and are appreciated for their community activities. Their motivation is also clearly displayed in their objectives, which, increasingly, are the development of the neighbourhood. Socially, people recognise themselves in the activities of the Sports and Culture Associations, which help defuse potential conflicts and maintain social peace in the district.

The expansion of the associative movement launched by young people has been a strong signal to the municipal authorities for a new governance that considers not only the requirements of their living environment and their economic involvement but also the requirements of their involvement in the decision-making concerning the affairs of the city (Niang). Aware of the scale and weight of associations, municipal authorities were directed towards local governance based on consultation and local participation carried out by community leaders and an institutionalisation of the neighbourhood scale through the creation of District Councils. In addition to the leadership of the Sports and Culture Associations, the District Councils appear mainly in the form of a federation of all the

Table III.4.1 Structures of the District Councils

Structures	Description	Missions	Features
Sports and Culture Associatio1CSA)	Citizens' Movement	Strengthening the capacity of youth socio-economic integration	Instrument of animation and mobilisation Recognition and ownership by the population Team spirit and challenges
Cultural Associations	Cultural Movement	Promotion of culture and solidarity among groups Strengthening social links	Ability to mobilise every individual Ability to respond satisfactorily in emergency situations
Religious Associations	Religious Movement	Promotion of spiritual values Support for members in difficult situations	Mobilisation Transmission of information
Social Associations	Citizens' Movement	Preservation of the physical and moral integrity of individuals	Support for marginalised or neglected individuals
Community Health Associations	Citizens' Movement	Promotion of different aspects of health	Ability to communicate Proximity to healthcare users Ability to induce change
Women Promotion Group	Women's Movement	Improving women's income Social inclusion Access to financial resources Socio-economic promotion Fight against poverty	Ability to mobilise women's economic resource
Young Women Organisation	Women's Movement	Promotion of young women	Focus on gender
Economic Interest Group	Economic Operators	Economic development of members	Job creation Generating income
Popular Bank	Financial Operators	Savings development	Capitalisation for low-income households
Togetherness Group/Friends' Group	Citizens' Movement	Bringing together individuals with a common goal	Volunteers Fundraising capacity
Neighbourhood Leader	Local Authority	Representing the municipality/state in public management Social mediator Administrative facilitator	Administrative legality Popular legitimacy
Local Residents in Neighbourhoods	Local Authority	Representing the interests of their political parties	Ability to discuss and share issues
Students' and Parents' Association	Citizens' Movement	Support for school management	Mobilisation of children's education
School and the Environment	Educational Structure	Implementation of research and action	Support for issues related to climate change

Source: original by the author.

associative structures working towards the development of the district. Local power-uniting populations have arisen quickly.

The Neighbourhood Council is defined in its constitution as an apolitical, non-denominational and non-corporate associative structure. Among its missions, it has the role of co-ordinating the various actions occurring in the neighbourhood, serving as a consultation framework office, mediating between different actors for any intervention needed to ensure community development through the mobilisation of all appropriate resources, and promoting citizenship and participatory democracy (Niang). Neighbourhood Councils have become the formal frameworks of management problems at the local level, the main contact for any work related to development activities in the district. This allows for a better rapprochement between local authorities and the people who now have a better reading of the roles of the municipal institution. It also allows the latter to strengthen the implementation of their programme through local management.

However, the Neighbourhood Councils do not work directly with the city; they are related to the Municipal Development Agency, the technical arm of the commune, which aided their establishment and continues to supervise their activities. This participation dynamic eventually established a governance framework uniting "decentralised" and "community" powers in urban areas. It should be noted that decentralised power is the one exercised by officials elected by universal suffrage, deconcentrated power is the one devolved to the territorial administration, and community power is related to civic commitment.

At the institutional and organisational level, this situation led to a consultation framework through multidisciplinary committees composed of "elected people", representatives of the decentralised services of the state, and members of the civil society, making District Councils appear as relays for the local authorities. This facilitates the mobilisation of all actors in disaster situations and presents a new turn in the different parties' actions. The set of associations, groups and individuals is actually represented in the various committees of the District Councils for inclusive participation.

IV Urban governance risk

The nature of the structures and the level of co-operation between actors complicate their mutual understanding and the ratio of power in the urban metabolism. The actors' play informs us about the nature of alliances and sources of actual or potential conflicts that limit formulated answers. In Saint-Louis, where the Sports and Culture Association is at the heart of potential alliances, there is joint control between several actors, as shown in their close distances and considering the differences and similarities between them.

Interactions between strategic actors indicate that "a possible prioritization of this diversity of actors can develop in a distinction between institutional actors (territorial levels) and industry actors (operators who animate the scale of the territory)" (Debrie 109). The analysis at the level of the community of plans between actors indicates that local authorities, the Municipal Development

Agency and the District Council have convergent actions. However, the first two maintain stronger relationships in public decision-making. Neat distances between actors both confirm and draw a triangle of actors directly involved. It is clear from the analysis that the actors and power relations producing urban metabolism in Saint-Louis occur in a climate of significant co-operation between actors. The ambivalence and divergence among actors are very low as they co-operate on development projects.

The convergence of actors is a major asset for a city exposed to several hazards. We are then in the context of what has been defined as a concrete action system (Crozier and Friedberg), i.e. in a context where the rivalry between the actors does not harm the urban development project. At the neighbourhood level, the District Council is therefore presented as a timely entity to support the well-being of populations, thus establishing the mechanism of civic co-operation. An evaluation of different District Councils reveals a differentiation in the level of achievement of their objectives. The explanation lies in the capacities of the people who support these organisations and their level of relationships.

Governance of urban risk in Saint-Louis seems to be the domain of the commons (Ostrom), where public authority, local actors and technical and financial partners liaise, sharing strong civic involvement. Power relations depend on the position, role and importance of the actors involved in urban construction. In the case of Saint-Louis, the District Council is the primary level of implementation of community directives that are either disputed or accepted. The contribution of Sports and Culture Associations and Basic Community Organizations is significant, especially in public education or in the mitigation of impacts, being the first actors to respond in cases of flooding, for example. This contribution of non-institutional structures in risk management indicates the relevance of their integration in the implementation of policies of prevention and response at the local level. They complement the implementation chain of public decisions. Perrin assimilates this kind of practice to an "institutional transplant, inaugurating a process of institutional convergence of territories" (281).

From the perspective of the theory of regulation (Reynaud), the strategy developed by the municipality of Saint-Louis of integrating non-institutional actors is a key factor of stability. Centralisation orientates people towards conducting civic actions or supporting the efforts of political authorities, thus avoiding a situation of anomie. The ways in which the city is led fit well within a sustainability perspective. Different actors adhere through a consultation approach and good attendance that installs the exercise of power in a scheme of political dialogue, understood here as civic debate.

V Political dialogue and participation

Political dialogue is a civic action that can foster a productive exchange between actors of different levels, allowing them to analyse as well as systematise possible causes and consequences of vulnerability, consolidate participation in times of crisis, cope with the challenge of collecting information and optimising solutions, and mediate knowledge. Political dialogue is a negotiating tool acting on consultation. Its aim and purpose is the search for consensus in a context of ongoing discussion with concerned parties on governance for social peace and human security. It is an approach that allows one to reach the household, the thinnest entity of vulnerability in the context of climate change.

Participation in climate change actions involves programmes related to storm water, integrated coastal zone management and generic environmental management. Involvement is quite varied and can be expressed in activities requiring the roles of prime contractor, client and even partly responsible for social engineering or planning.

As regards risk governance, political dialogue facilitates the citizens' empowerment through the reappropriation of technical, organisational, social and political dimensions of their environment. This allows them to negotiate the rules of the game and strongly contribute to the development of a common project. The public policies that emerge, having the support of the largest numbers, become operative.

The District Council of Diaminar carried out the difficult project of relocating 50 families under the supervision of the Municipal Development Agency. It is known that constraints arise with this type of intervention, not only involving the displacement of a population that is not always accepted by the persons concerned but also involving fair compensation for the displaced ones.

This approach centred on consultation and participation responds somehow to the principles of mobility in alliances and the continuity that must animate the game of actors. It recalls the "palaver tree" in rural Africa. Political dialogue in urban neighbourhoods is a hybrid of the "palaver tree" practiced by different people from different backgrounds who identify with such an approach. This approach capitalises certain African cultural values and places them at the service of the community. It promotes consensus, giving the floor to all people who would like to decide on common issues and find solutions. More formalised, it favours decentralisation, the exercise of power, accountability and proximity communication (Kessy).

A focus on "some positive traits of African culture such as community spirit, the trust agreement, the sense of communication and commitment" (Kessy 50), is essential to placing people at the centre of all activity in modern management. The main concern is the ongoing search for symbiosis and synergy between the universal laws of management and the inclusion of the African socio-cultural environment. The trading is thus done around typical elements of African culture under the prism of "political dialogue", which describes itself as a participatory approach to the governance of vulnerabilities and is defined as a civic action in Saint-Louis, Senegal (Vedeld *et al.*). This approach has implications for the decision-making process, with information flow coming directly from the citizen. Thus, in Saint-Louis, one can observe a correspondence of key actors around the objectives of group relationships, ethics and participation. The analysis of neat distances between management objectives of the district of Saint-Louis places the link between actors.

The results of the political dialogue are an increase in the level of knowledge with a return of experience, an inventory of solutions, approaches and methods from the communities, and population mobilisation. These results help build a common vision and shared scenarios with communities through a causal risk analysis following a confrontation between scientific knowledge and local knowledge for decision-making support. Appropriation of local development guidelines through consultation helps mitigate rivalries and conflicts that may arise between actors. As part of climate risk management, it has been a tremendous asset, ensuring mobilisation of all parties, better empowerment through feedback, and capacity building of grassroots organisations in risk management.

VI Participation as key to a culture of wellbeing

Welfare is the result of mobilisation around the exercise of citizenship that meets a set of requirements, namely understanding, sense of belonging, need to know, and will to act. Based on the assumption that in any group of humans, relationships are paramount through "daily activities, research and development" (Coly 182), participation appears as a vector of wellbeing. Considering the objectives of the actors on climate change, wellbeing is built around health issues, economic development and the security of populations in Saint-Louis. The involvement of citizens through political dialogue gives a special dimension to the mitigation measures proposed by the municipality in risk management.

It must be emphasised that, here, the verticality of decision centres and horizontality of places of power make the governance of urban risk more complex, because the mastery of the game involves controlling the power of decision-making. According to Crozier, an actor has power because s/he mastered an area of uncertainty. If you are the person that controls such an area, then you will have power over those who are affected by the uncertainty you control. However, as there is rarely a situation of monopoly, everyone tries to influence others depending on the uncertainties that they control, hence the complexity of the viewpoints, acts and contradictions regarding the streamlining.

The involvement of actors in climate risk management indicates substantial power relations. Strategic interactions, co-operative play among actors, and weak differences between actors are the result of consultation and participation of all parties in the affairs of the city, ensuring a level of information and consideration for all. The political dialogue characterising this form of governance is an element of culture which works as a response of human beings in their environment, tending towards an intellectual and moral balance between human beings and their environment (Senghor). It is a social construct well suited to urban spaces and aimed at implementing public action around specific values such as autonomy, tolerance, solidarity, responsibility.

In Saint-Louis, the participation of all parties generates a better connection between the municipality and the people. They have a better reading of the roles played by the municipal institution to better implement its programmes via local management. As a part of risk management, actors can be mobilised in cases of disaster to bring relief to the victims, especially in small territorial neighbour-hood units. The neighbourhood is relevant since it is defined as the smallest scale of interpersonal grouping, the recognised spatial level of an identity experiencing coincidental but inevitable encounters, especially as the people who share the space know each other and share the same standard of living and feel the same nuisances (Tall).

VII Conclusion

The process of creating a culture of wellbeing in Saint-Louis is carried out by the District Council and is characterised by participation (exercise of citizen-ship). Within this process, negotiation through consultation (political dialogue) is at work and leads to the definition of lines of action to address the manifesta-tions of climate change (floods, in particular).

The palaver tree is an instrument used by different social groups in Africa adapted to the urban context. It occurs in the city, in the public square, in the mosque, in the meeting room of the neighbourhood or the home of a citizen in several formats. It is an informal public hearing that can be spontaneous, based on the event. The process is supported by the associative fabric of the leadership of associations, with the Sports and Culture Associations being the most sophist-icated organisations that create the majority of membership.

The Municipal Development Agency has sought to formalise the framework of interventions of parties by proposing the development of a district develop-ment plan that retains the mechanisms of consultation and articulates issues of development at different levels, from district to town. The District Council is thus the "palaver tree", the place for political dialogue that establishes the satis-factory levels of living in terms of physical, economic and social health. Ongoing negotiations around issues of wellbeing occur by activating traditional African mechanisms and tools of co-operation.

Political dialogue retains the properties of an organisation with very small power distances integrating accountability and delegation of authority. Participa-tion thus appears as a generator of wellbeing encompassing health, socio-economic development and human security. Participation is the result of mobilisation around the exercise of citizenship that meets a set of needs. Urban culture in Saint-Louis has gradually developed around correspondence object-ives, pursued by the neighbourhoods expressing their power of association. A culture of wellbeing is taking shape, based on a triple relation to groups, ethics and participation.

Bibliography

Coly, Adrien. *Le système fluvio-lacustre du Guiers: étude hydrologique et gestion quantitative intégrée*. Thèse de doctorat de 3ème cycle, Université Cheikh Anta Diop de Dakar, Département de Géographie, 1996.

Crozier, Michel, and Erhard Friedberg. *L'acteur et le système*. 1977. Seuil, 1990.

Debrie, Jean. "Une approche territoriale de la gouvernance ou le dialogue secteur-territoire: l'exemple des projets urbains fluviaux". *CIST 2011 – Fonder les sciences du territoire*. Proceedings du 1er colloque international, CIST, 2011, pp. 109–113.

Diangitukwa, Fweley. "La lointaine origine de la gouvernance en Afrique: l'arbre à pala-bres". *Gouvernance*, vol. 11, no. 1, 2014, pp. 1–20, www.revuegouvernance.ca/images/content/Spring2007/fweley.pdf.

Godet, Michel. *Manuel de prospective stratégique*. Dunod, 1997.

Godet, Michel. "Jeu d'Acteurs et de Stratégie: la Méthode MACTOR". *Stratégique*, no. 46, 1990.

Habermas, Jürgen. *The Structural Transformation of the Public Sphere: An Inquiry into a Category of Bourgeois Society*. Translated by Thomas Burger, with the assistance of Frederick Lawrence, MIT Press, 1989.

Habermas, Jürgen. *Theory of Communicative Action, Volume Two: Lifeworld and System: A Critique of Functionalist Reason*. 1981. Translated by Thomas A. McCarthy, Beacon Press, 1987.

Habermas, Jürgen. *Theory of Communicative Action, Volume One: Reason and the Rationalization of Society*. 1981. Translated by Thomas A. McCarthy. Beacon Press, 1984.

Kessy, Marcel Z. "Des idées qui ont marqué le secteur de l'eau africain". *L'Afrique et l'Eau*, edited by Claude Jamati, Alphares, 2014, pp. 43–62.

Niang, Demba. *Gouvernance locale, maitrise d'ouvrage communale et stratégies de développement local au Sénégal: l'expérience de la ville de Saint-Louis*. Thèse de Géographie et Aménagement, Université de Toulouse le Mirail, 2007.

Ostrom, Elinor. *Governing the Commons: The Evolution of Institutions for Collective Actions*. Cambridge University Press, 1990.

Perrin, Pierre. "Réplication des Institutions et Convergence des Territoires". *Revue d'Économie Régionale and Urbaine*, 2006, pp. 281–301, doi 10.3917/reru.062.0281.

Reynaud, Jean-Daniel. *Les Règles du Jeu: L'Action collective et la Régulation sociale*. Armand Colin, 1997.

Sall Moussa. *Crue et élévation du niveau marin à Saint-Louis du Sénégal: impacts poten-tiels et mesures d'adaptation*, thèse de doctorat, Espace géographiques et sociétés, Université du Maines, 2006.

Senghor, Léopold S. *Négritude et humanisme*. Seuil, 1964.

Tall, Serigne M. "La Décentralisation et le Destin des Délégués de Quartier à Dakar (Sénégal). Plaidoyer pour les délégués de quartier de Dakar après la loi de décentralisa-tion de 1996". *Bulletin de l'APAD*, no. 15, 1998, http://apad.revues.org/567. Accessed 17 March 2017.

Vedeld, Trond, Wilbard Kombe, Adrien Coly, N. M. Ndour, Clara Kweka, and Siri Hellevik. *Urban governance, climate extremes and resilience in Dar es Salaam and Saint Louis*. Final report, CLUVA, Delivrable D3.1, 2013.

Weets, Guy, Adrien Coly, Emmanuel Tonye, Hamidou Toure, Kumelachew Yeshitela, and Wilbard Kombe. *Synthesis of the Results Obtained by the Case Studies*. CLUVA, Deliverable D5.8., 2014, www.cluva.eu/deliverables/CLUVA_D5.8.pdf. Accessed 28 March 2017.

III.5 Is ethno-tourism a strategy for sustainable wellbeing?

A focus on Mapuche entrepreneurs[1]

Gonzalo Valdivieso, Andrés Ried, and Sofía Rojo

I Introduction

Ethno-tourism has been studied from different perspectives. It has attracted the attention of scholars mainly owing to the possible benefits and negative effects it can have on the indigenous communities involved and the promotional role played by national or regional governmental agencies. However, limited attention has been devoted to the views and goals pursued by indigenous entrepreneurs themselves as they engage in productive activities related to serving tourists (Campelo *et al.*).

This chapter explores the motivations of Mapuche entrepreneurs who have developed and commercialised indigenous tourism. In particular, we have observed how the Mapuche value their own culture and social encounters with people from different backgrounds. Communities and families change when they engage in this kind of tourism enterprise, which are supposed to complement traditional economic endeavours.

This study was conducted in Araucanía, a region in the south of Chile, where most of the Mapuche population is located and a significant number of national and international tourists are received every year. Interviewees have showed two contrasting attitudes. Some believe that tourism can help them preserve and value their natural and cultural heritage. Others see opportunities for profit and employment, complementing traditional farm activities and allowing them to stay connected with their land.

The first part of this chapter introduces ethno-tourism with a specific focus on the experiences of providers. The second part is an overview to the context of the Mapuche people in Araucanía, detailing their status in relation to national policies over the last few decades. The third section presents methodological aspects, conclusions and final remarks.

II Tourism versus ethno-tourism

In recent years there has been an increasing interest in the contribution of ethno-tourism to local development and general wellbeing of indigenous peoples, with an eye to the preservation of cultural heritage (Pratt *et al.*; Yang; Yang and

Wall). Researchers have assessed the benefits of tourism on native communities as well as the possible negative effects (Dyer *et al.*; Tew and Barbieri; Trau and Bushell). Others have examined the role and impact of government policies on ethno-tourism via national, regional and local agencies (de la Maza; Ruhanen *et al.*; Yang and Wall). There are also some interesting papers about the motivations behind what is known as agritourism (Hernandez-Maestro and Gonzalez-Benito; McGehee and Kim). However, the motivations and perceptions of the indigenous entrepreneurs themselves and how these notions have evolved over time require closer inspection.

There are many different definitions of ethno-tourism (or indigenous tourism). It can be understood as a specialisation of cultural tourism, where the central activity focuses on an ethnic community, usually rural, whether living or historic, which expresses itself through daily activities and rituals, either material or spiritual. Pierre van den Berghe states that ethno-tourism exists where tourists actively search for ethnic exoticism: "The native is not merely a host, a provider of creature comforts, a servant, but becomes, quite literally, the spectacle" (9). It is generally accepted that ethno-tourism is an activity where travellers are looking for an encounter with an ethnic group whose culture is different from their own and, in some way, belongs to, or originates in, that place or village. "Ethnic tourism is drawn to those groups that are most clearly bounded and culturally different" (Picard and Wood 8). Ethno-tourism, as part of cultural tourism, takes place when the main attraction is the non-Western culture of a community, regardless of whether it is original or transplanted (de la Maza).

Even if ethno-tourism is related to one particular culture, expressed by native or aboriginal people, there is no wide consensus as to whether it should be owned and managed directly by them. In fact, there are many ethnic theme parks worldwide that are not managed by indigenous communities and yet have developed cultural attractions labelled as ethnic (Ryan and Huyton; Yang). Other authors like Butler and Hinch have stressed that indigenous people should be deeply involved in these activities: "Indigenous tourism refers to tourism activities in which indigenous people are directly involved either through control and/ or by having their culture serve as the essence of the attraction" (5).

There are many activities related to ethno-tourism which include cultural events, arts, performances, festivals, museums, villages, dance, trips to the natural heritage, craft production, revival of religious ceremonies, lodging in traditional houses, and gastronomy (Dyer *et al.*; Yang and Wall). Robert Wood in "Touristic Ethnicity" argued that tourism tends to make people much more self-conscious and reflexive about the "cultural stuff" they may have taken more or less for granted before (225).

One particular issue that comes to light in previous research is the relationship between tourism in rural areas and ethno-tourism. Even if they are not identical, they show common features. Indigenous communities that participate in tourism tend to be located in rural areas and in many cases they continue to engage in their own traditional agriculture. Another cause for this rural-ethnic connection is that many governments promote tourism in rural areas as a strategy to alleviate poverty or social inequalities within those communities, without regard to whether they are indigenous or not (Briedenhann and Wickens; Hall; Hall *et al*.).

II.1 Motivations in tourism

Motivations represent the "why" behind a certain behaviour and are related to several factors. Some are described as intrinsic (*push*) and understood as psychosocial forces stemming from past experiences. They are based on personal needs and tend to promote participation in specific activities. There are other motivations that attract (*pull*). These are more related to the nature of the activity and the impulse to choose it (Goytia, "Ocio Y Calidad de Vida"; Mannell and Kleiber).

Motivations are also strongly determined by social and environmental factors. Even if a person is innately predisposed towards a certain behaviour, it may not be expressed unless the social environment leads to it (Iso-Ahola). The effort by authors like Iso-Ahola, Mannell and Kleiber to include this social dimension should be complemented with a broader understanding of the social, cultural and anthropological aspects that create motivation. Motivations are a complex set of impulses, reasons and needs driven by some demand and/or their physiological, social or cultural satisfaction. They can lead to a particular behaviour which, when executed, provides feedback on the concrete results and the personal satisfaction achieved. As a consequence, participants decide whether to continue, reinforcing their initial motivation, or not.

It is clear that the motivational process takes individual differences into account. What motivates people, when and how much, can vary significantly. In any case, motivations energise, activate and direct our behaviour, even if they are socially determined. In this chapter, we will try to understand the main motivations of entrepreneurs in their decision-making process with regards to the use of their time and resources in tourist initiatives.

It is also important to consider a motivational approach in which tourism is perceived as a psychosocial phenomenon and, therefore, as a leisure experience. In this way, it can be understood as being autotelic, such that motivation comes naturally. Autotelic experience, being an end in itself, comes from the notion that leisure activity is driven by an intrinsic desire reinforced by the satisfaction that participation provides, rather than by some external reward (or punishment). This is very different from extrinsic motivation, wherein the reason to do something comes from elements external to the activity itself (Deci and Ryan 1035).

The idea that leisure and tourism experiences are, by definition, intrinsically motivated is widely accepted (Cuenca, *Ocio Humanista*; Iso-Ahola; Mannell and Kleiber; Neulinger; Reeve). According to Manuel Cuenca, "authentic leisure is a vital experience, a sphere of human development which, starting from a certain attitude towards the object of the action, rests on three essential pillars: perception of free choice, autotelism and a rewarding feeling" (*Ocio Humanista* 96). We could say that people are inherently endowed with a natural tendency to learn, develop and freely participate in actions driven by their internal world of needs, impulses and experiences, more so than by the external world of rein-forcement (Ryan and Deci).

The notion of tourist services providing a leisure experience is founded on the "valuable leisure" approach. This approach suggests that people and com-munities engage freely and voluntarily in initiatives linked to the rescue of their culture and traditions. They do this in an intrinsically motivated way, driven by the pleasure or satisfaction that comes with assigning value to that experience (Cuenca, *Ocio Valioso*).

When considering the motivations of entrepreneurs in tourism, there are two fundamental approaches. The first one is the Opportunity Entrepreneurship Activity (OEA), which includes ventures associated with the exploitation of an opportunity or necessity by people who create a company as a career option. The second one is the Necessity Entrepreneurship Activity (NEA), which is chosen by people not because they want to be independent, but because they cannot find another way to subsist (Minniti *et al.*).

Another important aspect behind the motives for starting a business are the incentives provided by governmental and non-governmental institutions. Some authors insist that even if incentives are important, the deciding factor in the process of entrepreneurship operates at the individual level. This has to do with the potential entrepreneur's internal aspirations, the external viability associated with the social image of the activity and the personal support of the people closest to them (Marulanda Valencia *et al.*).

The specific motivation for entrepreneurship in Mapuche tourism has rarely been observed or researched. One study proposes that Mapuche entrepreneurship is mainly an activity of economic of subsistence that re-values Mapuche culture (Landriscini). The author highlights the fact that, for the managers of these asso-ciative enterprises, opportunities for strengthening reciprocity between local groups is important. They are looking for ways to reinforce local organisations and give prestige to cultural values and resources. As such, Mapuche entrepre-neurship is an alternative to the current economic model. At the same time, it strengthens the Mapuche cultural identity, both individually and collectively, by associating the indigenous people with the formation of the first legitimate nation of Araucanía.

II.2 Expectations of tourism

Tourism has become one of the main economic activities in rural areas in recent years. Among other parameters, it is interesting to consider that the international movement of visitors to Chile in 2012 reached a record increase of 13 per cent, much higher than the estimation of 4 per cent given by the World Tourism Organization (WTO) to the continent. In terms of GDP, the results are even more encouraging. The contribution of the tourist sector to the national GDP is expected to double by 2020, with an annual growth of the number of tourists to more than 5.4 million and an increase in their spending by 50 per cent (Gobierno de Chile 17).

It is important to mention that the National Tourism Strategy has chosen "Original Nature" as a slogan which combines the environmental experience of beautiful landscapes and impressive vistas (lakes, woods, volcanoes) with cultural heritage, mainly represented by the Mapuche as the native peoples of the region, in order to attract long-distance travellers. In fact, 65 per cent of these tourists consider nature as the most important reason for visiting the country (*Chile. Por un Turismo Sustentable*).

In terms of rural tourism, its expansion derives from the diminishing productivity of traditional agricultural activities and low earnings as well as from governmental agency programmes oriented to diversify and increase the sources of economic income for rural families. However, rural tourism in the south of Chile shows many weaknesses that demand public and private investment interventions to make it a sustainable economic activity. The main weaknesses are low associativity, deficient management, low quality in services and limited offer. In order to face these gaps three kinds of assets have been envisioned: first, "personal abilities" focusing on responsibility, motivation and friendliness; second, the development of abilities in communication, teamwork and decision-making; third, the development of technical skills in customer services, tourism, management and social leadership (Szmulewicz *et al.* 1025).

Many rural places present some kind of cohesion formed by rural activities, rural life and tourism. These services and agritourism experiences partly incorporate the fundamental aspects of the Mapuche culture and traditions.

III The Mapuche people in Araucanía

The historical context of the Mapuche in Araucanía is significant. At the end of the nineteenth century, most of the land occupied by the Mapuche people was confiscated by the national military force. Then, Mapuche families (and communities) were relocated to different parts of the region with land entitlement (reservations). After this process, a national policy of assimilation was implemented, which forced the Mapuche people to set their own traditions aside and become part of the national identity of the country. This included the learning of Spanish and a ban on the use of the Mapuche language (Mapudungun) in public schools. The Mapuche communities were also absorbed into the national health system; they were given the right to vote and were obliged to abide by national laws (Boccara and Seguel-Boccara). For many, this process triggered the loss of ancestral traditions, cultural heritage and language. Most of the communities living in the reservations broke apart. Their members migrated to cities, took menial jobs and became part of the national Chilean society (Bengoa).

Today, the Mapuche people are approximately 11 per cent of the national population and 31 per cent in Araucanía. As an ethnic group, they have higher poverty rates than the rest of the country. They are located in the agriculturally less productive areas, have been almost completely assimilated by the national system and attend the same schools as the non-Mapuche people. They are part of the national health system, are entitled to the same governmental benefits, with some exceptions regarding specific ethnic subsidies, and go to national or regional higher education institutions. In conclusion, this means they no longer have their own institutions.

The transformation of the Mapuche society has been promoted and driven by the state through periods of agreement and discord, but in general with great contempt for their original culture and traditions. This entire situation almost caused the disappearance of their language and culture, especially during the twentieth century. Cultural values and heritage survive only in some specific places or among the older generations within communities.

In the 1990s, after the return of democracy following the end of Pinochet's dictatorship from 1973 to 1990, there was a change in this assimilation process. The comprehensive Ley Indigena was approved by the National Congress of Chile in 1993 and the Corporación Nacional de Desarrollo Indígena – CONADI (National Corporation for Indigenous Development) was created as an intermediary institution between the state and its first nations.

Some multicultural public policies were introduced, such as the bilingual education programme in some rural schools and the intercultural health programmes in some hospitals. The participation of traditional Mapuche healers, known as *machi*, was encouraged in some family health centres (Boccara and Seguel-Boccara). Slowly, the value of indigenous cultural heritage started to be acknowledged at the national level and, along with that, some social transformations began, including new public policies on poverty alleviation and local economic growth programmes. However, this did not solve all the Mapuche demands, and the so-called Mapuche conflict[2] arose. This is still an ongoing issue, especially in Araucanía. There have been acts of violence and attacks on private property. The Mapuche have demanded the return of their traditional land, along with legal autonomy and recognition as a nation.

In economic terms, the Mapuche people in Araucanía currently live in both cities and rural areas. They are involved in all sorts of economic activities, but more associated with traditional agriculture and handicraft production. They are generally of lower income status.

III.1 Ethno-tourism in Araucanía

The region of Araucanía in the south of Chile has an extraordinary natural heritage of lakes, volcanoes (some active), pristine temperate forest, hot springs, a vast ocean coast, national parks and glaciers. Since the 1940s the landscape and its natural resources have attracted national and international travellers (de la Maza; Gedda).

Being the land originally occupied by the Mapuche people, it still has the highest concentration of Mapuche communities. Since 1990, and increasingly so after the turn of the century, the idea of developing indigenous tourism has been promoted by the national and regional governmental agencies, as well as national and international NGOs.

One of the most interesting examples of this process was the establishment of the Mesa de Turismo Mapuche (Board of Mapuche Tourism) involving regional institutions and Mapuche entrepreneurs. The Board has defined Mapuche tourism as a concept which refers to ethno-tourism in the territory and identified some of the main characteristics this type of activity should have (de la Maza). The National Tourism Service (SERNATUR), a government agency in charge of developing tourism as an industry at the national level, formalised Mapuche tourism as part of its promotional strategy for 2010–2014. Araucanía has been designated as "Original Nature" including natural and cultural heritage, and the *ruka*, the traditional dwelling of the Mapuche people, has been chosen as the logo. Mapuche tourism has achieved recognition as an integral part of the national strategy for economic development in this area.

It is, however, important to be aware of the tension between this "commodification" of indigenous tourism and the authenticity of the tourist activities under development. One of the ways in which the Board of Mapuche tourism tries to manage this risk is by stressing the notion that only the Mapuche people, possessing a vast knowledge of their culture and values, can develop this kind of tourism.

Formal institutional consensus has been reached in Araucanía about qualifying indigenous tourism as Mapuche tourism. This is defined as an economically sustainable activity, performed by Mapuche entrepreneurs who preserve their own worldview and traditions, use their own language, and make their livelihoods in harmony with the environment, which in turn is both protected and creates value. This framework aims to offer an authentic cultural experience to national and international tourists (*Fundamentos del Turismo Mapuche*).

IV Methodology

A qualitative methodological approach has been adopted to identify and understand the views, motivations and meanings that Mapuche tourism providers attach to this activity. Sociological statistics have been covered in other studies. It is understood that experiences are better communicated if they are told directly by the protagonists in their own words, using their own symbols and metaphors (Henderson; Ruiz Olabuénaga).

This study is based on semi-structured interviews with 13 entrepreneurs of Mapuche tourism. The information was collected using 25 open-ended questions, each of which was transcribed and the results arranged in a matrix in order to analyse the main content of each response, identifying the issues most frequently cited by those who were interviewed. Analysis was performed by establishing hermeneutic codes (Glaser and Strauss) and then identifying interactions, similarities and concepts. We have examined and integrated a large data set by applying a comparative method, using the Atlas TI software and making connections among the talks of the participants.

The set of questions is divided into five parts: the beginning of the initiative; initial and present motivations; perceptions about Mapuche tourism; relationships between providers and visitors; the effects of this entrepreneurship on the community.

V Findings and discussion

The findings are presented below in two sections illustrating the perceptions and motivational relationships that Mapuche entrepreneurs have developed with regard to ethno-tourism. Each of them is preceded by a general statement and exemplified through quotations from the interviewees. All quotations are from Mapuche entrepreneurs who have been active in tourism for at least five years. They are translated from the original Spanish by the authors.

V.1 Perceptions regarding tourism

Important differences were observed in the perceptions about Mapuche tourism and tourism in general. Since all participants place their culture at the centre of their activity, Mapuche tourism is practiced to validate Mapuche culture and help tourists to appreciate it. At the same time, it is also quite clear that they see tourism as a business activity. In general, they regard it as the main component of a sustainable livelihood, rather than an end in itself.

V.1.a Intercultural encounter

A major idea expressed by the interviewees is that Mapuche tourism generates a cultural experience, a meeting between two worlds, a specific form of tourism in a place with an identity of its own, capable of showcasing local customs, eating habits, experiences, and traditions:

> for me tourism is ... about sharing experiences, or when other people really feel welcomed by ... a culture, by people, which is what they're really looking for. People want to travel, ... they need to find something that they don't have where they're from, so what they want is to know, to share and to live an experience, and that experience gets you to really show part of your culture, part of your identity, and in some way, tourism also helps you to strengthen your culture.

This is also related to the idea of tourism as leisure, based on a rewarding subjective experience, in which emotional, symbolic and motivational dimensions are combined and projected onto social and cultural areas (Goytia, *El Turismo como Experiencia de Ocio*). These experiences become memories that give meaning to the participants' lives.

V.1.b Conservation and transmission of worldview

Mapuche tourism is a way of transmitting their worldview, from beliefs and values to eating habits and traditions. Their hope is that a better understanding of Mapuche conceptions will generate a sense of value and respect.

I want our culture to be known, because how do they respect you? It's by knowing you, I mean, they have to know who you are, what you think, and then they're going to respect you. Any other way, anyone can use you maybe, but they're not going to know what you think, what you defend, what you're protecting at the end of the day.

We bring out tourists not to make money, but to teach them, in all areas, hot springs, all that it is to climb up a volcano, hike up in the mountains, in the waters. I mean, to teach more than anything else.

V.1.c Interpersonal experience

Cultural tourism creates a space for personal interaction. Social encounters do not only involve the providers but also the human beings with whom attitudes can be shared. This interaction occurs in close connection with the place itself, which offers a unique context.

Open their doors to … receive tourists, open their hearts as well to produce new friendships, … in reality, there are people, there are people who are willing to really receive you in their home without any problem, and in my family that's how I got to know about it.

We considered Mapuche tourism to be telling others about our experience, for example, how we see the world. We see the world from the right and the people who come see the world from the left, the clock runs from the left and our clock runs from the right, for example, we have a 1,000-year old cosmology and there are people who are interested in learning how people, our people, lived in the past. As we have inherited stories, inherited knowledge, experience, it's about sharing that experience with others.

V.1.d An initiative of the Mapuche people

Mapuche tourism ought to be managed by the Mapuche people themselves and, whenever possible, by the whole community: this claim was to be expected although, as explained earlier, it is not considered as a normal prerequisite for ethno-tourism. As far as the Mapuche providers are concerned, tourism is a community endeavour.

Mapuche tourism is an activity that has to be done by the Mapuche or by the community, ideally by the community, because when you perform an activity like music or dance, the family isn't capable of taking care of all the things that have to be done in an activity or an experience of Mapuche tourism. For this reason, one of the positive things is that it is integrating, it is an integrating force for the community.

This idea is also associated with an increase in the size of the family and growth of self-esteem: "to be more in contact with the exterior world, in the case of communities that develop tourism, makes them feel more valued".

V.1.e A way to improve environmental awareness

Finally, environmental sustainability is a significant factor, as people think of tourism and environmental protection in a synergic way:

> Here tourism is seen as a complementary job, a new concept ... but, at the same time, in certain ways, tourism helps a lot. It helps a lot with regard to environmental issues, for example, in economic terms obviously, but also in environmental terms because, for example, before people began to work in tourism, they did not know how to help the environment.

V.2 Motivations

One of our main aims is to identify the reasons why entrepreneurs dedicate a significant part of their time and energies to tourism activities. We asked them about their motivations in different periods, such as when and why they started and how they found out about this possibility. They were also asked about their current motivations, why they continue and what inspires them to continue. In the final part, they were asked about their perception of why tourists visit them. We wanted to know what they understood about tourists' motivations for visiting.

The main reason for pursuing tourism is cultural. Preserving ethnic identity is the main goal. The idea that tourism allows entrepreneurs to cultivate their traditions, customs, language and environment is central.

There is also a very clear economic motivation, though mentioned as a necessity and not as an end in itself. In some interviews, it is seen as an income source which allows them to practice natural and cultural conservation. We did notice an internal conflict between the payment required for tourism services and the notion of a gratuitous personal encounter. Money is mentioned as a possible cause for conflict within the community but is generally overridden by the potential benefits.

In conclusion, the reasons why entrepreneurs start are: supplemental income; ethnic identity and awareness of local culture; external social interaction; demand; environmental protection; a desire to live on their land. The reasons why they continue are: the will to promote cultural identity and awareness of their traditions; environmental protection; pride, self-esteem and confidence; extra income; a desire to live on their land with their families; social interaction, intercultural relationships. The reasons why tourists come are the desire to get to know new cultures and to escape from the city routine by returning to the countryside.

V.2.a Promotion of Mapuche culture, income and sustainability

One of the main motivations shared by all the people involved is that this activity validates Mapuche culture by promoting awareness among outsiders, which is seen as contributing to a social objective beyond tourism.

> The motivation that spurred us to take on this entrepreneurship is that we always wanted to, we had been looking for ways in which we would let people get to know us, we could show everyone our customs, our roots, teach people about them, and we said: here is the opportunity.

Some people refer to this as an opportunity to show that the Mapuche identity is much more than the conflict.

It is interesting to notice the number of different traditional activities with tourist potential according to the interviewees. Tasks such as animal husbandry, identification of plants and basic cultivation seem to be activities that can attract visitors. This is consistent with notions widely circulating in other areas where rural tourism is practiced. The "promise of tourism" is that it will bring visitors and thus provide resources with which to keep this natural heritage alive.

> We have a mission regarding the forest, and that is … like our line of work, right now we are doing tourism, but our tourism has to do with the relationship we have with the forest and how everyone can have that same relationship.

The interviewees have highlighted the link between economic aims and sustainable environmental gains. The synergy between tourism and environment can bring many potential benefits to a community or a region (Meinking *et al.*). Mapuche tourism allows communities to preserve their traditional activities. It enables them to live there rather than migrate, looking for jobs. It provides them with supplemental income and a way to preserve natural heritage. The benefits reported are closely related to sustainability, as they overcome a strictly environmentalist approach and indicate a holistic view. Being the original inhabitants of Araucanía, the Mapuche people are an integral part of the local ecological balance, and, as such, their permanence in the region can be understood as an indicator of environmental sustainability. Even though it is difficult to foresee whether this promise can become a long-term achievement, it is what informs the lives of the people involved in this study.

V.2.b The role of visitors

Another important external force that promotes and motivates local people to develop entrepreneurships is that they have the opportunity to do so, as ethnotourism is fashionable and driven by visitors. In many cases, people describe how at the beginning visitors asked them to provide these activities. They wanted to know more about them, their traditions, their food, their beliefs, their worldview and way of life. This was perceived as an opportunity to recover traditions and customs and to reaffirm their value. Later it became tourism.

> People came and asked if they could take a picture of the *ruka*, or if they could see what the *ruka* was like on the inside, and afterwards my kids started to bring their classmates to the countryside at the end of the year. That's where we got the idea.
>
> I made cheese all my life, butter, I sold eggs, so the tourists who went to the hot springs passed by to see if they could buy some cheese, so that was what brought me into this. I said, well, so many people come through here, maybe it would be good if I could offer something different.

Finally, there is one more motivation deriving from the Mapuche worldview itself: what I "must do" or what "my ancestors are telling me to do". This is a very personal motivation, a much more intimate, almost spiritual drive.

> You start to miss your field, your land, then one day I was bored being so far from my family, I said, I made a prayer and I said that I wanted God to show me where I had to be, because I had travelled all over and I wanted to establish myself in a home, I dreamed that I had to come home and spoke to my father and he told me, he said, there's a little piece of land that's all alone there, he said, there's nobody there and that's where I would go, to that little piece of land.

VI Final remarks

The aim of this study is to investigate the motivations and understandings of ethno-tourism among the Mapuche entrepreneurs of Araucanía in the south of Chile. Concerns for culture and the environment play a primary role. Moreover, when economic motivations are mentioned, they are intrinsically connected with the idea of conservation and validation of their own traditions.

By offering opportunities to interact with different people and explore cultural differences, tourism helps Mapuche providers to value and appreciate their own traditions as well as allows them to communicate with their visitors, creating awareness and good will. This dynamic also boosts general self-esteem across the community through an increased interest in local customs, stories, food and worldview. The outcome is a conscious construction of social capital, both within the community and among visitors. In this way, Mapuche ethno-tourism helps create intangible capital goods such as communication and trust, which are at the core of any sustainable development process.

Nature and environmental awareness also play a key role. This implies a traditional relationship with the place, the land, water, animals and native vegetation. Clearly the economic component is also fundamental, as ethno-tourism allows the Mapuche communities to live on their land and earn supplemental income by supporting traditional activities as well as gain extra resources for implementing environmental conservation strategies and methods.

Sustainability and communal wellbeing are goals that must be examined in a practical way with regards to how they materialise in real places with real people. Ethnicity, authenticity and identity construction, which fuel the contemporary debate on ethno-tourism, invite further investigation. It has now become clear that Mapuche ethno-tourism can generate wellbeing and sustainability by preserving culture and the environment through socio-economic development.

Notes

1 This study has been conducted with the support of the Centre for Intercultural and Indigenous Research (CIIR) CONICYT/FONDAP/15110006.
2 The so-called Mapuche conflict is a term used by the government and the media to identify the claim regarding the land, recognition and autonomy which the Mapuche people have laid through actions such as road-blocks, forestry machinery burning, sit-ins and protests in Araucanía and neighbouring regions.

Bibliography

Bengoa, José. *Historia de un Conflicto. El Estado y los Mapuches en el Siglo XX*. Octubre 1999. 2nd edn, Editorial Planeta, 2002.

Boccara, Guillaume, and Ingrid Seguel-Boccara. "Indigenous Policy in Chile During the Nineteenth and Twentieth Centuries: From Assimilation to Pluralism. The Case of the Mapuche". *Revista De Indias*, vol. 59, no. 217, 1999, pp. 741–774.

Briedenhann, Jenny, and Eugenia Wickens. "Tourism Routes as a Tool for the Economic Development of Rural Areas – Vibrant Hope or Impossible Dream?" *Tourism Management*, vol. 25, no. 1, 2004, pp. 71–79.

Butler, Richard, and Thomas Hinch. *Tourism and Indigenous Peoples: Issues and Implications*. Routledge, 2007.

Campelo, Adriana, Robert Aitken, Jürgen Gnoth, and Maree Thyne. "Sense of Place: The Importance for Destination Branding". *Journal of Travel Research*, vol. 53, no. 2, 2013, pp. 154–166.

Cuenca, Manuel. *Ocio Valioso*. Documentos De Estudios De Ocio, Deusto Digital, 2014.

Cuenca, Manuel. *Ocio Humanista. Dimensiones y manifestaciones actuales del ocio*. Vol. 16. Universidad de Deusto, Bilbao, 2003.

de la Maza, Francisca. "State Conceptions of Indigenous Tourism in Chile". *Annals of Tourism Research*, vol. 56, 2016, pp. 80–95.

Deci, Edward L., and Richard M. Ryan. "The Support of Autonomy and the Control of Behavior". *Journal of Personality and Social Psychology*, vol. 53, no. 6, 1987, pp. 1024–1037.

Dyer, Pam, Lucinda Aberdeen, and Sigrid Schuler. "Tourism Impacts on an Australian Indigenous Community: A Djabugay Case Study". *Tourism Management*, vol. 24, no. 1, 2003, pp. 83–95.

Gedda, Manuel. *Patrimonio de la Araucanía – Chile. Manual de Interpretación y Puesta en Valor*. Ediciones Sede Regional Villarrica, Pontificia Universidad Católica de Chile, 2010.

Glaser, Barney G., and Anselm L. Strauss. *The Discovery of Grounded Theory: Strategies for Qualitative Research*. Transaction, 1967.

Gobierno de Chile. *Estrategia Nacional de Turismo 2012–2020*. 2012.

Goytia, Ana. "Ocio y Calidad de Vida". *AGATHOS: Atención Sociosanitaria y Bienestar*, vol. 8, no. 2, 2008, pp. 4–13.

Goytia, Ana. *El Turismo como Experiencia de Ocio: Introducción a una Nueva Perspectiva Psicosocial*. Universidad de Deusto – Deustuko Unibertsitatea, 2003.

Hall, C. Michael. "Pro-Poor Tourism: Do 'Tourism Exchanges Benefit Primarily the Countries of the South'?" *Current Issues in Tourism*, vol. 10, no. 2–3, 2007, pp. 111–118.

Hall, Ruth, Poul Wisborg, Shirhami Shirinda, and Phillan Zamchiya. "Farm Workers and Farm Dwellers in Limpopo Province, South Africa". *Journal of Agrarian Change*, vol. 13, no. 1, 2013, pp. 47–70.

Henderson, Karla. *Dimensions of Choice: Qualitative Approaches to Research in Parks, Recreation, Tourism, Sport, and Leisure*. Venture Publishing, 2006.

Hernandez-Maestro, Rosa M., and Óscar Gonzalez-Benito. "Rural Lodging Establishments as Drivers of Rural Development". *Journal of Travel Research*, vol. 53, no. 1, 2013, pp. 83–95.

Iso-Ahola, Seppo E. *Social Psychological Perspectives on Leisure and Recreation*. Charles C. Thomas, 1980.

Landriscini, Graciela. "Economía Social y Solidaria en la Patagonia Norte: Experiencias, Saberes y Prácticas. Casos y Reflexiones". *Revista Pilquen*, vol. 16, no. 2, 2013, pp. 1–15.

Mannell, Roger C., and Douglas A. Kleiber. *A Social Psychology of Leisure*. Venture Publishing Inc., 1997.

Marulanda Valencia, Flor Ángela, Iván Alonso Montoya Restrepo, and Juan Manuel Vélez Restrepo. "Teorías Motivacionales en el Estudio del Emprendimiento". *Pensamiento and Gestión*, vol. 36, 2014, pp. 206–238.

McGehee, Nancy G., and Kyungmi Kim. "Motivation for Agri-Tourism Entrepreneurship". *Journal of Travel Research*, vol. 43, no. 2, 2004, pp. 161–170.

Meinking, Adriana, Alexander Schiavetti, and Salvador Dal Pozzo Trevisan. "Distorsiones entre el Concepto y la Práctica del Ecoturismo. El Caso de Itacaré, Bahía-Brasil". *Estudios y Perspectivas en Turismo*, vol. 14, no. 3, 2005, pp. 243–262.

Minniti, Maria, William C. Bygrave, and Erkko Autio. *Gem Global Entrepreneurship Monitor: 2005 Executive Report*. London Business School, 2006.

Neulinger, John. *The Psychology of Leisure*. Charles C. Thomas, 1974.

Picard, Michel, and Robert Everett Wood. *Tourism, Ethnicity, and the State in Asian and Pacific Societies*. University of Hawai'i Press, 1997.

Pratt, Stephen, Dawn Gibson, and Apisalome Movono. "Tribal Tourism in Fiji: An Application and Extension of Smith's 4hs of Indigenous Tourism". *Asia Pacific Journal of Tourism Research*, vol. 18, no. 8, 2013, pp. 894–912.

Reeve, Johnmarshall. *Understanding Motivation and Emotion*. 5th edn, McGraw-Hill, 2015.

Ruhanen, L. M., C.-L. J. McLennan, and B. D. Moyle . "Strategic Issues in the Australian Tourism Industry: A 10-Year Analysis of National Strategies and Plans". *Asia Pacific Journal of Tourism Research*, vol. 18, no. 3, 2013, pp. 220–240.

Ruiz Olabuénaga, José Ignacio. *Metodologia de la Investigación Cualitativa*. 5th edn, Universidad de Deusto, 2012.

Ryan, Chris, and Jeremy Huyton. "Tourists and Aboriginal People". *Annals of Tourism Research*, vol. 29, no. 3, 2002, pp. 631–647.

Ryan, Richard M., and Edward L. Deci. "Promoting Self-Determined School Engagement". *Handbook of Motivation at School*, edited by Kathryn R. Wentzel and Allan Wigfield, Taylor & Francis, 2009, pp. 171–195.

Servicio Nacional de Turismo – SERNATUR. *Fundamentos del Turismo Mapuche y Orientaciones para su Desarrollo. Región de la Araucanía.* SERNATUR, Gobierno de Chile, 2013.

Servicio Nacional de Turismo – SERNATUR. *Chile. Por un Turismo Sustentable. Manual de Buenas Prácticas. Municipalidades.* SERNATUR, Gobierno de Chile, 2012.

Szmulewicz E., Pablo, Cecilia Gutiérrez, and Karen Winkler. "Asociatividad y Agroturismo: Evaluación de las Habilidades Asociativas en Redes de Agroturismo del Sur de Chile". *Estudios y Perspectivas en Turismo*, vol. 21, no. 4, 2012, pp. 1013–1034.

Tew, Christine, and Carla Barbieri. "The Perceived Benefits of Agritourism: The Provider's Perspective". *Tourism Management*, vol. 33, no. 1, 2012, pp. 215–224.

Trau, Adam, and Robyn Bushell. "Tourism and Indigenous Peoples". *Tourism, Recreation and Sustainability: Linking Culture and the Environment*, edited by S. F. McCool and R. Moisey, 2nd edn, CABI International, 2008, pp. 260–282.

Van den Berghe, Pierre L. *The Quest for the Other: Ethnic Tourism in San Cristóbal, Mexico.* University of Washington Press, 1994.

Wood, Robert E. "Touristic Ethnicity: A Brief Itinerary". *Ethnic and Racial Studies*, vol. 21, no. 2, 1998, pp. 218–241.

Yang, Li. "Ethnic Tourism and Cultural Representation". *Annals of Tourism Research*, vol. 38, no. 2, 2011, pp. 561–585.

Yang, Li, and Geoffrey Wall. "Ethnic Tourism: A Framework and an Application". *Tourism Management*, vol. 30, no. 4, 2009, pp. 559–570.

III.6 Japanese castle towns as models for contemporary urban planning

Shigeru Satoh

I Japanese castle towns

Almost all cities in Japan were originally constructed as castle towns. The apparently chaotic spatial composition of these castle towns, built in the late sixteenth and early seventeenth centuries, makes it difficult to identify the design principles on which these towns as a whole were organised. On the other hand, it is well known that these castle towns offer a variety of scenic vantage points, such as local streets, bridges and towers, from where people can enjoy the picturesque urban landscape in harmony with the natural setting. The purpose of this chapter is to clarify some urban design principles inherent in Japanese urbanism that were applied to the design of the castle towns.

From the late sixteenth to the beginning of the seventeenth century the castles built on the Japanese islands were symbols of the sovereign right of the feudal Lords, and these historical castles became the foundation of the modern castle towns. Subsequently, a castle town city today is defined as a city built according to the principles of a castle town. The relationship between this urban typology and its outskirts shows how urban design relates to the natural topography and climate. This kind of city is commonly seen not only in Japan but also in the regions of East and Southeast Asia that are affected by the monsoon season, including south-eastern China, Korea, Taiwan and Vietnam.

The various forms of castle towns show that they cannot be defined as a mono-urban type at the first glance, since each castle town is intricately connected to its site and outskirts, showing creativity in historical design. Castle towns were built in Japan as the capitals of feudal domains. A castle town, *jo-ka-machi* in Japanese, means "the town below the castle". Prior to the Edo period, most Japanese castles were located on high ground, with the associated urban area of administrative, commercial and residential districts being built on lower ground. During the Edo period, castle towns had two main components: the castle area and the town itself. Thus the expression "castle town" referred to the whole town, which included the central "castle area" and the urban area, separated from the castle area by double or triple moats. The fact that the castle was located on high ground was very important both for defence – as the military base of the clan – and as a symbol of the local government.

A castle town consisted of four zones: the castle, residential areas for warriors ranked by class, areas for "urban commoners", and temple and shrine sites. The castle functioned as a military fortification and as a symbol of the political power of the territorial lord. Following the fashion of building in periods of warfare, castles were situated at various locations such as hilltops and edges of plateaux, surrounded by double or triple moats. The influence of the spatial configuration of European or Middle-Eastern castles is recognisable. The areas for the "urban commoners" were commercial centres, and had a modified grid street pattern, modelled after Kyoto, which was the most sophisticated grid-pattern city.

The origin of the residential areas for warriors was the warrior settlements located around the castle in the medieval era, and the areas were zoned according to the distinctive class ranks. The castle was surrounded by the warriors' residential areas that had a complex street layout, with dead-end alleyways and irregularly angled streets, to provide better protection against attacks. The warriors' residential areas ranged from high-class warrior residences occupying large plots to terraced blocks for lower-class warriors. Temple districts were located at the edge of a plateau or on the outskirts of the castle town. They consisted of large sites and open spaces.

I.1 Two typical castle towns – Tsuruoka and Morioka

The castle town had a unique spatial system comprising the urban structure of the castle with major streets planned according to topographical and natural parameters. Consequently, the castle town as a whole was organised as a combination of these components, as exemplified by Tsuruoka (Figure III.6.1) and Morioka (Figure III.6.2).

The laying-out of streets oriented toward the tops of surrounding mountains was a traditional approach to urban design in Japan. In many cases, the major streets function as primary lines of vision (vistas) towards the tops of the surrounding mountains. In Tsuruoka the moat to the west of the castle lies exactly on the line of vision connecting the top of Mount Chokai, a major mountain located in the far north of the city, with Mount Kinbo, an object of religious belief located in the south. The major streets running north and south were laid out as lines of vision towards the tops of two mountains located in the south, and consequently various vantage points in the town provided differing views of the mountains. In Morioka the layout, which unifies the entire urban structure, was dictated by the different lines of vision towards the castle towers and turrets. The irregular five-sided street pattern is laid out in line with the surrounding mountain tops.

I.2 Geometrical relationships between urban landscape features

Major landmarks, street intersections and topographical points in castle towns were located with respect to the geometrical relationships between them. In Tsuruoka (Figure III.6.1) major temples and shrines, street corners of the

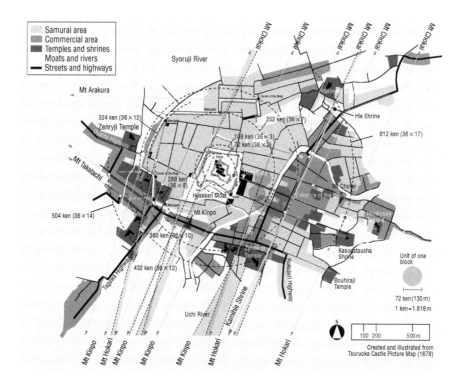

Figure III.6.1 Urban design of Tsuruoka.

Source: Satoh, Shigeru, editor. *Shinpan Zusetu, Joukamachi-toshi* [*Illustrations of Japanese Castle-town Cities*]. Kajima Institute Publishing, 2015, p. 34.

commercial areas and the line of moats were laid out in a concentric manner centred on the castle keep. It is obvious that in Morioka (Figure III.6.2) major landmarks in the castle town were located mainly on the points of two different concentric circles. The centre of this is *Iwakura*, a huge sacred rock which was discovered at the time the Castle was constructed. In addition, Oote-do street in front of the castle district is laid out on the vista line linking *Iwakura* with the top of the sacred mountain Atago-Yama. In Morioka the location of major facilities was determined according to the five-sided relationship between the castle tower and the surrounding mountain tops – features of the core natural topography.

I.3 Are castle towns military towns?

There are a number of answers to the question, "What is a castle town?", the most common being, "Castle towns were military towns" with various stratagems to fortify their defense. Probably this was an influence from the military science of the Edo period. Even so there were unbalances in the structure of the

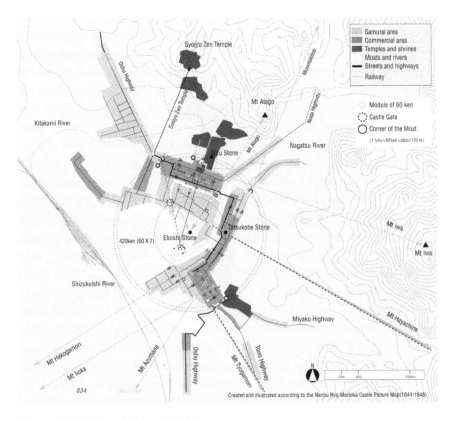

Figure III.6.2 Urban design of Morioka.

Source: Satoh, Shigeru, editor. *Shinpan Zusetu, Joukamachi-toshi* [*Illustrations of Japanese Castle-town Cities*]. Kajima Institute Publishing, 2015, p. 48.

town and its streets, as is also described above. If the castle town was solely for military purposes, the most obvious and simplest way would be to build ramparts around the town, as in China and Europe. The construction of Huế in Vietnam in the nineteenth century was influenced by China and France and was surrounded by thick ramparts to make it a closed citadel. Castle towns in Japan, in contrast, were structured so they would be connected to the surrounding farmlands.

Of course, these castle towns were fortified for defensive purposes with wooden fortresses, moats and *masugata* (area between the inner and outer gates), but military defense was only one of the purposes of the castle town structure; they were not built for military purposes only. In fact, the castle town was constructed as the centre of the feudal domain to fulfil the expectations and needs of a town of that time, and urban functions were incorporated and planned purposefully.

In order to fulfil the needs of the town, the castle town inevitably comprised various spatial systems. There were commercial areas with systematic rectangular plots; *samurai* areas with gardens and also residential areas for *ashigaru* (foot soldiers) where people lived in high-density housing. A variety of residential and architectural styles were laid out systematically in the castle town. In addition, the layout was aimed at connecting man-made buildings with the natural landscape, for example a river running behind the high-density commercial streets, so as to allow the people to live close to nature. The castle town brought together different spatial systems and social classes in one place, and its urban structure also exhibited a wide variety of residential styles. These various spatial systems were unified in the castle town by means of a method of design based on aesthetic landscaping compatible with the environment and climate, and on the superb techniques of town construction and environmental management that were the result of long experience.

Castle towns have played a variety of important roles in each region, as centres of local administration, culture, trade and industry. Since the Meiji Period, local government centres – the seats of each prefectural office – have been established in many of these castle towns, which have expanded in population, area and urban function. In some cases the spatial structure has gradually changed, while others have maintained their original configuration as well as their social and spatial context.

II The history of emerging castle towns up to the beginning of the pre-modern period

As was previously mentioned, one example of a typical Japanese city is Kyoto, which became the capital of Japan after Nara. The city had the same grid-pattern street structure as is seen in the US, but it also had symbolic places and a framework that included palaces and the Suzaku-Oji street, which had a clear hierarchical structure. Consequently, the layout differed from that of cities in the US in that the relationship between the city and the surrounding natural environment was elaborately designed. This design was introduced from China. Apart from the grid pattern, the design included coexistence with the environment, as well as Feng Shui principles of a design commonly seen in East Asian countries with a monsoonal climate.

The early-modern castle towns were formed through transformations that took place in the Middle Ages and the Period of Warring States. Various forms of dwelling were developed throughout Japan, and the castle town was designed as a centre of local commerce, representing the integration of these transformations. In the castle town, water, wind and topography were controlled; towns were constructed for the purpose of defense and commerce and castles were designed with aesthetic sensibility, as symbols of power.

The development of civil engineering had a long tradition. Excavations in Asuka-kyo revealed that sophisticated environmental management techniques had been applied in this capital. The waterways and ponds were connected to the underground water system and were used to control it. The construction of fosses and waterway networks was designed so as to control the environment, taking into consideration the topography, geography and underground water system. The castle town incorporated many charming features, not only the townscape and aesthetics but also an array of technologies which enabled the people to live in harmony with the natural environment. In other words, the technology was environmentally friendly, and this was the principle on which the castle town was built.

More than 100 major castle towns constructed at around the same time have a refined process of construction that makes them typical of the "Castle Town of the Edo Period". During this period civil engineering techniques were developed on the ancient principle of coexistence with the natural environment, and these techniques survived until just before the development of the destructive modern construction techniques of the late nineteenth and twentieth centuries. Castle towns were constructed using delicate engineering techniques, and laid out with sophisticated land works. Neither high-level nor rough engineering techniques were utilised, but rather the appropriate technology that was adapted to the ecological system.

Up to the late sixteenth century, various types of castles emerged throughout Japan for the purpose of preparing for warfare and providing protection against enemy attack. Fully-fledged urban communities formed around the castles, with the enforced resettlement of warriors and the growth of commercial areas. Each area was independent, forming its own spatial system. The castle towns were developed with a new urban form to integrate whole feudal communities, by concentrating different types of settlements around the castles, each with its own spatial system. Consequently, a castle town did not have a single distinctive layout, because it was a complex of differently laid-out elements in association with the natural setting.

However, there was a set of implicit urban design principles that gave a sense of formal unity to the urban landscape of the castle towns, each of which is a complex of various spatial systems. The location and configuration of the castle town was determined with respect to the overall natural setting – the topography and major natural features such as mountain ridges, hills, plateaux, plains, rivers and lakes. Optimum use of those natural settings was made in developing the castle town in accordance with a clear zoning system based upon the different urban functions and social classes. This is now well known, and has been demonstrated through a range of studies that have looked at the process by which these towns were transformed into zones. However, few studies have sought to clarify the design principle which integrated the entire castle town, with the exception of some studies focusing on the specific aspects of spatial features in castle towns.

II.1 Castle towns and the origin of the modern cities of Japan

The design principles of the castle town described above were formed during the time of castle town construction in the late sixteenth to early seventeenth centuries through a process of trial and error based on philosophy and techniques developed in Japan. Heian-kyo, laid out according to *jobosei*, a grid pattern on the Chinese model, certainly influenced the layout of the castle towns. However, the design principles of the castle towns that we have described previously differ substantially from the ancient capitals laid out on the *jobosei* principles first imported from China, which were first applied at Fujiwara-kyo and reached their perfection at Heian-kyo.

After that, a more flexible Japanese method was developed that integrated the rational Chinese method. This method was brought to final fruition in the castle towns. Yoshihiko Amino's theories explore the idea that the diversified flexible method was applied in the medieval period of Japan. A very interesting theory is that the Ritsuryo system of law, the layout of the capital and the construction of straight roads throughout the country were introduced from China, but that these were at odds with the original Japanese culture. The construction of the castle towns was the result of a concentration of various experiences over a long period of history, and consequently it was in this way that the Japanese method was developed. This method reached some degree of skill through the creation of stable residences adapted to the environment and climate of monsoonal East and Southeast Asia, and then these residences were refined by being connected to each other by topographical conditions and variations.

Furthermore, modern society can also be regarded as the "New Medieval Period". The vigorous world of the medieval period described by Amino overlaps current Japanese society and the future of its cities. In short, a uniform model cannot be applied to cities by a single power or authority; instead there are various authorities and organisations that must act together in order to undertake urbanisation and spatial planning. The large frameworks, such as geographical features, natural environmental conditions, scenery, etc., provide a guide for urbanisation and spatial planning. The design of the castle towns brought together complex structures based on these large frameworks, and this has something in common with modern-day urban planning and design.

III The castle town as a garden city – its landscape, spiritual design and appearance

III.1 Design principles and characteristics of castle towns – Shan Shui and Feng Shui

The majority of the castle towns were designed from the very beginning in the late sixteenth and early seventeenth centuries. Although there had been earlier villages on a few of these sites, almost all of the castle towns were constructed or expanded using new planning and design principles. Urban design and construction technology were considerably well developed by that time. One of the morphological characteristics of castle towns is that their configuration was adapted to the surrounding natural environment, which brought organic order to the man-made urban space. This principle was not only applied to Japanese castle towns, but also introduced in other East Asian countries. The principle of urban design and civil engineering at that time was one of harmony with, rather than conflict with or domination over, the natural climate and landscape.

The design of a castle town needed to meet diverse needs. The military function was, obviously, the most important requirement. Therefore, it was necessary for each castle town to take advantage of its particular location and geographical conditions. Second, residential areas had to be strictly zoned according to social class. Third, a rational and functional urban structure was needed to demonstrate the rule of the political and administrative power over the whole feudal domain. Fourth, the construction of the castle town was an expression not only of the power of the Daimyo, the landholding military lords, but also of their aesthetic values.

The principle known in China as Feng Shui, that is, of harmony with the natural environment, was introduced into city and garden design in the early modern period, and also had an effect on the design of the castle town. When cartographic evidence of the spatial structure of the castle towns is examined, it can be seen that this philosophy is a major design principle. The physical planning of the castle town appears to incorporate four main principles that reflect Feng Shui, as follows; the existing landscape and topography was essential in determining the location and the overall physical form of a new castle town; the spatial composition of the castle town reflected the features of the surrounding natural setting by conserving the distinctive views of those features from the town; advantage was taken of the micro-landscape in the zoning and location of residential areas according to the social hierarchy; and last, the physical urban space was arranged according to geometrical rules and in modular units. By the beginning of the seventeenth century, when Japan was in the process of shifting from a long period of war to the peaceful Edo era, the innovative planning methods used in the castle towns had spread widely, and been elaborated on.

During times of war the Daimyo engaged in battles and mastered the methods of surveying topography and controlling water streams, which was essential in planning battlefield strategy and in managing their territory for agricultural production. At the same time, in stark contrast, they occupied themselves with the tea ceremony and sophisticated garden design. Oda Nobunaga and Toyotomi Hideyoshi unified the whole of Japan and, finished with war, constructed the marvellous Azuchi Castle and castle-town and Osaka Castle, respectively. These became the original models for later constructions. Kobori Enshu, the architect of Katsura no Rikyu (Katsura Detached Palace), is famous as a tea master and as a designer of many sophisticated gardens and castle towns. He is one of the most important artists from the period of wartime to the peaceful Pre-Modern and Edo Period, and was involved in the design and construction of castle towns for a number of Daimyo. The Daimyo also played a major role in the design and construction of a series of castles and castle towns. It appears that each one tried to make use of his knowledge, aesthetic sense and environmental management skills in the design of the peacetime capital of his own territory. Many Daimyo emerged as successful builders of castles and castle towns, designing them as a garden in microcosm.

Thus it was that the castle town was designed as a garden with aesthetic sensibility. Examples include Tsuruoka City, together with Mt. Chokai, Gassan, Kinpo, and Hokari, which are the pride of the people of Shonai. The city is in harmony with the surrounding mountains and landscape, and was indeed designed to symbolise the integration of the whole *han* territory. The vista lines, running toward the sacred mountains, are not merely used in planning street and road demarcation, but demonstrate that the city of Tsuruoka unifies the landscape of the whole territory.

The city was designed to have sequential scenery from the human viewpoint and to represent the "integration of scenery". Each castle town was designed on the basis of its spatial form, with various city components and regional conditions integrated into its own unique design. There was no basic model, no mechanical grid pattern, no design guidelines like Vitruvius' *De Architectura* or the Laws of the Indies that set out plans for the construction of Spanish colonial cities in the Americas, or the Feng Shui principles of ancient Chinese capital construction; in the castle town, the immutable climate, landscape, topography, geology and resources of the location became the foundation on which the town was laid out.

III.2 Methods and techniques for planning the castle town

Various techniques were combined and applied in the planning of a castle town. These approaches to the construction of a modern castle town are summarised in Figure III.6.3.

The spatial presentation of elements, such as bridge piers, curved streets, perspectives, symbolic structures such as turrets, etc., were variously composed and combined to create the town as a whole.

First, the geographical conditions were examined, then the disposition of the main components and framework of the castle town were explored. At this time, probably the large components – the geographical features and symbolic scenery, vistas towards the mountains, and adjustments to the existing land – were determined. Next, a basic town was laid out using a system of modules measuring 50, 75, or 30 ken (a traditional unit of measurement in Japan, equivalent to 1.82 m).

Most of the commercial areas were formed by means of a combination of several grid patterns with the town framework. The grid patterns that were applied in the commercial areas were somewhat irregular. On the other hand, the *samurai* areas were laid out in a variety of grid patterns: T-shaped, curved, and *masugata*. The large framework of the town as a whole was co-ordinated with small-town planning. Basically, the composition of the castle town was flexible, being divided into different areas; and of course some areas fulfiled a military purpose.

The castle town design was created by bringing together various planning and design techniques and combining them through the process of going back and forth between small-scale and large-scale methods. The overall town design emerged after this process. The design of the town was not prescribed extensively

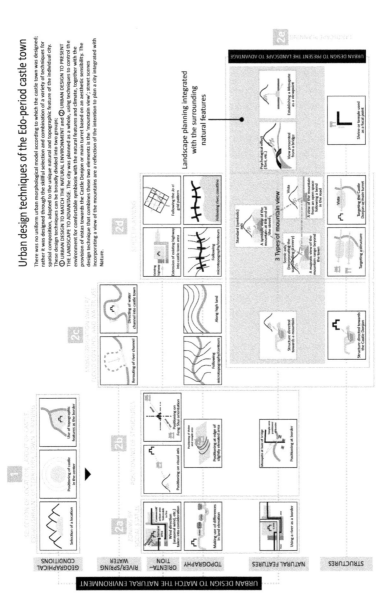

Figure III.6.3 Compilation of design elements and techniques incorporating the natural environment into the structure of a castle town.

Source: Satoh, Shigeru, editor. *Shinpan Zusetu. Joukamachi-toshi [Illustrations of Japanese Castle-town Cities]*. Kajima Institute Publishing, 2015, p. 26.

according to any fixed regular law; rather it was constructed by making full use of various techniques developed through an understanding of the local conditions. If beautiful mountains had religious objects located around them, the mountains were included into the design naturally; in addition, rivers were utilised as moats and a landscape-axis towards the turrets of a flatland castle would also be included in the plan. Moreover, as we have mentioned, about 300 castle towns were constructed in a short period of time. Thus, probably, it will be appropriate to consider the castle towns to have been designed using common, shared techniques, with a combination of various techniques according to the purpose and features of the location. Accordingly, these unique castle towns could not be completed through the application of a uniform model.

Analysis of the spatial composition of various castle towns throughout Japan has revealed a common set of design principles and design methodology, and also suggests the following conclusions. The castle towns were composed of different zones each with its own spatial system, such as the castle, urban administrative area, business areas, residential areas and temple areas, in the manner of a mosaic. The spatial system of each zone is based upon the following fundamental principles.

First, the structure of the castle town is organised with vistas towards the ridges of the surrounding mountains, and with the castle tower and major turrets as landmarks visible from viewpoints on and near the major linear spatial elements such as streets, rivers and moats.

Second, the components of the castle town such as the castle tower, major turrets, shrines and temples, street corners in commercial areas, and natural landscape elements represented by mountain ridges are located with respect to the distinctive geometrical relationships between them.

Third, each castle town is spatially articulated and ordered within its own modular structure of approximately 30 ken.

The basic principle of aesthetic sensibility in the design of the castle town was to unite the town with the climate and surrounding nature; this shares a common direction with contemporary city planning in its emphasis of the sense of place – what is called "city identity".

In the castle town, a moat surrounded the castle, and in order to maintain the moat the town was designed and constructed taking into account the rational relationship between the flow of the river and other water courses, and the direction of the wind. Residential areas were planned in harmony with the topography and flow of water, and this topography was also utilised for its military implications. Furthermore, the symbolic structures of the castles and its turrets together with the surrounding mountains were included in the design of the urban landscape.

It is quite possible that this basic principle could be applied to our contemporary cities when we consider their ideal situation utilising their regional assets. There are many design elements found throughout the castle town that could be applied to modern cities. Castle towns were constructed as the quintessence of landscape culture, and they could provide the foundation for modern city design. They were neglected in modernisation, whereby the symbiotic relationships were lost through the filling in of the fosses and moats, the covering over of rivers with concrete, and the disconnection from the underground water system. Recent environmental regeneration plans have included the restoration of fosses and moats and the renaturation of rivers. The principles of castle town construction have been recognised as valuable assets for every city, and the trend is towards the regeneration of ecological principles such as the flow of wind and water. Consequently, the principle of castle town design, in which the most is made of the natural environment, is being revived in the design of ecological cities in the twenty-first century.

Bibliography

Amino, Yoshihiko. *Nihon no rekishi wo yominaosu* [*Reinterpreting Japanese History*]. Chikuma Shobo, 1991.

Itoh, Hirohisa. *Kinsei toshi-kuukan no genkei* [*Original Landscapes of Urban Space in the Modern Era*]. Chuo Kouron Bijyutsu, 2004.

Miyamoto, Masaaki. "Kinsei shoki joukamachi no vista ni motozuku toshisekkei" ["Urban Design Based on the Vista of Castle Towns in the Early Modern Era"]. *Kenchiku-shi-gaku*, no. 6, 1986, pp. 72–103.

Miyamoto, Masaaki. *Toshi kuukan no kinsei-shi kenkyu* [*A Study of the History of the Modern Era of Urban Space*]. Chuo Kouron Bijyutsu, 2005.

Satoh, Shigeru, editor. *Shinpan Zusetu joukamachi-toshi* [*Illustrations of Japanese Castle-town Cities, New Edition*]. Kajima Institute Publishing, 2015.

Satoh, Shigeru. "Urban Design and Change in Japanese Castle Towns". *Built Environment*, vol. 24, no. 4, 1999, pp. 217–234.

Satoh, Shigeru. "The Morphological Transformation of Japanese Castle-town Cities". *Urban Morphology*, vol. 1, no. 1, 1997, pp. 11–18.

Satoh, Shigeru. *Joukamachi no kindai toshi dukuri* [*Castle-town Cities in Japan and Modern Urban Morphology*]. Kajima Institue Publishing, 1995.

Sugano, Keisuke, Ryutaro Okitsu, and Shigeru Satoh. "Medieval Castles and Pre-Modern Castle Towns Planned with Nature as the Heritages for Landscape Design Today: A Case Study of Nanbu Region in Tohoku". *History, Urbanism, Resilience: Planning and Heritage. International Planning History Society Proceedings*, vol. 4, 2016, pp. 274–284.

Takahashi, Yasuo, Yoshida Nobuyuki, Miyamoto Masaaki, and Ito Takeshi, editors. *Zushu Nihon Toshi-shi* [*An Illustrated History of the Japanese City*]. Tokyo University Press, 1993.

Takahashi, Yasuo, and Yoshida Nobuyuki, editors. *Nihon Toshi-shi Nyumon* [*Introduction to Japanese Urban History*], vols. 2–3. Tokyo University Press, 1990.

Takahashi, Yasuo, and Yoshida Nobuyuki. *Nihon Toshi-shi Nyumon* [*Introduction to Japanese Urban History*], vol. 1. Tokyo University Press, 1989.

Tatsumi, Kazuo. *Machiya-gata syugou jutaku* [*Machiya (Town House)-style Housing*]. Gakugei Shuppansha, 1999.

Yamori, Kazyo, editor. *Jouka machi no chiiki kouzou* [*Regional Structure of Castle Towns*]. Meicho, 1987.

Yamori, Kazyo. *Toshi puran no kenkyu* [*On the Regional Structure of Japanese Castle Towns*]. Daimeido, 1970.

III.7 Vietnam's pathway towards sustainability

Stories half-told

Nhai Pham and Yen Dan Tong

I Sustainable development in Vietnam: policies and institutions at a glance

Vietnam has developed an early awareness of sustainability in its contemporary development process, considering sustainable development (SD) the inevitable path and "an important strategic goal that the Communist Party, Government and People of Vietnam are determined to attain" (Government of Vietnam, *Implementation of SD* 5). Sensible SD policies have been promulgated in the country. However, detailed regulatory provisions and active enforcement of these policies have not yet been efficiently implemented.

Vietnam's strong commitment was induced by critical environmental problems in the country during the 1980s. Such commitment was then evidenced by the country's pioneer participation in the global movement towards sustainability in that decade.[1] Vietnam reunified the country in 1975 and henceforth witnessed significant environmental degradation of its natural resources, including forests, watersheds, marine resources, land and air due to years of war (Government of Vietnam, *The Plan* 5). The country was additionally under severe pressure from speedy post-war population increase, which created substantial stress on the environment (Government of Vietnam, *The Plan* 5). In 1986, Vietnam initiated the *Doi moi* (*Economic Reform*) process which transitioned the country from a centrally-planned economy to a socialist-oriented market economy. In this new era, economy has had to recover from the aftermath of a long war and natural resources have been heavily deployed to propagate strong economic growth. An evaluation of Vietnam's environmental conditions conducted by the International Union for Conservation of Nature and the World Wide Fund concluded that "what Vietnam faces today is a grave ecological crisis" (quoted in Government of Vietnam, *The Plan* 24) and that the nation's forests would be completely eliminated by the early part of the twenty-first century if current trends in environmental degradation were to continue.

Ultimately, extensive policies and institutional frameworks have been constituted in Vietnam in support of SD. In 1992 Vietnam approved the *National Plan for Environment and Sustainable Development 1991–2000: Framework for Action* (the Plan).[2] The Plan was the first document on SD at national scope, rectified in preparation for the Earth Summit in Rio De Janeiro. It was primarily established to outline a national framework for action in the area of environmental protection and sustainable development. It is comprised of three major components: (1) the development of an institutional, legislative and policy framework; (2) action programmes; (3) support activities. The Plan simultaneously identified specific short-term actions to address priority problems regarding sustainability for the period 1990–1995 and 1996–2000. Among them is the urgent establishment of an environment authority at central level and the development of environment law and legislations. Some of the actions recommended in this plan are well-advanced compared to Vietnam settings at that time. Examples include the implementation of the Environmental Impact Assessment, the proposal to formulate Energy Efficiency Standards or the internalisation of pollution costs in which the polluter has to pay to clean up the environment so that resource prices are regulated by the market. The Plan is considered to be the blueprint for action taken by the government of Vietnam at a later stage in order to protect the environment and advance toward sustainability.

Vietnam's enduring commitment to develop sustainably is reaffirmed through various manoeuvres. In 1992, it was a delegate at the Rio Summit and later a signatory to the Climate Change Convention and Convention on Biological Diversity on 16 November 1994. In 1998, it issued Directive No. 36-CT/TW dated 25 June on enhancing environment protection which again highlights that environment protection is fundamental and indivisible in Vietnam's socio-economic development towards sustainability. This philosophy is expressed in the Resolution of the ninth, tenth and eleventh Party Congress and mainstreamed into development plans at central and local levels.

Apart from releasing important legal documents in the field, SD policies in Vietnam stagnated until the introduction of the *Strategic Orientation for Sustainable Development*, also known as *Vietnam Agenda 21*, in 2004. This framework strategy created the legal foundation for ministries, sectors, local governments, relevant organisations and individuals to implement and co-ordinate actions aimed at ensuring sustainable development in the twenty-first century. Ministries and localities have designed and enacted their respective *Agenda 21*s.

Vietnam Agenda 21 recognises eight principles of SD and 19 areas of priority in socio-economic development (see Appendix 1). However, no specific SD indicators were proposed there. Some of the SD indicators were later identified in the Vietnam Socio-economic Development Plan 2006–2010 and 2011–2015. It was not until 2012–2013 that Vietnam, for the first time, released the *Indicators for Monitoring and Evaluating Sustainable Development* at national and local level. Local indicators include one composite indicator (the Human Development Index-HDI), seven economic indicators, 11 social indicators, nine natural resources and environment indicators, and 15 region-specific indicators (see Appendix 2). Whether Vietnam meets the indicators' target or not remains to be extensively evaluated, but it is very likely that many of the indicators have not been accomplished. For example, whilst the rate of solid waste collection and treatment is set at 85 per cent in 2015, the actual rate is only 40 per cent for hazardous waste nationwide and 40 to 55 per cent for ordinary waste in rural areas (Government of Vietnam, *State of the Environment* 62).

Still, the introduction of the *Vietnam Agenda 21* marks a period of stronger actions in SD, which has been actively mainstreamed into socio-economic development. Subsequent legislation was built up to support the agenda. The *Vietnam Socio-Economic Development Strategy for 2011–2020* restates the philosophy that "rapid development in close linkage with sustainable development represents an all-thru requirement in the Strategy" (1). In 2012, the *Vietnam Sustainable Development Strategy for the Period from 2011–2020* was approved. In 2013, the Prime Minister approved the *National Action Plan for Sustainable Development 2013–2015* under Decision No. 160/QĐ-TTg. These legal documents have definitely expedited SD activities in Vietnam after a period of stagnancy.

Despite extensive regulations, SD activities in Vietnam were not soundly monitored until a later stage. Although an institutional framework was proposed in the 1992 Plan, it was constituted only in 2005 in order to guide and overlook SD policy implementation. Such framework is led by the National Council for Sustainable Development (NCSD)[3] and a Sustainable Development Office (SDO)[4] in the Ministry of Planning and Investment which serves as the standing office of the NCSD. The NCSD monitors and assesses the implementation of SD objectives in the country. It was at first set up as a bulky system of 44 members (Khoa and Sinh) and reduced to 30 members in 2009.

A Steering Committee/Council for Sustainable Development and an Office for Sustainable Development have also been founded at ministerial and local levels. In addition, Vietnam has established the Vietnam Business Council for Sustainable Development in response to the UNDP's call for a more active contribution of the business sector in achieving SD goals. The Council aims to build up a Vietnamese business community that is sustainable, energetic and successfully integrating into the world of the twenty-first century, as well as contributing to the nation's sustainable development (Government of Vietnam, *Implementation of SD* 26). Such institutional framework maintains the tight vertical co-ordination among SD offices at different levels while still allowing for a certain horizontal co-ordination owing to the inclusion of ministers and leaders from various ministries into the NCSD.

Despite a lengthy formulation, the current comprehensive system of legal documents and institutional framework has co-ordinated agencies at different levels, thus enabling Vietnam to pursue their SD commitments. Some remarkable achievements have been recorded in the country over 30 years of economic reform. However, challenges remain. In the next section, a flash evaluation of SD policy implementation in Vietnam is discussed to provide an overview of its outcomes in the pathway towards sustainability.

II A flash evaluation of SD policy implementation in Vietnam

Vietnam has been heralded for prominent achievements in socio-economic development during 30 years of economic reform. Rapid economic growth has been attained along with a dramatic reduction in poverty. However, such strong economic growth has not been accomplished without considerable costs for the environment.

Vietnam has experienced convincing performances in socio-economic development ever since it enacted the *Doi moi* process and the sustainable development process a few years later. Its GDP increased seven-folds, from US$26.33 billion in 1986 to US$193.6 billion in 2015. The GPD growth rate from 2005 to 2015 is 6.25 per cent per annum: from a poor country with chronic food shortages, in 2016 Vietnam became a lower-middle income country with GPD per capita of US$2,111 (Figure III.7.1) and the second largest rice exporter in the world. Poverty rate (at US$1.9 per day) decreased from 49.2 per cent in 1992, the year that Vietnam released its national framework for sustainable development, to 3.06 per cent in 2014 (World Bank, *Vietnam-World*) (Figure III.7.2).

Vietnam attained the target of primary education universalisation while gradually improving education quality (Government of Vietnam, *Implementation of SD* 14). Clean water has been supplied to 83 per cent of households in the country since 2010 (Government of Vietnam, *Implementation of SD* 37). Owing to these remarkable achievements, a country that has reconstructed its economy over the course of 30 years after a long and damaging war has been able to achieve five of the eight Millennium Development Goals (Thang).[5]

These achievements are primarily the result of growing industrialisation (ADB and UNEP, *Vietnam National* 8). The share of industry in GDP increases from 22.7 per cent in 1990 to 33.25 per cent in 2015 (GSO, *Statistical Yearbook 2015*).

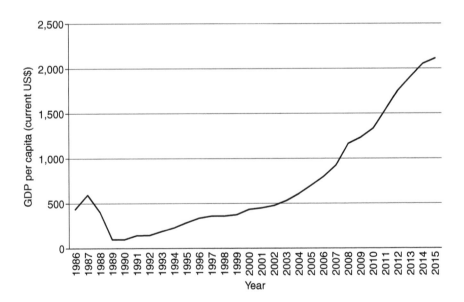

Figure III.7.1 GDP per capita in Vietnam, 1986–2015.

Source: World Bank.

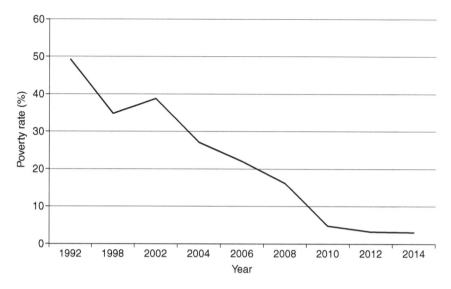

Figure III.7.2 Poverty rate in Vietnam, 1992–2014.
Source: World Bank.

However, its industrial sector is in the initial stages of development, with production still being dominated by small-scale units which utilise technologically
simpler processes producing more waste and adverse environmental impacts
than larger, more modern, facilities (ADB and UNEP, *Vietnam National* 8).
More importantly, the Environmental Impact Assessment (EIA) of investments
is undertaken too late in the project planning cycle, resulting in EIAs that react
to development proposals instead of anticipating them (Clausen *et al.* 137). Even
before EIAs are approved, agreement for the development of a particular project
at a particular location is sought and often granted from national and local
authorities (Clausen *et al.* 138). An example of the non-serious enforcement of
EIA regulations is the case of JA Solar, a US$280 million project by a Chinese
corporation in the Bac Giang province. JA Solar was granted permission by the
local government to start its construction process on 27 November 2016, long
before the submission of their EIA to the relevant authority in early February
2017 (Truong and Hoang). Post-EIA activities are also not adequately adminis-
tered once EIAs are approved, resulting in EIAs being adjusted without notice
(Kinh). Therefore, rapid economic growth has been obtained at some considerable costs for the environment. Industrialisation process has been associated with
prolonged and severe environment problems, as recognised by the Government
of Vietnam (*Implementation of SD*):

- The efficiency of the economy remains low. Economic growth has been
  reliant on exploitation of raw natural resources.

- Economic growth has not been harmoniously combined with the protection of natural resources and the environment. In many economic aspects it has mainly relied on export of raw natural resources.
- Natural resources have been subjected to wasteful exploitation and inefficient use: environmental problems such as water and air pollution, biodiversity reduction, unfulfilled solid waste collection rates and rampant mineral exploitation have all caused resentment in the population.

As a resource-based economy, Vietnam over the last 30 years has been reliant on heavy exploitation of natural resources to stimulate economic growth. To assess whether growth is sustainable or not, the economics literature examines changes in the total national wealth and its components. Based on that, growth can be classified in terms of "weak sustainability" or "strong sustainability". The notion of "weak sustainability" presupposes that natural capital and physical capital are perfect substitutes: as long as the total capital stock is maintained constant, or any depleted natural capital is substituted with physical capital, economic development will be sustainable (Pearce and Atkinson; Neumayer). Thus, natural capital should be managed efficiently so that welfare losses from environmental damages are minimised and resource rents arising from the exploitation of natural resources, accounting for environmental externalities, are maximised (Barbier). The notion of "strong sustainability", on the other hand, defines natural capital and physical capital as complements, not substitutes. Thus, at least a portion of a nation's natural resources must be preserved at all times.

Vietnam appears to be on a weak sustainable development path with an adjusted net saving[6] of 15 per cent of Gross National Income – GNI (OECD, *Towards Green Growth* 98). However, the high rate of natural capital depletion in Vietnam casts doubts on the sustainability of the current development model (Figure III.7.3). By 2012, Vietnam's natural capital depletion rate was 8.82 per cent of GNI, which is second to only Brunei Darussalam in South East Asia and is significantly higher than countries such as China (3.86 per cent) or India (3.68 per cent). Such rate of depletion of natural resources is considered to be worrying (OECD, *Towards Green Growth* 98) because it could reach the threshold level that triggers irreversible changes in the environment, leading to large adverse long-term consequences for the economy and human welfare. In addition, the deterioration of the environment questions the "weak sustainability" outcome that Vietnam seems to have accomplished.

First, there is a considerable decline in biodiversity. Vietnam is at risk of biodiversity loss with around 300 threatened species. Forest, which hosts biodiversity, is decreasing in quality although coverage is maintained at about 40–42 per cent. About 50 per cent of primary forest in Vietnam has been lost in the period from 2000 to 2010 and reforestation cannot compensate for the environment damage caused by the loss of primary forest (OECD, *Towards Green Growth* 102). Newly planted forests do not provide the same ecosystem services as a primary forest since they absorb less carbon. Additionally, primary forests are usually a more suitable habitat for a wider range of native

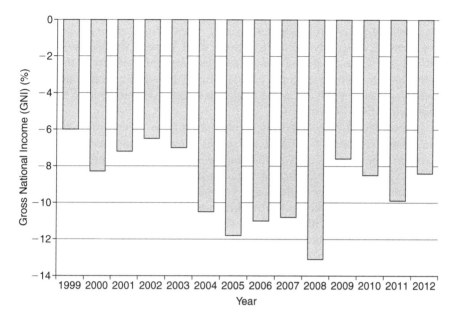

Figure III.7.3 Natural capital depletion in Vietnam. Unit: percentage of GNI.

Source: OECD, *Towards Green Growth* 98.

forest species than plantation forests (Brockerhoff *et al.*). At the same time, mangrove forest, an important habitat for biodiversity and coastal protection, has persistently declined, largely due to intense shrimp farming. Vietnam has lost 67 per cent of its mangrove forest since 1943 (Government of Vietnam, *State of the Environment* 150). In terms of fish stock, overfishing and the use of destructive fishing methods have resulted in fewer catches in spite of a better technology. Furthermore, Vietnam is among the countries in the world most vulnerable to climate change impacts on fisheries (Allison *et al.*). Ulti-mately, biodiversity has been severely degraded (Government of Vietnam, *State of the Environment* 143).

The increasing or persistent level of pollution in many other components of the environment, including air, water, and soil is another major issue of sustainability in Vietnam. A sharp increase in the number of motor vehicles, rising industrial production and reliance on fossil fuels are all contributing to higher rates of air pollution (OECD, *Towards Green Growth* 154). Air quality, assessed through the air quality index (AQI)[7] is shown to be poor in large urban areas during a large proportion of the year. Statistics from Gia Lam air station in Hanoi, the capital city of Vietnam, shows that air quality was poor or worse in 189 days in 2014. There are even days when the air is harmful to human health. *Towards Green Growth* by the OECD states that air pollution is the cause of about 30,000 deaths in Vietnam annually (155). It is of even greater concern in industrial areas. The level of TSP, a measure of the mass concentration of particulate matter in the air, exceeds the Vietnam standard in all industrial parks throughout the country. Even in rural areas the figures do not improve, especially around traditional craft villages. For the period from 2011 to 2015, the air around these areas compares unfavourably to the period from 2006 to 2010 due to the popular use of low-quality coal, out-of-date technology and insufficient waste treatment systems (Government of Vietnam, *State of the Environment* 114).

In a similar fashion, water, including both surface water, groundwater and sea water, is severely polluted and poses threats to human well-being in Vietnam. The levels of BOD_5 (biochemical oxygen demand) and COD (chemical oxygen demand) in surface water in urban areas well surpass national standards. This is due to the large volume of untreated sewage water discharged to lakes, rivers or canals. The proportion of sewage water treated is only 10–11 per cent with only 40 out of 787 urban areas possessing the necessary water treatment facilities to meet the national standard. As far as industrial wastewater is concerned, only a few of the total 625,000 enterprises have got operational wastewater treatment facilities, and the discharge of BOP reagent constantly increased from 1995 to 2003 (ADB and UNEP, *Vietnam National* 10). At the same time, most river basins in Vietnam are contaminated, especially in areas with a high concentration of industrial activities (Government of Vietnam, *State of the Environment*). Underground water quality is decreasing with ammonia content exceeding the national standard many times over many areas in Vietnam. Sea water is also at risk in areas such as large ports or industrial parks. In 2016, extremely high phenol and cyanide levels in the discharge from Formosa Ha Tinh Steel Corporation have led to mass fish deaths and coral reef bleach in four provinces in central Vietnam (Government of Vietnam, *State of the Environment* 100). This is an apparent example of the trade-off between economic growth and environment degradation and of the short-sighted "grow-now-clean-up-later" model.

Soil pollution also proves the argument about Vietnam's weak sustainability. The *State of the Environment Report* confirms that soil pollution is on the rise due to the overuse of fertilisers and pesticides in agriculture (129). Only 40 to 50 per cent of fertilisers is absorbed by plants. The rest accumulates in the environment, leading to soil and underground water pollution. Soil pollution near landfill areas is particularly severe and heavy metal contents in soil well surpass the national standard.

It is clear that although significant socio-economic goals have been achieved, Vietnam is under great risk of severe environment pollution and degradation. This implies that the extensive policy and institutional framework on sustainability management are unable to protect the country from the pathway of the grow-now-clean-up-later model. Significant economic growth has come at the cost of the environment which may result in substantial damage to future prosperity.

III (Un)sustainable development in the Mekong delta

In the previous sections, development policies towards sustainability have been discussed. Despite strong and early commitment, Vietnam has not successfully implemented relevant SD policies at macro-level, evidenced by the poor state of its environment. At micro-level, SD policies are not always well-planned, which can lead to irreversible and long-term consequences to the economy and the natural surroundings. An example is the plan to develop the Mekong delta, the largest delta and rice bowl in Vietnam. The original research in this section analyses the environmental impacts of the Mekong Delta Development Plan (MDDP) and again highlights the fact that economic ends have been reached at a high cost to the environment.

Vietnam's strategy in the Mekong delta is to develop a system of large infrastructure projects to enable rice intensification in the area. This strategy was also recognised within the *National Plan for Environment and Sustainable Development 1991–2000: Framework for Action*, which states that "agricultural practices should emphasize intensified cultivation" (8). Vietnam's target in uplifting rice production in the delta has been largely achieved; yet, whether such development is sustainable or not remains a thorny question.

III.1 *A brief introduction to the Vietnam Mekong delta (VMD)*

The VMD is the largest delta and rice bowl in Vietnam and among the largest mega deltas in the world. It has been formed by the deposition of sediments from the Mekong River over thousands of years (Tri). The area of the VMD in Vietnam covers $39,000 \, km^2$, equivalent to 5 per cent of the Mekong Basin area. The VMD is a rich soil area, characterised by a dense canal system, covering 12 provinces located in south-western Vietnam. The population of this region reached 17.3 million by 2013 (Thong *et al*. 38). The delta is a flat landscape, except for some mountains in the Long Xuyen Quadrangle. The average elevation of the region is only 0.8 m above sea level. Tides occur semi-diurnally with an amplitude of oscillation between 2.5–3.0 m, while in the southern sea they occur daily with a lower amplitude, oscillating between 0.4 and 1.2 m. The intertidal zone constitutes an area of 480,000 hectares, of which approximately 300,000 hectares are suitable for saltwater and brackish water aquaculture. The total area of the inland wetland is about $36,000 \, km^2$, equal to 90 per cent of the inland area. Two-thirds of the wetland area have been used for agricultural production and aquaculture (Thong *et al*. 38).

Given the physical location of the VMD within the Mekong River Basin, the Mekong River plays a very important role in the agricultural and aquaculture production of Vietnam. This river and its tributaries bring alluvium-rich water to the VMD. The ongoing sediment deposition extends the Mekong Delta continuously towards the South China Sea, improves the fertility of fields, provides a well-stocked fishery and creates very favourable conditions for aquaculture. The flood flow also improves water quality, particularly by flushing acid sulfate soils and agricultural pests. In addition, this freshwater source is important for agriculture, aquaculture, and livelihoods, and provides a vast freshwater ecosystem for the plain (Tri). Consequently, the VMD is considered to be the "rice bowl" of Vietnam and is, by far, the largest cultivated region in Vietnam. This region's land use for growing cereals provides 47 per cent of the national total, more than northern and central Vietnam combined. Most land is used for rice cultivation, which provides almost 56.7 per cent of the total output. In 2016, Vietnam was the world's second largest rice exporter. Around 6.3 million tonnes of rice were exported, of which the Mekong Delta supplied approximately 90 per cent (Sang). The Delta also produces 57 per cent of the total national fishery catch and 79 per cent of its aquacultural production (GSO, *Statistical Yearbook 2013*).

Several natural phenomena significantly influence life and production in the VMD, including floods, rising and falling tides, and saline water intrusion from the sea (Tri; Käkönen). The annual flood pulse in response to the western North Pacific monsoon during the months of July to October is the key hydrological characteristic of the Mekong River (Tuan *et al.*; Manh *et al.*). Flood in the wet season and saline intrusion in the dry season are natural in the VMD. Water has always had a double-sided impact here. On the one hand, the great variation in the water flow of the Mekong River drives productivity of the agro- and aqua-ecological systems of the VMD. On the other hand, the permanent threat of water disasters in the form of high floods and severe saline intrusion also limits, to some extent, the intensive production systems (Käkönen; Tuan *et al.*; Tri).

In the VMD, livelihoods and farming systems used to be characterised by adaptation to as well as control of the changing environmental conditions. Over a long period of time on the floodplain the rice farming system had been designed to well suit the annual floods and the tidal regime. Floating rice grew along with rising floodwaters, and was thus fed by it. Then it was harvested once the floods had receded. In areas near the coast, the brackish water period was avoided, and rain-fed rice was sown when the rains began. Then the rice would be harvested prior to the intrusion of brackish water (Nguyen). In the last few decades, attitudes towards the changing flow pattern of the Mekong and the delta's environment have significantly changed from farm-level adaptation to centralised control. Such shifts have derived from the VMD's development plan which stresses the need to intensify agriculture and aquaculture, the intensive use of agro-chemicals and the construction of large-scale water control infrastructure (NCNST).

The MDDP was created in the 1960s and water control emerged as pivotal in order to enable the intensive farming and multi-cropping systems. The objective of the MDDP was to control both flooding and salinity intrusion so as to enable the development of agriculture. By the 1980s, secondary-scale units containing regulated intakes and offtakes were created. However, it was not until the 1990s that a series of large-scale watercontrol infrastructures were built. These projects were in line with the national policies of the 1980s and early 1990s, emphasising rice intensification to improve food security, as confirmed by the *National Plan for Environment and Sustainable Development*

1991–2000: Framework for Action. Large-scale watercontrol infrastructures have targeted mainly the floods in the upper part of the delta and saline intrusion in the coastal areas (Käkönen 205).

In the upper section or the floodplain, the MDDP operated most intensively in the An Giang province. The three rice crops have been made possible by a very effective water-control system erected mainly during 1997–2000 (Käkönen). During 2000–2012, the three rice crop areas located in four provinces on the floodplain increased sevenfold. Specifically, the area increased from 53,500 to 403,500 hectares within the span of 12 years (Duong *et al.*). Nowadays, the floodplains are described as "orthogonal, wide and shallow cross sections separated from the channels by dykes and connected to the channels by control structures such as sluice gates" (Manh *et al.* 3034).

In coastal areas, the salinity-control projects had played an important part in the plans developed during the 1960s. However, similarly to the floodplain areas, it was only in the 1990s that the coastal delta went through an intensive phase of construction projects. The aims of salinity control were both to protect formerly freshwater systems from more salinity and transform artificially former brackish areas into freshwater areas. Since the beginning of the year 2000, following shrimp farmers' protestations, the government has allowed for re-entry of seawater to the formerly salinity-control area (Käkönen). Presently, almost all irrigation systems in the coastal zone have been closed at points of inflow and outflow to control saline intrusion for both agriculture and aquaculture, except when the canal systems are opened and linked together in the Ca Mau Peninsula for both irrigation and navigation. The conflicts generated by water used for rice production and shrimp culture still systematically persist (Tri).

The expansion of the MDDP is reflected in huge structural modifications in the VMD. Over 91,000 km of canals, 13,300 km of flood control structures and 800 km of dykes have been constructed. During the period 1995–2010, the Vietnamese government has invested in building 15 large-scale water control systems across the Mekong delta with nearly 916,000 hectares of service areas.

III.2 Impacts of the MDDP on the VMD

III.2.1 Significant increases

Apart from rice cultivation, which is the key economic activity, aquaculture is also a major business in the VMD. Production has increased significantly owing to the large-scale construction and intensification of the infrastructure. Such increases are considered as a great success achieved by the MDDP.

III.2.2 Significant decreases

The role of the VMD in supplying natural products is also very important. For example, about 85 per cent of the local people use plants as construction materials. Large amounts of pinene and terpinene, which are important in the pharmaceutical industry, are contained in Melaleuca oil. Also, materials for many kinds of handicraft products for domestic use and export are provided by forests. Hundreds of species of medical plants, wild vegetables, spices, birds, mammals, reptiles and amphibian species are the pharmaceutical materials and main food source for the community.

These natural products, however, have declined as a consequence of the MDDP (NCNST, *Downstream Mekong River*). This is due to the significant reduction of natural wetland areas and their higher intensification in order to implement the MDDP. The VMD contains vast wetland areas. These were covered mainly by mangroves and Melaleuca forests, from which most of the timber stock and fuel materials are derived. Currently, high-quality mangrove forests are located only in reserves and national parks and they cannot provide enough timber for the future. Some popular medical plants have become rare, especially those in the mangrove forests. Similarly, though Melaleuca has been largely replanted, the declining natural area of Melaleuca has led to a fall in the oil stock by about 13 per cent over the past 36 years. Changes in natural habitats have also caused changes in the composition and quantity of natural food items. A significant number of them have become scarce, hence their use as a food source is endangered (NCNST, *Downstream Mekong River*; Thong *et al.*).

In short, the MDDP has increased provisioning services (i.e. rice, shrimp products) at the expense of other provisioning services (i.e. timber, natural food). Noticeably, the MDDP has also endangered the regulating services provided by VMD ecosystems, in that it has reduced the areas of water storage habitats. Lakes, ponds, Melaleuca forests, and flooded grasslands are natural storage sites for water, which they regulate in the dry season before irrigation begins. The dramatic decline in these habitats in recent years has deeply compromised water regulatory capacity and made people highly dependent on rain and irrigation (NCNST, *Downstream Mekong River*). Most recently, the MDDP has further threatened such regulating services by attempting to totally prevent the flood from VMD's floodplain.

III.2.3 Threats to natural ecosystems and their services

The natural products listed above are only some of the benefits that the VMD has foregone due to the implementation of the MDDP. There are, however, other more severe and long-term losses, particularly due to the disruption of natural ecosystems caused by the MDDP.

The significant ecological impacts of the MDDP has been summarised by Hashimoto. At a larger scale, the entire delta may be viewed as a component of the Mekong River catchment system. In this context, the delta is a sink and a transfer zone for matter derived from the upstream parts of the catchment and transported downstream. At the delta scale, the diverse biophysical environments are all components of a network of temporary sources and sinks for matter such as sediment, water, carbon and sulphur, linked by numerous transfer pathways. For example, the floodplain acts as a temporary storage for flood waters travelling from the catchment to the ocean. During the storage, much of the suspended sediment transported from the catchment is deposited, such that the floodplain also functions as a sediment sink. Given the inherent role of deltas as sediment sinks and the rapid rates of geomorphic processes driven by a large river discharge and sediment load, the VMD is a highly dynamic biophysical environmental system. As such, the delta is in a constant state of evolution.

Deltaic ecosystems have generally developed through the opportunistic occupation of ecological niches, which are created through the appearance, development and disappearance of diverse environments within the delta during the course of its evolution. Feedback effects are often significant in the evolution of deltaic ecosystems. As the physical environments of the delta evolve, the ecosystems undergo concurrent changes which follow a relatively predictable pathway. At a general level, the pattern of ecosystem succession is driven by a progressive increase in the substrate elevation and a decrease in environmental salinity levels.

The MDDP, however, has disrupted the natural evolutionary trends of the biophysical environment and changed the pattern of ecosystem succession by fragmenting biophysical environments within the delta, simplifying its ecosystems and triggering cumulative effects. From an ecological point of view, these all have had detrimental consequences on the VMD's ecosystem and thus reduced the adaptability of ecosystems to future changes for different reasons.

First, the fragmentation of biophysical environments within the delta has produced a reduction of available refuge areas. For example, the widespread proliferation of dykes and canals around mangrove areas will hinder the landward relocation of the mangrove habitat under a rising sea level, such that mangroves will eventually die out locally unless they are able to maintain their substrate through sufficiently rapid sedimentation. In the absence of such structural obstructions, areas to the landward side of the mangroves would have served as a refuge during a sea level rise.

Second, fragmentation also increases the level of background environmental stress placed on ecosystems, as it facilitates the convergence of negative impacts of activities in the surrounding areas, and limits the potential for reproduction, feeding and symbiosis through a low background population of individual species. For example, under natural conditions the diverse biophysical environments within the delta would be generally separated from each other by gradational boundaries reflecting the natural environmental gradients in parameters such as salinity levels, sedimentation rates and energy of the river flow. Water-control structures, such as dykes and sluice gates, have converted these gradients into abrupt boundaries across which the transfer of matter is hindered. Such erratic environmental conditions are likely to reduce ecological diversity as only the more adaptable species are allowed to survive. Other species, if not eliminated, suffer from elevated environmental stress. A future environmental change would merely serve to further increase the level of stress placed on these ecosystems, increasing the risk of collapse.

Third, simplification is another factor reducing the capacity for future change. When natural or traditional rural ecosystems are modified through human activity, species composition and age/physical structure of ecosystems become simplified. The MDDP has strongly encouraged such a tendency, directly through the replacement of stratified, highly diverse ecosystems with uniformly structure monocultures (e.g. irrigated rice, intensive shrimp aquaculture), or indirectly through degradation of natural ecosystems. Adaptability to changing conditions is reduced in ecosystems where the collective resilience to environmental adversities, such as weather, pests, diseases and human exploitation, is weakened.

Fourth, further threat to the survival of ecosystems through future environmental changes derives from the high background environmental stress generated by the cumulative effects of the MDDP, such as increased agro-chemical pollution from intensified rice culture and eutrophication caused by effluent discharge from shrimp ponds. Stressed ecosystems are typically more susceptible to damages or collapse under adverse or changing environmental conditions.

Such a reduction in the ability of the environment to withstand future change is not limited to the biosphere. Geomorphic and geochemical processes have widely been altered through the effects of the MDDP (i.e. existing infrastructure development interventions within the VMD), such that the threshold of tolerance of the physical environment to changing conditions has been lowered.

With regard to human utilisation of the delta environment, this is the first time that the rate of ecosystem modification has been as rapid as it has been observed since the MDDP started (Hashimoto). Since the MDDP has decreased the resilience of VMD ecosystems, its sustainability is clearly questionable.

III.3 Is the MDDP sustainable?

It is clear from the above sub-section that the MDDP has significantly and adversely altered ecosystems in the VMD due to Vietnam's policy in rice intensification. A strong uptake in rice production has been witnessed in the delta. However, a question remains: is the MDDP economically sustainable? In the following part, we quantitatively answer this question by discussing some relevant findings from our survey conducted in the VMD from 2011 to 2012.[8]

III.3.1 Assessing a highly intensified agricultural system

This sub-section is devoted to discussing whether or not newly intensified agricultural systems set up under the MDDP are better alternatives to existing less intensified systems. Increasing the provisioning service (such as the rice output) is an overarching goal of the MDDP. However, the feedback costs on rice output are not adequately recognised. This analysis hence can be considered as a way to examine the sustainability of this programme from the perspective of rice farmers.

First, it is necessary to define the terms "balanced cropping" and "intensive cropping". The system within low dykes is based on an integrated system of two rice crops and one natural flood-food capture. This natural flood "crop" provides free goods such as wild fish, other aquatic animals, and aquatic vegetables for local people, especially the poor and landless. Since most Mekong fish species depend on flooded areas for food and reproduction, rice fields are important aquatic habitats that significantly determine the inland fish catch. For example, 40 per cent of the total wild fish catch in the An Giang province is from rice fields (Phan and Pham). The Mekong Delta in Vietnam alone produces about 700,000 tonnes of inland fish per year. Since this accounts for one-third of the overall Mekong fish catch, which is exceptionally important by global standards (Baran), this flood crop is substantial. In addition, by letting the rice fields be flooded during the flood season, the two-crop system makes use of the flood for an efficient and environmentally sound production of rice. For example, the floodwater brings alluvial sediments to rejuvenate the fields, by adding both macronutrients (e.g. N, P, K, Ca, Mg, S) and micronutrients (e.g. B, Cu, Fe, Cl, Mn, Mo, Zn). It also provides additional organic matter to the soil. The amount of sediment deposited on the fields ranges from a few to ten tonnes per hectare (Duong *et al.*). This helps to replenish the soil and maintain fertility for rice cultivation. It was found that annually deposited nutrients (N, P, K), associated with the sediment deposition, provide on average more than 50 per cent of mineral fertilisers that are typically applied to rice crops in the non-high dyke floodplain (Manh *et al.*). The flood season also provides the natural pest control mechanisms for rice fields through the role of flood-reliant aquatic species. For example, natural fish can act as a bio-control agent in rice field ecosystem (Xie *et al.*). On the other hand, three rice crops seem to be against good agricultural practices since they limit the possibilities of, for example, integrated pest management which relies on crop rotation and longer fallow time. Studies in the Mekong Delta have found that farmers practising two crops have higher rice yields per crop than farmers practising three crops (Berg; Huynh). The negative impacts of high rice cropping intensity on rice productivity are further confirmed by a long-term continuous cropping experiment in the Philippines (Dobermann *et al.*). Cumulatively, Tong shows that yields decreased by 38–58 per cent within the 24-year period of growing three rice crops a year. The average yield reduction ranged from 1.4–1.6 per cent of each crop per year. The two-crop system, therefore, is more "balanced" than the alternative intensive three-crop system. Henceforth, in this study the term "balanced cropping" refers to the two-crop, natural flood capture system associated with low dykes. "Intensive cropping" refers to the three-crop system associated with high dykes.

Second, we summarise some key findings from our survey of agricultural outcomes on balanced and intensive system farmers in the Mekong Delta. The survey was conducted on intensive and balanced cropping farmers from two sites in the An Giang province from November 2011 to October 2012. The assumption is that rice production had homogenous characteristics between the two sites before the dyke heightening. In addition, one more balanced cropping site in another province, Dong Thap, was added. The inclusion of this third site, originally dissimilar to the first two sites, was to ascertain the character of balanced cropping in the presence of site-specific conditions. All these three sites are situated in the major rice producing areas of the VMD floodplain and all once experienced the same flooding levels preceding the use of high dykes.

Differences among the three cropping sites were investigated using analysis of variance (ANOVA) comparisons of means across categories. In case the ANOVA test proved inconclusive, differences were further analysed by comparing medians using a Wilcoxon rank-sum test. We conducted all analysis using the software package STATA 10. The results, published by Tong in 2017, are summarised here. The case for intensive rather than balanced farming requires that the value of the additional third rice crop exceeds the sum of the value lost in the first two crops, the increased labour costs incurred, and the value of the foregone fish output. Results from this study indicate that this is unlikely. The value of adding one additional crop does not compensate for the lost net income of the first two crops in using intensive cropping. The net benefit per crop is significantly lower in intensive cropping than balanced cropping. Moreover, the annual net income from intensive cropping with one additional crop is not significantly different from balanced cropping, even ignoring extra imputed labour costs and the value of foregone fish outputs. Also, spillovers all point towards an even greater level of production cost resulting in lower annual net income from intensive cropping. Considering the imputed family labour costs, the intensive cropping further underperforms owing to more family labour days needed on the extra crop but also on the first two crops. This result would be strengthened if adjustments were made reflecting the foregone value of natural fish value, which is substantial.

The study shows that rice farmers have a high adaptation capacity in dealing with negative intensive cropping impacts by changing the character of their farming systems: they increase the input use and change rice variety. However, these changes are capable of offsetting possible rice yield losses at the expense of value in rice output. Farmers do not improve their net income by switching to intensive cropping. They are, instead, constrained to work harder but less effectively and cannot revert to the balanced cropping which has now become impossible owing to the irreversible investment in high dykes. As the exceptional importance of the Mekong fisheries is matched only by their economic role in rural livelihoods and food security (Baran *et al.*), the adverse effects on fishers can cause socio-economic tension by increasing poverty and reducing community self-sufficiency.

Ultimately, the long-term sustainability of recent policies favouring farming conversion on the floodplain is questionable. Instead of focusing on increasing the rice output to achieve rapid economic development based on exports, the government should incorporate a greater appreciation of the Mekong Delta as an environmental system which provides multiple highly-valued ecosystem services. In this regard, the use of low dykes are preferable to reliance on full-flood protection high dykes. In contrast to intensive rice cropping, "a sustainable food production in the Mekong Delta should aim to reduce the resource use, avoid overuse of agrochemicals and improve production efficiency through increased recycling of nutrients and matter" (Berg 95). In that context, supporting balanced cropping facilitates an economically as well as ecologically viable option.

III.3.2 Assessing large-scale water-control structures

The above section has shown that the switch from two rice crops to three rice crops on the floodplain makes no economic sense *even if* intensive rice farmers bear no dyke costs and the analysis does not account for the loss of natural fish crop from fishers. In a socially optimal approach, a comprehensive cost-benefit analysis of an investment needs to incorporate all external costs into the assessment. With specific regard to the MDDP, economic evaluation of the dyke construction must take into account dyke capital costs and some negative externalities to the environment, such as the external cost of chemicals used in rice production. Inevitably, the case for rice intensification under the MDDP must worsen when these factors are accounted for. Ultimately, dyke heightening cannot be a socially optimal technology. As expected, the findings in Tong's "Rice Intensive Cropping" broadly confirm previous results that pointed to the economic inefficiency of the MDDP.

In order to collect the data needed for calculating the costs of dyke heightening on rice productivity and on pesticide-use externalities, the team conducted surveys among rice farmers in the high-dyke and nearby low-dyke areas, with the assumption that these areas used to have the same natural and social conditions before dyke heightening. In order to estimate the decrease in profit from agriculture production due to dyke heightening, a dyke-heightening dummy variable was introduced into the rice profit function. The pesticide environmental accounting (PEA) methodology was applied to calculate the increase in external cost of pesticide use. The result in details was published by Tong in 2015.

Again, our findings are consistent with regard to the inefficiency of the MDDP. High dykes, by preventing flooding, provide the opportunity for cultivating three rather than two rice crops. However, our research shows that dyke heightening has caused society to lose VND7,165 billion (US$344 million). The An Giang province alone also lost VND816 billion (US$39 million) from this project. It is equivalent to more than one-third of the provincial GDP in 2012 (AGSO, *Statistical Yearbook*) and hence is a major cost. Taking into account the uncertainty with respect to the various assumptions made, sensitivity analyses have showed the high robustness of our cost-benefit analysis conclusions.

As regards the benefits for different stakeholders, intensive cropping farmers receive only a very low positive net benefit when they cultivate the third crop. If they choose not to follow intensive cropping, they will further underperform. Meanwhile, fishing people, balanced crop farmers, and the other local people, in general, are disadvantaged by dyke heightening and the disadvantages affecting these groups exceed the advantages brought to the intensive farmers. The conclusion that the MDDP is both economically and environmentally unsustainable is therefore soundly supported.

IV Conclusion and policy implications

This chapter discusses the case study of Vietnam, a fast-growing, resource-based developing economy facing significant environmental challenges in its development process. Our study of the legal and institutional framework for sustainable development reveals that the country is characterised by an early awareness of and strong commitment to sustainability. However, the active implementation of policies has been very slow and there is still a considerable gap between policies and practices.

At the macro-level, Vietnam is attached to the grow-now-clean-up-later model, even though sustainability is considered as an inevitable path. This is evident in the degradation of the environment during the industrialisation process. The "weak" sustainability criteria with which Vietnam seems to comply are still questionable owing to the high rate of natural resources depletion and the poor state of the environment. Whether depleted natural wealth is sufficiently reinvested in other product assets remains unevaluated in Vietnam. Whilst international experience shows that a boom in non-renewable natural resource revenues can become a "curse" by depressing economic growth, worsening poverty and increasing political instability (ODI, *Meeting the Challenge* v), Vietnam needs to be cautious not to fall into this trap.

At the micro-level, our focus on the sustainability of the Mekong Delta Development Plan reveals that, although a short-term economic benefit is achieved through rice intensification, the long-term damages of the MDDP to the delta are significant and, to a large extent, irreversible. Apart from massive negative impacts of the plan on the VMD's ecosystems, substantial economic losses to society have been documented when benefits of all stakeholders in the delta are taken into account.

Whilst moving towards sustainability, it is critical that Vietnam focuses on the enforcement and monitoring of SD policies, particularly as regards the environment. Pro-active prevention also needs to be strengthened by ensuring that the environmental impact assessment is conducted at an earlier stage in feasibility studies and is monitored afterwards. A comprehensive cost-benefit analysis, in which negative externalities are accounted for, must be conducted to ensure the sustainability of development projects and to avoid irreversible damages to the environment.

Appendix 1

Vietnam's principles and priority areas for sustainable development

Eight principles for SD

1   Human beings are at the centre of sustainable development.
2   Economic development is considered as the central task of the next development periods.

3 Protection and improvement of environment quality are to be considered an inseparable factor of the development process.

4 The development process must satisfy the needs of present generations without causing obstacles to the life of future generations.

5 Science and technology are the foundations and momentum for the country's industrialisation and rapid, strong and sustainable development.

6 Sustainable development is a goal shared by the whole Party, governments at all levels, ministries, sectors and localities, agencies, businesses, social organisations, communities and the whole people.

7 Development of an independent and autonomous economy must be linked with international economic integration in order to ensure sustainable development of the country.

8 Socio-economic development and environmental protection should be closely tied with a guarantee of national defence and security as well as social safety and order.

Nineteen priority areas for sustainable development

1 Maintaining rapid and sustainable economic growth rate;
2 Switching to environment-friendly production and consumption models;
3 Implementing "clean" industrialisation;
4 Ensuring sustainable agricultural and rural development;
5 Ensuring sustainable development of regions and localities;
6 Making focused efforts to eliminate hunger and reduce poverty, and furthering efforts to achieve social progress and equity;
7 Continuing to reduce population growth rate and creating jobs for the workforce;
8 Orienting the urbanisation and migration process in order to sustainably develop urban areas and reasonably distribute population and labourers at regional level;
9 Improving education quality in order to raise the education level, professional skills and qualifications of the population and meet the requirements for the development of the country;
10 Developing healthcare services, improving working conditions and healthy living environments;
11 Preventing soil degradation and promoting the use of land resources in an efficient and sustainable manner;
12 Protecting water bodies and using water resources in a sustainable manner;
13 Ensuring rational exploitation and sustainable use of mineral resources;
14 Protecting marine, coastal and islands environments and developing marine resources;
15 Protecting and developing forests;
16 Reducing air pollution in urban and industrial zones;
17 Managing solid waste and hazardous waste;
18 Conserving biodiversity;

19 Adopting measures for mitigating climate change and limiting its negative impacts, preventing and combating natural disasters.

Appendix 2

Compulsory indicators for monitoring and evaluating local sustainable development for the period 2013–2020

1 General indicators

Composite indicator: Human Development Index (HDI)

Economic indicators

1 Investment per local GDP ratio
2 Incremental Capital Output Ratio (ICOR)
3 Social productivity
4 Local budget revenue – expenditure ratio
5 Land for rice production area.

Social indicators

1 Poverty rate
2 Unemployment rate
3 Percentage of trained workers
4 Gini coefficient
5 Gender at birth ratio
6 Percentage of residents with social insurance, unemployment insurance, and medical insurance
7 Percentage of local public budget for cultural and sports activities
8 Percentage of communes meeting "new rural area" criteria
9 Under-five-year-old mortality rate
10 Road toll number
11 Percentage of children starting school at the right age.

Environmental and natural resources indicators

1 Percentage of population with access to clean water
2 Percentage of biodiversity-protected area
3 Land erosion percentage
4 Percentage of industrial and processing zones with solid waste and waste water treatment meeting environmental requirements
5 Forest coverage
6 Percentage of solid waste collected and treated
7 Number of natural disasters and damages.

2 Local-specific indicators

For mountainous areas: Number of forest fires and illegal forest harvest
For deltaic areas: Percentage of irrigated area
For first-tier urban areas: (1) accommodation area per head; and (2) the decrease of surface and underground water
For rural areas: (1) Revenue per cultivated hectare; (2) Percentage of rural population with access to clean water; and (3) Percentage of rural solid waste treated.

Notes

1 Khoa and Sinh further argue that sustainable practices in economic and social development have been traditionally implemented in Vietnam through a widespread use of renewable energy such as wind and solar power in production and recycling factors of production.
2 The Plan was developed with the assistance of UNDP, SIDA and UNEP under project document VIE/89/021.
3 The NCSD is chaired by a Deputy Prime Minister and involves ministers and leaders from various ministries, including the Ministry of Planning and Investment; the Ministries of Labour, Invalids and Social Affairs; the Ministry of Natural Resources and Environment; the Ministry of Foreign Affairs; the Ministry of Finance; the Ministry of Agriculture and Rural Development.
4 The SDO is also known as the *Vietnam Agenda 21* Office.
5 According to Thang's *Millennium Development Goals Lessons*, Vietnam has been able to: (1) eradicate extreme poverty and hunger; (2) provide universal primary education; (3) promote gender equality and empower women; (4) reduce child mortality; (5) improve maternal health. It has been partially able to: (1) combat HIV/AIDS, malaria and other diseases; (2) ensure environment sustainability; (3) develop global partnership for development.
6 Net saving is calculated by (1) deducting capital consumption from gross national savings to obtain net national savings; (2) adding current expenditure on education as a proxy for human capital accumulation; (3) subtracting estimates of the depletion of different kinds of natural resources; (4) subtracting estimates of pollution damage in the form of health costs due to particulate emissions (OECD, *Towards Green Growth*).
7 Vietnam AQI is assessed according to 5 levels: 0–50 is good (no damage to human health); 51–100 is average (sensitive people should limit outdoor time); 101–200 is poor (sensitive groups, i.e. children, old people and people with respiratory diseases, should limit outdoor time); 200–300 is very poor (sensitive groups should avoid outdoor activities, others should limit outdoor time); over 300 is harmful (everyone should stay indoors).
8 More details about the survey can be found in Tong (*A Cost-Benefit Analysis* and "Rice Intensive Cropping").

Bibliography

AGSO. *Statistical Yearbook of An Giang 2013* (in Vietnamese). An Giang Statistical Publishing House, 2013.
Allison, Edward, Allison L. Perry, Marie-Caroline Badjeck, W. Neil Adger, Katrina Brown, Declan Conway, Ashley S. Halls, Graham M. Pilling, John D. Reynolds, Neil L. Andrew, and Nicholas K. Dulry. "Vulnerability of National Economies to the

Impacts of Climate Change on Fisheries". *Fish and Fisheries*, vol. 10, no. 2, 2009, pp. 173–196.

Asian Development Bank and United Nations Environment Program (ADB and UNEP). *Vietnam National Environmental Performance Assessment (EPA) Report*. GMS Environment Operations Center, 2008, www.gms-eoc.org/uploads/resources/24/attachment/Viet%20Nam%20EPA%20Report.pdf. Accessed 26 March 2017.

Baran, Eric. *Mekong Fisheries and Mainstream Dams*. International Centre for Environmental Management, 2010.

Baran, Eric, Teemu Jantunen, and Chiew Kieok Chong. *Value of Inland Fisheries in the Mekong River Basin*. World Fish Center, 2007.

Barbier, Edward B. *Natural Resources and Economic Development*. Cambridge University Press, 2007.

Berg, Hakan, "Rice Monoculture and Integrated Rice-fish Farming in the Mekong Delta, Vietnam – Economic and Ecological Considerations". *Ecological Economics*, vol. 41, no. 1, 2002, pp. 95–107.

Brockerhoff, Eckehard G., Christopher P. Quine, Jeffrey Sayer, and David L. Hawksworth. "Plantation Forests and Biodiversity: Oxymoron or Opportunity?" *Biodiversity and Conservation*, vol. 17, no. 5, 2008, pp. 925–951.

Clausen, Alison, Hoang Hoa Vu, and Miguel Pedrono. "An Evaluation of the Environmental Impact Assessment System in Vietnam: The Gap between Theory and Practice". *Environmental Impact Assessment Review*, vol. 31, no. 2, 2011, pp. 136–143.

Dobermann, Achim, David Dawe, Reimund P. Roetter, and Kenneth G. Cassman. "Reversal of Rice Yield Decline in a Long-term Continuous Cropping Experiment". *Agronomy Journal*, vol. 92, no. 4, 2000, pp. 633–643.

Duong, Hoang Thai Vu, Van Trinh Cong, Franz Nestmann, Peter Oberle, and Nam Nguyen Trung. "Land Use Based Flood Hazards Analysis for the Mekong Delta". *Proceedings of the 19th Congress of the Asia and Pacific Division of the International Association for Hydro Environment Engineering and Research. IAHR – APD*, 2014, www.iahr.org/site/cms/contentdocumentview.asp?chapter=49&documentid=1790&category=326&article=876. Accessed 22 March 2017.

Government of Vietnam. *State of the Environment Report 2011–2015*. Government Printing Office, 2015.

Government of Vietnam. *Implementation of Sustainable Development, National Report at the United Nations Conference on Sustainable Development (RIO+20)*. Government Printing Office, 2012.

Government of Vietnam. *National Plan for Environment and Sustainable Development 1991–2000: Framework for Action*. Government Printing Office, 1991.

GSO. *Statistical Yearbook of Vietnam 2015* (in Vietnamese). Statistical Publishing House, 2015.

GSO. *Statistical Yearbook of Vietnam 2013* (in Vietnamese). Statistical Publishing House, 2013.

Hashimoto, Takehiko. "Riko". *Environmental Issues and Recent Infrastructure Development in the Mekong Delta: Review, Analysis and Recommendations with Particularly Reference to Large-scale Water Control Projects and the Development of Coastal Areas. Working paper 4*. Australian Mekong Resource Centre, The University of Sydney, 2001.

Huynh, Viet Khai. *The Costs of Industrial Water Pollution on Rice Production in Vietnam*. Economy and Environment Program for Southeast Asia, 2011, www.eepsea.org/pub/tr/Huynh-Viet-Khai-Technical-Report-Sep2011.pdf. Accessed 23 March 2017.

Käkönen, Mira. "Mekong Delta at the Crossroads: More Control or Adaptation?" *AMBIO: A Journal of the Human Environment*, vol. 37, no. 3, 2008, pp. 205–212.

Khoa, Lê Văn, and Sinh, Nguyễn Ngọc. "Sustainable Development in Vietnam: Status-quo, Challenges, and Resolutions" (in Vietnamese). *Vietnam Association for Conservation of Nature and Environment*, www.vacne.org.vn/phat-trien-ben-vung-o-viet-nam-hien-trang-thach-thuc-va-giai-phap/2149.html. Accessed 5 January 2017.

Kinh, Nguyễn Khắc. "Issues in Vietnam Environmental Impact Assessment and Important Points in Environmental Impact Assessment under the Law on Environment Protection 2014 and Other Legal Provisions" (in Vietnamese). *People and Nature Reconciliation*, http://nature.org.vn/vn/wp-content/uploads/2015/09/20152209_batcaptrongDTM.pdf. Accessed 1 February 2017.

Manh, Nguyen Van, Nguyen Viet Dung, Nguyen Nghia Hung, Bruno Merz, and Heiko Apel. "Large-scale Suspended Sediment Transport and Sediment Deposition in the Mekong Delta". *Hydrology and Earth System Sciences*, vol. 18, 2014, pp. 3033–3053.

National Centre for Natural Science and Technology (NCNST). *Downstream Mekong River Wetlands Ecosystem Assessment*. Institute of Geography, NCNST, 2005.

Neumayer, Eric. *Weak Versus Strong Sustainability: Exploring the Limits of two Opposing Paradigms*. 2nd edn, Edward Elgar, 2003.

Nguyen, Bao Ve. "Factors Affecting the Sustainability of Three Rice Crops System in the Mekong Delta". Conference on *Improving the Three Rice Crop System*, 18 October 2009, An Giang Department of Agriculture, An Giang Province. Keynote Address. Accessed 4 January 2017.

Organisation for Economic Co-operation and Development (OECD). *Towards Green Growth in Southeast Asia*. OECD Publishing, 2014, http://dx.doi.org/10.1787/9789264224100-en. Accessed 5 January 2017.

Overseas Development Institute (ODI). *Meeting the Challenge of the Resource Curse: International Experiences in Managing the Risks and Realizing the Opportunities of Non-Renewable Natural Resource Revenue Management*. ODI, 2006.

Pearce, David W., and Giles D. Atkinson. "Capital Theory and the Measurement of Sustainable Development: An Indicator of 'Weak' Sustainability". *Ecological Economics*, vol. 8, no. 2, 1993, pp. 103–108.

Phan, Thanh Lam, and Mai Phuong Pham. *Preliminary Results of the Involvement in Fisheries from the Baseline Survey in An Giang Province, Vietnam*. AMFP Technical Report, Mekong River Commission, 1999.

Prime Minister of Vietnam. *Vietnam Socio-Economic Development Strategy for 2011-2020*. Decision 432/QĐ-TTg. Hanoi, 2012. Print.

Sang, Thanh. "Rice Export Faces Challenges", *BNews*, 25 September 2015, http://bnews. Vn/xuat-khau-gao-gap-kho/1524.html. Accessed 10 August 2016.

Thang, Nguyen Trung. *Millennium Development Goals Lessons for Sustainable Transition: Vietnam Experience*. ISPONRE, 2016, www.asef.org/images/docs/03_Vietnam_SDGs_implementation.pdf. Accessed 26 March 2017.

Thong, Mai Trong, Hoang Luu Thu Thuy, Vo Trong Hoang. "Ecosystem Assessment of Cuu Long River Delta Wetland, Vietnam". *Journal of Environmental Science and Management*, vol. 16, no. 2, 2013, pp. 36–45.

Tong, Yen Dan. "Rice Intensive Cropping and Balanced Cropping in the Mekong Delta, Vietnam – Economic and Ecological Considerations". *Ecological Economics*, vol. 132, no. 1, 2017, pp. 205–212.

Tong, Yen Dan. *A Cost-Benefit Analysis of Dike Heightening in the Mekong Delta*. WorldFish, 2015.

Tri, Vo Khac. "Hydrology and Hydraulic Infrastructure Systems in the Mekong Delta, Vietnam". *The Mekong Delta System. Interdisciplinary Analyses of a River Delta*, edited by Fabrice G. Renaud and Claudia Kuenzer, Springer, 2012, pp. 49–81.

Truong, Nguyen, and Tran Hoang. "A Hundred Million Dollar Project 'Eludes' the Environmental Impact Assessment" (in Vietnamese). *Tien Phong*, 11 March 2017, www.tienphong.vn/dia-oc/du-an-tram-trieu-do-tron-danh-gia-tac-dong-moi-truong-1129073.tpo. Accessed 13 March 2017.

Tuan, L. A., Chus Thai Hoanh, and Chiew Kieok Chong. "Chapter 1: Flood and Salinity Management in the Mekong Delta, Vietnam". *Challenges to Sustainable Development in the Mekong Delta: Regional and National Policy Issues and Research Needs*, edited by Tran Thanh Be, Bach Tan Sinh, and Fiona Miller, SUMERNET, 2007, pp. 15–68.

World Bank. *Vietnam-World Bank Data*. The World Bank, http://data.worldbank.org/country/vietnam. Accessed 26 March 2017.

Xie, Jian, Liangliang Hu, Jianjun Tang, Xue Wu, Nana Li, Yongge Yuan, Haishui Yang, Jiaen Zhang, Shiming Luo, and Xin Chen. "Ecological Mechanisms Underlying the Sustainability of the Agricultural Heritage Rice Fish Coculture System". *Proceedings of the National Academy of Sciences of the USA*, vol. 108, no. 50, 2011, pp. 1381–1387, doi: 10.1073/pnas.1111043108.

Xuan, Vo Tong, and Sheigo Matsui, editors. *Development of Farming Systems in the Mekong Delta of Vietnam*. Ho Chi Minh City Publishing House, 1998.

Index

Page numbers in *italics* denote tables, those in **bold** denote figures.

accountability 140, 141, 143, 250, 252
adaptability *27*, 40, 86, 293, 294
agency 50–1, 55, 59, 64, 70, 86, 89
Agenda 21 47, 50, 55–7, 58, 138, 141
Albrizio, Silvia 113
Aldiss, Brian 93, 96, 97
Amino, Yoshihiko 275
Ammon, Ulrich 188
apocalypse 85–104; *fin-de-siècle* apocalypse
   86–92; nuclear weapons and World War
   III 92–3, 97–8; and palingenesis 85, 90,
   92, 99, 102; risk and resilience in the
   twenty-first century 98–103; science
   fiction 93–7; steampunk fiction 3,
   99–100; twentieth century as
   apocalypse 92–8; in utopia as a literary
   genre 85–6
Appadurai, Arjun 2
Approaches to a Cultural Footprint.
   Proposal for the Concept and Ways to
   Measure it (Baltà Portolés and Roig
   Madorran) 11, 12
Aristotle 24, 150–1, 162–3
Arntz, Melanie 8
Asia–Europe Foundation (ASEF) 10–11
Attitudes of Europeans towards the Issue
   of Biodiversity. Analytical Report (EC)
   170

Baker, Sarah E. 217
Ballard, J. G. 93–5
Baltà Portolés, Jordi 11, 12
Barro, Robert J. 126
BEAGLE (Biodiversity Education and
   Awareness to Grow a Living
   Environment) project 177
biodiversity 7, 20, 109, 167–84;

awareness of biodiversity and the
   Natura 2000 network 170–4, 181;
   Decade on Biodiversity 2011–2020
   176–7, 180–1, 181–2; education and
   communication in the convention on
   biological diversity 178–80;
   extinctions 168, 169, 173; information
   and education for the protection of
   biodiversity 174–8; and senior citizens
   177; state of biodiversity in Europe
   168–70; and sustainable development
   167–8; Vietnam 286–7, **287**
Book of Dave, The (Self) 101–2
Book of Revelation 85, 86
Boyce, James K. 9
Bradbury, Ray 92
Brandolini, Andrea 139
Brave New World (Huxley) 92
Brazil *see* settlements and urban river
   basins
Brundtland Report *see Our Common*
   Future (World Commission on
   Environment and Development)
Builes Vélez, Ana Elena 222–31
Burford, Gemma 54
Butler, Judith 63
Butler, Richard 255

capacities for human flourishing 20,
   23–45, 40–1; adaptability *27*; adaptation
   and limitation 40; affinity and
   reciprocity 35; analysis approach 30;
   capability domains 30–1; care and trust
   35; circles of social life approach 25, **26**,
   28–9, 30, 31, 43–4, **43**; commitment and
   purpose 41; communication and
   dialogue 35, 37; conditions, capabilities

capacities for human flourishing *continued*
   and capacities for social life 25–33, *27*;
   constructive grounding of the approach
   29; conviviality and hospitality 36; core
   capacities 23–5; dignity and recognition
   33–4; emotion and feeling 33; endurance
   and patience 40–1; enquiry and vision 39;
   faith and love 36; freedom 28–9; health
   and wellbeing 33; imagination and
   creativity 39; innovation and change
   39–40; integrity and consonance 34;
   justice and truth 35–6; key considerations
   for structuring and choosing capacities
   32; knowing and comprehending 38–9;
   learning *27*, 37–8; liveability 25, *27*;
   negotiation 36; practical consciousness
   (pragmatics) 38; practicality and
   technique 39; productivity 33, 36–40, *42*,
   43; receptiveness and responsiveness 40;
   reconciliation 27, *27*, 36; reflective
   consciousness (reflection) 38–9; reflexive
   knowing (reflexivity) 39, 40; relationality
   27, *27*, 32, 34–6, *42*, **43**; resilience *27*,
   40; security and safety 34; sensory
   experience (feeling) 38; sensuality and
   sexuality 34, 37; social capacities 24–5;
   stability and continuity 41; stewardship
   and custodianship 41; strength and vigour
   33; sustainability *27*, 33, 40–1, *42*, **43**;
   teaching 37–8; vitality 32, 33–4, *42*, **43**;
   vocation 39
Capital in the Twenty-First Century
   (Piketty) 139
capitalism 48, 67, 77, 80, 81, 101, 116
Capitalism, Socialism and Democracy
   (Schumpeter) 81
Caradonna, Jeremy L. 46
Carpenter, Edward 65, 68–71
Carson, Rachel 48
Cassilha, Gilda Amaral 207–21
castle towns *see* contemporary urban
   planning, castle towns
catastrophe 21, 64, 85–104, 163
Centre for Environmental Studies and
   Sustainable Development (UCBS)
   175–7
Chang, Ha-Joon 156
Chapman, Richard 185–99
Children of the Dust (Lawrence) 96–7
Chile, ethno-tourism 205, 254–68
circles of social life concept 24–33, **26**
cities 108–9; autonomous cities 160–2, **162**,
   163–4; climate effectiveness of 160;
   climate governance 146–66; and climate

policy architecture 149–50; decentralised
   power vs deconcentrated tasks of cities
   146; effective climate policy 146–8;
   emission reduction 149, 150; governance
   choice based only on effectiveness and
   efficiency 150; from hierarchy to
   autonomy and networks 157–9; historical
   developments in governance 150–1; and
   institutionalist climate governance 156–7,
   160; as part of public governance in
   climate policy 148–52; planning and
   control governance 152; planning and
   control instrumentation for effective
   climate policy 152–3; planning for wider
   welfare optimisation 153–4; proactive
   decentral action 147; reasons for taking
   active responsibility for emission
   reduction 159–60; role under planning
   and control climate policy 154–5; taking
   centralised responsibility 157–60; *see
   also* climate change and urban life;
   contemporary urban planning, castle
   towns; settlements and urban river basins;
   social urbanism
citizen participation 139, 140, 158–9, 209,
   225–7, 241, 244, 246–9, *247*; as key to a
   culture of wellbeing 251–2; and political
   dialogue 249–51
Clarke, I. F. 87–8, 89
climate change and urban life 205, 244–53,
   247; district councils in Saint-Louis du
   Sénégal 246–8, *247*, 249; flooding
   245–6, **245**; neighbourhood organisation
   246–8, *247*; participation as key to a
   culture of wellbeing 251–2; political
   dialogue and participation 249–51;
   sports and culture associations 244, 246,
   247, 249; urban governance risk 248–9
climate governance 146–66; autonomous
   cities 160–2, **162**, 163–4; cities and
   effective climate policy 146–8; cities as
   part of public governance in climate
   policy 148–52; city role under planning
   and control climate policy 154–5;
   climate effectiveness of cities 160;
   climate policy 108–9; climate policy
   architecture and cities 149–50;
   co-ordination by hierarchy or
   incentivised autonomy and networks
   151–2; decentralisation, democracy and
   legitimacy 162–3; decentralised power
   vs deconcentrated tasks of cities 146;
   emission reduction task of cities 149;
   governance choice based only on

effectiveness and efficiency 150; from hierarchy to autonomy and networks 157–9; historical developments in governance 150–1; institutionalist climate governance and cities 156–7, 160; interactions between strategic actors 248–9, 251–2; planning and control governance 152; planning for wider welfare optimisation 153–4; proactive decentral action 147; reasons for taking active responsibility 159–60; taking centralised responsibility 157–60
Cloud Atlas (Mitchell) 100–1
Club of Rome 48
"Co-Designing in Love: Towards the Emergence and Conservation of Human Sustainable Communities" (Salazar) 54
Coal Question, The (Jevons) 64
Colebrook, Claire 63
Coly, Adrien 244–53
communication 2, 6, 10, 11, 12, 20, 35, 37, 53, 93, 139, 168, 175, 178–80, 185–200, 225, 227, 244, 250, 258, 265; *see also* English (language)
community 19, 22, 25, 34, 41, 65, 66, 68, 70, 86, 90, 97, 140, 178, 194, 208, 209, 212, 217–19, 224–30, 236, 240, 246–51, 255, 260, 262–6, 283, 292
consumption 72–84, 128; demand, economic development and consumption 78–9; economic activities, inputs and outputs 74–5; efficiency 77–8, *77*, 79; environmental impacts 82; Keynes' predictions on consumption 80–2, 82–3; physical goods 75–6; productive efficiency 75; reduction in focus on 143; sobriety and voluntary simplicity 140; structural change and economic development 72–82, *77*; structure of an economic system 73–4, 75, 76; stylised facts and economic development 74; trajectories of economic development 76–8, *77*; variety (creativity) 74, 75, 76, 77, *77*, 79; working hours reduction 79, 80–1, 81–2
contemporary urban planning, castle towns 205–6, 269–80; background information 269–70; castle towns as military towns 271–3; civil engineering 273–4; design principles and characteristics 275–7; Feng Shui principle 276; geometrical relationships between urban landscape features 270–1; history of castle towns to the pre-modern period 273–5;

Japanese modern city origins and castle towns 274–5; Kyoto 270, 273; Morioka 270–1, **272**; planning methods and techniques 277–80, **278**; street layout 270; Tsuruoka 270–1, **271**, 277
Convention on Biodiversity (CBD) 168, 178
convergence 112–13; in human development 125–8, 130
conviviality 36, 66, 140, 143
Costanza, Robert 49
Covenant of Mayors for Climate and Energy in the EU 155
creativity 4, 12, 20–1, 36–7, 39, 65, 66, 74, 75, *77*, 147, 151, 190, 194, 224–7, 230, 269
Crozier, Michel 251
Crutzen, Paul 10
Crystal Age, A (Hudson) 86, 88, 89–90
Crystal, David 186, 187
Cuenca, Manuel 256
culture 1–13, 65, 70, 92, 93, 97, 135, 192, 194, 205, 209, 222, 250–2, 254–66; cultural footprints of sustainability and wellbeing 9–12; *see also* tourism
Curitiba *see* settlements and urban river basins
Cutter, Susan L. 212–13

Daly, Herman E. 49
Darwinism 100
Davies, Alan 189
Davison, Aidan 71
De Abaitua, Matthew 102
Decade on Biodiversity 2011–2020 176–7, 180–1, 181–2
degrowth 108, 135–45, 197; accountability system based on a plurality of variables 143; conviviality and reciprocity 140; democracy 139–40; equity 138–9; minimum and maximum wages and guarantees 142; participation and transparency 140–1; pillars of 135–6; principles and initiatives 140–1; redefinition of wellbeing and accountability system 141; reduction in focus on consumption 143; research gaps, future developments and policy implications 141–3; sobriety and voluntary simplicity 140; subsidiary and sustainable production 140; and sustainability 136–7; sustainable production activities, support for 143; and wellbeing 137–8, *138*; work sharing and working time reduction 142

democracy 23, 136, 139–40, 141, 143, 151–2, 155, 162–3, 223, 244, 248
Demystifying Economic Evaluation (Ozdemiroglu and Hails) 5
Dias, Braulio Ferreira de Souza 167
Diatta, Mohamed B. C. 244–53
dignity 33–4
Drowned World, The (Ballard) 94
Duhau, Emilio 228

Earth Summit Declarations 1972–2012 20, 46–62; contradictory rationales in the global sustainability agenda 51–2, *52*; institutional sustainability 55; land ethic concept 48; Rio de Janeiro 1992 55, 56, 168, 282; socio-economic rationale 56; sustainability as a mask for global neoliberalism: local effects 57–9; sustainability, incongruities of: epistemological change or progress 47–9; sustainability practice and localisation 50–1; sustainability trend based on a socio-ecological epistemology 55–7; sustainability trend based on the epistemology of progress 52–5; three pillars of sustainability concept 52–5, 58
ecological footprint paradigm 11–12
economics: consumption, demand and economic development 78–9; and convergence analysis 126–8; creative economy 10–11; economic and policy implications of sustainability 128–30; economic development and structural change 72, *77*, 82; economic growth in Vietnam 283–4, 285–6; efficiency 77–8, *77*, 79; inputs and outputs 74; Keynes' predictions on consumption 80–2, 82–3; productive efficiency 75; socio-economic discourses of sustainable wellbeing 6–9; structure of an economic system 73–4, 75, 76; stylised facts and economic development 74; sustainability and economic development 112–13; trajectories of economic development 76–8, *77*; variety (creativity) 74, 75, 76, *77*, *77*, 79; and wellbeing 4–6; working hours reduction 79, 80–1, 81–2
ecosystems 11, 53, 57, 68, 82, 167–84, 281–304
education 2, 3, 8, 11, 78, 107, 109, 125, 167–84; learning 37–40
Edwards, Rosalind 217

efficiency 72–84, *77*
Ekins, Paul 114
"Elemental Apocalypse Quartet" (Ballard) 94
emissions 122–4, **123**, 129, 146; reduction in cities 149, 150, 154–5, 157–60
English (language) 109–10, 185–99; context-dependency 196–7; de-culturing 191–2, 194; discourse 194–5; dominance of 185; homogeneity 188–9; as a lingua franca 189–92; linguistic interactions 193; measuring global English 186–7; native speakers 189; pragmatic aspects 192–5; roles of 187–8; tensions in the ELF model 190–1; translanguaging 195–6; as a universal language 195–7
Enshu, Kobori 276
environmental factors: and consumption 82; ecological taxes and environmental policies 113–14; environmental policies 113–14; environmental awareness 3; environmental policy stringency and ecological taxes 119–22, **120**, **121**, **122**, 131n11; environmentalism 47–9
Environmental Kuznets Curve 82, 130n4, 136–9
equity 2, 135–45, 223–6
ethno-tourism *see* tourism
EU Biodiversity Strategy to 2020 (EC) 173–5
eudaimonia concept 24, 223
Europe: awareness of biodiversity and the *Natura 2000* network 170–4, 181; state of biodiversity 168–70
European Commission: Strategic Planning and Programming Cycle 153
European Environment – State and Outlook 2005, The 113
European Environment Agency 113, 169

Feng Shui principle 205, 275–7
Ferreira, Leila C. 208
flooding 210–11, 212, 213, 219–20, 245–6, **245**, 290, 295
Forum d'Avignon 10
Foster, John 48–9
Franz, Gianfranco 9, 232–43
Fraser, Nancy 33–4
freedom 23, 28–9, 197
Funtowicz, Silvio 139

Gabardo, Marta Maria Bertan Sella 207–21

Garrett, Peter 193
Geddes, Patrick 50
Geologia stratigrafica (Stoppani) 13n6, 9–10
"Geology of Mankind" (Crutzen) 10
Giglia, Angela 228
Gilli, Marianna 111–34
Gilligan, Carol 35
Global Biodiversity Outlook 4 178
global carbon fund (CGF) 124
globalisation 11, 12, 63, 70, 123, 144, 148
Godard, Jean-Louis 93
Godet, Michel 244
Gottdiener, Mark 220
Graddol, David 187, 196
gross domestic product (GDP) 8, 111–12, 116–19, 122, 126, 128, 130n2, 130n4, 130n13, *138*, 141, 142, 233, 246, 257, 284–5, 284, **284**
Grubb, Michael 153
Guidobono, María Florencia 222–31

Habermas, Jürgen 244
Hails, Rosie 5
Haraway, Donna 51
Hay, William Delisle 86, 87
Helliconia novels (Aldiss) 96, 97
Hinch, Thomas 255
Honneth, Axel 33–4
Hourcade, Jean-Charles 153
Hudson, William Henry 86, 88, 89
Huele, Ruben 146–66
human development 108, 111, 115, 226, 256; convergence in human development 125–8, **127**, 130
Human Development Index 30, 31, 43–4, 154, 208, 282, 300
Hungry Cities Chronicle, The (Reeves) 99–100
Huppes, Gjalt 146–66
Huxley, Aldous 92

Ice (Kavan) 95–6
identity 28, 34, 81, 191, 192, 193, 194, 197, 229, 238, 240, 241, 244, 246, 257, 258, 261, 263, 264, 266, 279
If Then (De Abaitua) 102
Industrial Revolution 20, 72–84, 86, 90
inhabiting 227–30
innovation 39–40, 72–3, 113–14, 128–9; and co-operation 158–9; creative social innovation for the development of sustainable community 230; institutionalism for decentral innovation

156; and invention 115–19; and knowledge transfers 122–5, **123**; patents filed 115–16, **115**, 116–17, **118**, 119, **119**; research and development expenditure 116, **117**, **118**; transformation through creative social innovation or social urbanism 226–7; urban development through creative social innovation 224–6
institutionalism 158, 160, 162; for decentral innovation 156; and international governance 161, **162**; and monopoly 156–7
International Year of Biodiversity 2010 175, 176
investment 112–13, 128
Italy *see* post-disaster planning

Jackson, Tim 9
James, Paul 23–45, 209
Japan *see* contemporary urban planning, castle towns
Jefferies, Richard 86, 88–9, 90
Jenkins, Jennifer 186–7, 190, 191, 192
Jevons, W. S. 64, 78, 136
Johannesburg Declaration 2002 52, 56, 57
Jung, Carl 96
justice 35–6, 53, 66, 157, 224

Kachru, Braj B. 186, 187, 188
Kalinowska, Anna 167–84
Kavan, Anna 95–6
Keough, Noel 70
Ketterer, David 93
Keynes, John Maynard 21, 72, 80–3, 113
Kingsley, Charles 94–5
knowledge 7, 50–1, 58, 76, 92, 97, 102, 128–9; innovation and knowledge transfers 122–5, **123**; knowing and comprehending 38–9
Kuznets curves 82, 130n4, 136–9
Kuznets theory 138–9

land ethic concept 48
language 109–10; English 185–99
Lawrence, Louise 96–7
learning 37–40
Lefèbvre, Henri 36–40, 216
legitimacy 108–9, 148, 155, 162–3, 225
Leitão, Sylvia Ramos 207–21
Leopold, Aldo 48
Libellus vere aureus (More) 86
Limits to Growth, The (Meadows) 48, 136
Lindell, Michael K. 240

liveability 25, 27, *27*
localisation 49, 50–1, 59
Louit, Robert 93–4
Luhmann, Niklas 35

MacDuffie, Allen 64, 65
McHale, Brian 94
McHarg, Ian 48
MacKenzie, Ian 189
MACTOR analysis 244–5
Malthus, Thomas Robert 4, 72
Many Faces of Biodiversity, The
    (campaign) 175–6
Mapuche people *see* tourism
Marândola, Eduardo, J. R. 212
Marcovitch, Jacques 212
Marx, Karl 4, 37, 77, 79
Mazzanti, Massimiliano 1–16, 107–10,
    111–34, 203–6
Medellín *see* social urbanism
Medovoi, Leerom 63, 67
Mekong Delta *see* sustainable development
Menoni, Scira 239
migration, intra-urban migrations 227–30
Mill, John Stuart 4, 64
Mitchell, David 100–1
monopoly 156–7
Montgomery, Scott L. 185, 187, 196
Moorcock, Michael 93
More, Thomas 86
Morgan, Jamie 9
Morin, Edgar 207
Morris, William 65–8, 87, 90–2
Mumford, Lewis 48

Naess, Arne 49, 51
*National Plan for Environment and
    Sustainable Development 1991–2000:
    Framework for Action* (Vietnam) 281–2,
    289, 290–1
Natura 2000 109, 170–6, 181
Nature for All (campaign) 179
neoliberalism 29, 48, 56, 57–9, 193–4, 197
Neuhoff, Karsten 153
*News from Nowhere: or, An Epoch of Rest
    Being Some Chapters from a Utopian
    Romance* (Morris) 66, 67–8, 70, 87, 90–2
Nussbaum, Martha 20, 25, 28, 29–30, 31,
    32, 33, 34, 35–6, 38–9

Orr, David W. 51
Ostler, Nicholas 189, 191, 196
Ostrom, Elinor 157
Our Common Future (World Commission

on Environment and Development) 44,
    47, 52, 53
Ozdemiroglu, Ece 5

Paquot, Thierry 208
Paris Agreement (2015) 146, 149, 150
Parkins, Wendy 63–71
patents 115–19, **115**, **118**, **119**, 156–7
Pattern of Expectation (1644–2001), The
    (Clarke) 87–8
Pham, Nhai 281–304
Pigou, Arthur Cecil 113
Piketty, Thomas 9, 139
Pinner, Richard S. 189
Plumwood, Val 68
pollution 112, 113, 120, 129, 136–7, 172,
    206, 208, 216, 282, 286, 287–8, 294
population growth 63, 72, 77, 78, 167
post-disaster planning 204, 232–43;
    businesses 240, 241; citizen
    participation 241; community
    transformation 238–9; cultural heritage
    239; effects of the disaster and early
    results of reconstruction 232–3; Friuli
    237, 239; long-term management and
    sustainability problems 236–7;
    management of the phase between
    reconstruction and new development
    237–40; operational plan 234–5;
    planning innovations 233–5; planning
    successes and limits 235; road networks
    241–2; San Francisco 204, 237; social
    value of property 239–40; special
    programme for the area 233–4; Umbria
    and Marche 237–8, 239; urban tax-free
    zones 235; "where it was, as it was"
    reconstruction method 238, 239, 240
Postmodernist Fiction (McHale) 94
poverty 53, 56, 138, 208, 218, 222–4,
    258–9, 284, **285**
productivity 33, 36–40, *42*, **43**; productive
    efficiency 75; sustainable production
    activities, support for 143

quality of life 1–2, 24, 140, 168, 180, 208,
    217–19, 223–4, 227, 229

Rasmussen, Douglas B. 24, 41
Ravetz, Jerry 139
real-estate 216–17
reality 4–5, 49, 143
reconciliation 27, *27*, 36
Red Men, The (De Abaitua) 102
Reeve, Philip 99–100

reflexivity 39, 40
relationality 27, *27*, 32, 34–6, *42*, **43**
religion 54
research and development (R&D) 116,
  117, **118**
resilience 1, 2, 4, 9, 12, 25, *27*, 40, 86,
  98–103
Ried, Andrés 254–68
Rio Declaration of 1992 52, 54, 55, 56
Rio+20 Declaration of 2012 54, 57, 58–9
risk 1, 2, 12, 21–2, 28, 86, 98–103, 163,
  212–13, 226, 245; interactions between
  strategic actors 248–9, 251–2
Robeyns, Ingrid 31
Rojo, Sofía 254–68
Rowbotham, Sheila 70
Ruskin, John 64

Saint-Louis du Sénégal, urban life and
  climate change 205, 244–53
Sala-i-Martin, Xavier 126
Salazar, Gonzalo 46–62
Sall, Fatimatou 244–53
Sarr, Chérif Samsédine 244–53
Satoh, Shigeru 269–80
Saviotti, Pier Paolo 72–84
Schumpeter, Joseph 81
Schussel, Zulma das Graças Lucena
  207–21
science fiction 93–7
Self, Will 101–2
Sen, Amartya 20, 25, 29, 32, 111, 153
Sendai Framework for Risk Reduction
  2015–2030 102
senior citizens 177, 238
Sennett, Richard 35
settlements and urban river basins 204,
  207–21; circles of sustainability concept
  207; circles of sustainability, the case of
  Belém River in Curitaba 209–10; coping
  with an unsustainable way of life
  219–20; Curitiba Preliminary Urban
  Plan 211; definitions of urban
  sustainability 208; drainage problems
  208–9; favelas 208, 210, 211–12, 213,
  218; Guaíra River sub-basin 214, **215**,
  216; hydrography 209–10; Pinheirinho
  River basin 214, **215**; quality of life and
  perception of wellbeing 217–19; real-
  estate prices 216–17; settlement process
  210–12; Social Life Questionnaire 218;
  urban and environmental legislation,
  failure to comply with 214–19;
  vulnerability and risks 212–13

Shang Shui principle 205–6, 275–7
Sigur Rós 98
skills, reskilling 107
slow living and sustainability 63–71;
  community, importance of 70;
  ecosocialist thinking 66–8; Edward
  Carpenter (1844–1929) 65, 68–71;
  idealism 69–70; resource depletion, risk
  of 64–5; William Morris (1834–1896)
  65–8
slums 226–7; favelas 207–21
Smith, Adam 4, 37
social media 139, 175, 179, 180
social urbanism 204, 222–31; conceptual
  approaches and definitions 222–4;
  creative social innovation for the
  development of sustainable community
  230; España Library Park 229–30;
  Medellín cómo Vamos MCV 223, 230n1;
  Moravia Cultural Centre project
  229–30; PUI strategy 225–6;
  transformation through creative social
  innovation or social urbanism 226–7;
  transformation, ways of inhabiting and
  urban migrations in Medellín 227–30;
  urban development through creative
  social innovation 224–6; wellbeing,
  theories of 222–3
Soja, Edward W. 220
*Special Eurobarometer 421. The European
  Year for Development – Citizen's Views
  on Development, Cooperation and Aid*
  (EC) 173–4
Speck, Stefan 6, 114
Spinozzi, Paola 1–16, 19–22, 85–104,
  203–6
State of Nature in the EU (EEA) 169
steampunk fiction 3, 99–100
Stoppani, Antonio 9–10, 13n6
*Study on Environmental Fiscal Reform
  Potential in 12 Member States*
  (European Commission) 114
sustainability 19, 24, *27*, 33, 222; and
  biodiversity 167–8; as capacity for
  human flourishing 40–1, *42*, **43**; circles
  of sustainability concept 207, 209–10;
  contextualising sustainability 111–34;
  convergence in human development
  125–8, **127**, 130; cultural footprints of
  9–12; definitions related to wellbeing 3;
  and degrowth 136–7; economic and
  policy implications 128–30; and
  economic development 112–13;
  environmental policies and ecological

sustainability *continued*
   taxes 113–14, 119–22, **120**, **121**, **122**,
   131n11; innovation and invention
   115–19, **115**, **117**, **118**;
   interdisciplinary projects 3–4; and
   knowledge transfers and innovation
   122–5, **123**; and slow living 63–71;
   (un)sustainable societies 204–5; urban
   sustainability 208, 224; within cultural
   discourses 2; *see also* Earth Summit
   Declarations 1972–2012
sustainable development 2, 44, 136;
   environmental impact assessments in
   Vietnam 285; highly intensified
   agricultural systems, assessment of
   294–7; impacts of the MDDP on the
   Mekong Delta 291–4; large-scale water-
   control structures, assessment of 297;
   Mekong delta 289–97; Mekong Delta
   Development Plan (MDDP) 289, 290–1;
   natural capital depletion in Vietnam
   286–7, **287**; policies and institutions in
   Vietnam 281–3; policy implementation
   in Vietnam 283–8, **284**, **285**; policy
   implications for Vietnam 298; pollution
   287–8; principles and priority areas for
   sustainable development in Vietnam
   298–300; production increases and
   decreases as a result of the MDDP
   291–2; and public policy in Vietnam
   206, 281–304; regulations and legal
   documents 283; sustainability of the
   MDDP 294–7; sustainable development
   indicators 143, 282–3, 300–1; threats to
   natural ecosystems and their services
   from the MDDP 292–4; weak
   sustainability in Vietnam 286
sustainable wellbeing 1–16, 203–4, 223;
   theory and methodology 1–4; socio-
   economic discourses of 6–9

taxation 113–14, 120–30, 131n12, 131n13;
   and environmental policy stringency
   119–22, **120**, **121**, **122**
teaching 37–40
technology 7–8, 54, 73, 76, 93, 96, 97,
   101, 107, 108, 111–34, 139, 142, 146,
   149, 150, 157, 205, 242, 274, 275–6,
   287–8, 297, 299
Teilhard de Chardin, Pierre 10
This is Tomorrow 93
*Three Hundred Years Hence; or, A Voice
   from Posterity* (Hay) 86, 87
Throsby, David 7

*Together: The Rituals, Pleasures and
   Politics of Cooperation* (Sennett) 35
Tong, Yen Dan 281–304
Torras, Mariano 9
tourism 3, 4, 238, 254–68; conservation
   and transmission of worldview 261–2;
   definitions of ethno-tourism 255;
   entrepreneurship 257; environmental
   awareness 263, 264; ethno-tourism in
   Araucanía, Chile 259–60; expectations
   of tourism 257–8; as an initiative of the
   Mapuche people 262–3; intercultural
   encounter 261; interpersonal experience
   262; Mapuche people in Araucanía,
   Chile 258–60; motivations in tourism
   256–7, 263–5; perceptions regarding
   Mapuche tourism 261–3; promotion of
   Mapuche culture, income and
   sustainability 264; rural-ethnic
   connection 255–6; rural tourism in Chile
   258; tourism versus ethno-tourism
   254–8; visitors, role of 264–5
training 8, 107, 114, 125, 176
transparency 139, 140–1
Truffaut, François 93

UNESCO Chair on Sustainability 226–7
United Nations Office for Risk Reduction
   86
*University Educators for Sustainable
   Development* (*UE4SD*) programme 179
urban areas *see* climate change and urban
   life; contemporary urban planning,
   castle towns; social urbanism
utopia as a literary genre 3, 21, 66, 67, 70;
   apocalypse in utopia 85–6

Valdivieso, Gonzalo 254–68
Van den Berghe, Pierre 255
variety 74, 75, 76, *77*, 79
Vernadski, Vladimir I. 10
Victor, Peter A. 9
Victorian period 63–71
Vietnam *see* sustainable development
Vietnam Agenda 21 282, 283
Villaça, Flávio 216
vitality 32, 33–4, *42*, **43**
vulnerability 64, 98, 102, 212–13, 218–19,
   240, 250

wages 74, 78, 79, 125, 128, 142, 211, 218
wants, theory of 78
Watene, Krushil 31
Water Babies, The (Kingsley) 94–5

Webb, Gary R. 240
wellbeing 1–16, 19, 24, 32, 47, 85, 114,
115, 125, 137–81, 141, 143, 150, 154,
162–3, 173, 174, 180–1, 203–4, 224–7,
229, 242, 244, 249; and citizen
participation 251–2; cultural footprints
of 9–12; and degrowth 135–8, *138*; and
economics 4–6; and health 33;
perception of 217–20; theories of 222–3

Wilson, Edward O. 167, 171–2
Winnicott, Donald 37
Wolff, Hans 191
Wood, Robert 255
working hours 79, 80–1, 81–2, 83, 142
World Commission on Environment and
Development 53, 136
World Trade Organization (WTO) 161
Wyndham, John 92, 93

For Product Safety Concerns and Information please contact our EU
representative GPSR@taylorandfrancis.com
Taylor & Francis Verlag GmbH, Kaufingerstraße 24, 80331 München, Germany